ATMOSPHERE

ATMOSPHERE

Editor

Erminio Gimondi

scitus
academics

Atmosphere

Edited by **Erminio Gimondi**

This Book is Printed on Acid-Free Paper.

ISBN: 9781681178936

Library of Congress Control Number: 2017948921

SCITUS Academics LLC,

616, Corporate Way, Suite 2, 4766,

Valley Cottage, NY 10989

www.scitusacademics.com

Notice

Printed in United States of America.

Preface

The earth is the only known planet, on which life exists. The composition of the atmosphere and the way its gases interact with electromagnetic radiation determine the atmosphere's effect on energy from the Sun—and vice versa. The present condition and properties of earth's atmosphere are one of the main reasons for earth to support life. Atmosphere is a gaseous layer surrounding the earth. In other words, we can say that our earth is surrounded by a thin layer of gases, called atmosphere. The atmosphere is special because it contains life-sustaining oxygen in large quantities. In fact, it took millions of years to reach the present condition by various processes. Along with its development, life came into existence and evolved. Today human actions are altering key dynamic balances in the atmosphere. Most importantly, humans are increasing greenhouse gas levels in the troposphere, which raises Earth's surface temperature by increasing the amount of heat radiated from the atmosphere back to the ground. Carbon dioxide is emitted by all living organisms as an end product of respiration. This carbon dioxide, in turn is used by producer organisms for the synthesis of food. This is another example how natural processes are inter-related and inter-dependant.

This book 'Atmosphere' discovers the fundamental processes that cause atmospheric circulation and create climate zones and weather patterns, and learn how carbon cycling between atmosphere, land, and ocean reservoirs helps to regulate Earth's climate. As a large amount of carbon dioxide is utilized by plants, on average, it comprises only 0.04 percent of dry air. Nevertheless, it plays significant role in keeping the atmosphere at temperatures that permit life. The concentration of carbon dioxide, nowadays, is increasing due to human activities. Today, man burns large quantities of fossil fuels for various purposes. As a result, huge amounts of carbon dioxide are emitted into the atmosphere. Its ever increasing concentration has already resulted in global warming and in some places, melting of ice.

Table of Contents

CHAPTER 1

The Atmospheric Boundary Layer during Wintertime Persistent Inversions in the Grenoble Valleys

Yann Largeron[1*] and Chantal Staquet[2]

[1]*Centre National de Recherches Météorologiques, UMR 3589, Météo-France/CNRS, Toulouse, France*

[2]*Université Grenoble Alpes, CNRS, LEGI, F-38000, Grenoble, France*

ABSTRACT

This study addresses the atmospheric boundary layer dynamics in the Grenoble valleys during persistent inversions, for 5 months during the 2006–2007 winter. During a persistent inversion, the boundary layer contains a layer with a positive vertical temperature gradient over a few days. Temperature data recorded on the valley sidewalls are first used. A bulk measure of the boundary layer stability, based upon the temperature difference between the valley top and the valley bottom, is introduced and a criterion is proposed to detect persistent inversions. We show that this criterion is equivalently expressed in terms of the heat deficit inside the boundary layer. Nine episodes are detected and coincide with the PM_{10}-polluted periods of the 2006–2007 winter. Secondly, the five strongest and longest persistent inversions are simulated using the MesoNH model. Focus is made on the stagnation stage of the episode, during which the inversion exhibits a diurnal cycle that does not significantly evolve from day to day. Whatever the episode, the inversion develops from the ground over a height of about 1200 m, with a nighttime temperature strength reaching 20 K. The boundary-layer dynamics within the inversion layer are fully decoupled from the (anticyclonic, weak) synoptic flow, independent from the synoptic-wind direction and similar whatever the episode. This implies that these dynamics are controlled by thermal winds and solely depends upon the geometry of the topography and upon the radiative cooling of the ground. Finally, a 2-day high-resolution simulation is made for the strongest case, representative of any persistent inversion. The flow pattern displays a well-defined spatial structure, with a vertical layering resulting from the superposition of the down-valley winds flowing from the different valleys surrounding Grenoble. This pattern persists all day long over a shallow convective layer of about 50 m forming above the ground during the reduced daytime period. Within this shallow layer, convection triggers a weak up-valley wind. Ventilation and stagnation areas in the surface layer are also computed, providing insights for air quality studies. The main characteristics of these persistent inversions are comparable to the most extreme wintertime inversions recorded in the Grand Canyon, Arizona.

1. INTRODUCTION

A temperature inversion consists in a layer of positive temperature gradient. Urbanized sites are very sensitive to temperature inversions forming in the atmospheric boundary-layer (ABL), as poor air quality is then induced (Figure 1A). Pollutants emitted by traffic, industry and biomass burning from wood combustion are indeed trapped into the inversion layer and induce well-documented health impact (e.g., Pope et al., 2002). Many world urban centers are located in complex terrain, such as valleys and basins (Fernando, 2010). In valleys, inversions are also called cold-air pools and form as a result of both cooling of the ground due to long-wave radiation and nocturnal down-slope winds (Whiteman, 2000). They usually extend from the bottom of the valley up to the boundary-layer top. In winter, inversions can last for a few days to several weeks, being triggered and maintained by anticyclonic conditions at synoptic scale (Reeves and Stensrud, 2009). We then talk about persistent inversions, as opposed to nocturnal inversions that form even in summer because of the boundary-layer diurnal cycle but are destroyed each day by thermal convection in the surface layer (f.i., Whiteman, 1982; Neff and King, 1989; Whiteman et al., 2008). During persistent inversions, a warm air mass surmounts the region at mid-altitude which results in a quasi-decoupling of the ABL within the valley or basin and the free troposphere, and the buildup of the inversion layer (Largeron and Staquet, 2016, and the present study). This accounts for the sustained increase of pollutants in urbanized complex terrain all along an inversion period (Silcox et al., 2012).

FIGURE 1. (A) Particulate Matter pollution during wintertime inversions over Grenoble (Photography: courtesy of Eric Chaxel). **(B)** View of the mountain ranges and Grenoble valleys discussed in the paper. **(C)** Locations of the ground-based Automatic Weather Stations (yellow) and air-quality stations (red, not used in the present study); the acronyms are defined in Table 1. Extracted and adapted from Largeron and Staquet (2016) with background maps taken from Google Earth.

The ABL in complex terrain during persistent inversions has therefore very specific features compared to the flat ABL, since slope and valley winds control the flow dynamics in the former case. As a result, data stemming from field campaigns or numerical experiments over flat terrain (see Mahrt, 2014; Sun et al., 2015, for a review) cannot be used to get insight into the ABL dynamics in complex terrain.

TABLE 1. Description of the ground-based automatic weather stations in the grenoble valleys.

Station	Acronym	Elevation	Elevation category
Le Versoud	LV	220 m	Low
Champ sur Drac	CD	230 m	Low
Pont de Claix	POC	237 m	Low
Champagnier	CP	363 m	Low
Peuil de Claix	PEC	935 m	Medium
Le Gua	LG	1600 m	High
Saint Hilaire	SH	1700 m	High
Chamrousse	CR	1730 m	High

The low-elevation stations are located at the valley bottom.

Several field campaigns dedicated to the study of persistent inversions in complex terrain have been conducted, with gradual focusing on smaller scales and growing importance of numerical studies for planning and conducting the campaigns and analysing the data. Campaigns were held in the Colorado basin (Wolyn and McKee, 1989; Whiteman et al., 1999a) or Columbia basin (Whiteman et al., 2001; Zhong et al., 2001) and first aimed at studying the formation, stagnation and destruction of the inversions (or cold air pools). Recent programs also include air quality concerns, such as PCAPS (Persistent Cold-Air Pool Study) in the Salt Lake valley (Silcox et al., 2012; Lareau et al., 2013; Wei et al., 2013). In Europe, the COLPEX program (COLd air Pooling and fog formation over complEX terrain), led by the Met Office in the UK, involved field and high-resolution numerical studies in a region of small hills in the UK (Price et al., 2011; Vosper et al., 2014). In the Alps, the first wintertime field campaign aiming at studying persistent inversions was held in the Chamonix-Mont-Blanc valley in January–February 2015, motivated by the anomalously high pollution level frequently recorded there; both meteorological and chemical data were collected during the campaign (Chemel et al., 2016).

In valleys, during persistent inversions, several mechanisms can significantly impact the concentration and the transport of pollutants, among which the slope and valley winds (Gohm et al., 2009). Down- and up-slope winds can drive the spatial distribution of pollutants which can be strongly heterogeneous inside the valley ABL (Schnitzhofer et al., 2007; Harnisch et al., 2009). Interactions between the cold-air pool and the valley winds are frequent (Banta et al., 2004; Cuxart et al., 2007) and the local specificities of a given site need to be investigated to know whether these valley-scale processes can explain the observed evolution of pollutant concentrations.

In France, the city of Grenoble is very critical in terms of pollution due to both its deep and landlocked orography and to the high-occupancy of the territory (Figure 1A): the city counts over half a million inhabitants and many industries. Moreover, the region of Grenoble is a strategic traffic crossroads, as a transit point and a necessary access to many ski resorts. Pollutant emissions, such as PM_{10} (particulate matter of diameter less than 10 μm) are then important and often concentrated on specific periods which coincides with periods when the boundary layer is the most stable. These emissions thus tend to result in acute episodes of local pollution, which makes Grenoble stand among the most polluted cities in France (Air Rhône-Alpes, 2011). Despite the strong impact of wintertime ABL dynamics on local air quality, field campaigns held in the Grenoble valleys up to now have focused on the summertime ABL to account for ozone pollution (Couach et al., 2003) and on the chemistry of aerosols from urban ground stations (Favez et al., 2010). No wintertime field campaign investigating the ABL dynamics has been performed yet. The present study, mostly numerical, is the first attempt to identify the circulation and thermal structure in the Grenoble valleys during wintertime persistent inversions, thus providing insights into the surface layer ventilation during such events. More details can be found in Largeron (2010). As we shall show, some conclusions of the present study apply to valleys with similar deep and complex orography, implying that they could be generalizable to many other world valleys. In the following, focus is made on the persistent stage of the inversion, between formation and destruction. Time series of 2-m air temperature data recorded at different altitudes around the city, and simulation results from high-resolution mesoscale simulations are used. In this study, we used the MesoNH numerical model (Lafore et al., 1998) that has already been successfully used to model stable ABL (e.g., Cuxart and Jiménez, 2007; Martínez et al., 2010) and valley or basin winds (e.g., Flamant et al., 2002; Cuxart et al., 2007; Drobinski et al., 2007).

In Section 2, we use the temperature data to identify the persistent inversion of the 2006–2007 winter in the Grenoble area. Nine such inversions are identified between November and February, which all coincide with very poor air quality. Numerical simulations of the five longest and strongest persistent inversions are then carried out, to investigate the general flow organization inside the valley and its possible coupling with the free troposphere (Section 3). A detailed study of one such inversion is next conducted, using a higher vertical resolution. Results are reported in Section 4 for the wind pattern and in Section 5 for the thermal structure of the persistent inversion. Conclusions are drawn in Section 6.

2. DETECTION OF PERSISTENT INVERSIONS

2.1. The Grenoble Valleys and the Ground-Based Observations

2.1.1. The Grenoble Valleys

The city of Grenoble is located in the French Alps and lies at the junction of three valleys, the Grésivaudan valley on the north-east, the Voreppe valley in the north-west and the Drac valley in the south (see Figure 1B). These valleys have a width comprised between 3 and 5 km and are of several tens of kilometers long. They separate three mountain ranges, the Vercors range on the west of the city, the Chartreuse range on the north and the Belledonne range on the east part. The highest altitude of the two former ranges is 2000 m ASL (above sea level) while the Belledonne range peaks at 3000 m ASL. The site also involves a fourth valley, the Romanche valley located south-east of Grenoble, which is narrower, deeper and higher in altitude than the other three valleys. The area comprising these four valleys and the city of Grenoble will be referred to as the *Grenoble valleys* hereafter. The flat part of the valley around and including Grenoble, will be referred to as the *Grenoble basin*; the largest width of this basin is about 6.5 km, and its altitude is 210 m ASL. The city of Grenoble has severe wintertime PM_{10} pollution issues. The number of days where the daily-average PM_{10} concentration exceeds the warning threshold of 50 µg m^{-3} has been of 50 in average per year over the 2007–2010 period, therefore exceeding the threshold of 35 per year set by the French legislation (Air Rhône-Alpes, 2011).

2.1.2. The Ground-Based Observations

The Grenoble valleys are equipped with ground-based Automatic Weather Stations measuring 2-m air temperature, and 10-m wind speed and direction every hour. These stations, which are either routinely operated by Meteo-France, the French weather forecast service, or by Air Rhône-Alpes, the air quality agency of Région Rhône-Alpes, are located at different places and altitudes in the Grenoble valleys. Time series of the temperature at eight stations have been used for the present study, whose name and elevations are displayed in Table 1, and locations indicated in Figure 1C. As can be seen from Table 1, the stations can be broadly categorized as low elevation, medium elevation and high elevation stations.

2.2. Methodology

We developed a simple method to detect persistent inversions of the 2006–2007 winter using the temperature records from the eight ground-based meteorological stations listed in Table 1. The principle of this method, detailed in Largeron and Staquet (2016), is to compute a bulk measure of the stability of the ABL from which persistent inversions are detected when this bulk measure exceeds a given threshold.

Figure 2A displays the 2-*m* air temperature recorded at the stations LV and CR during the month of December. Two points should be noticed: (i) the temperature difference between these two stations exhibits a high variability over the month; (ii) two main persistent inversion episodes occur, around mid-december and at the end of the month, during which the temperature at the valley bottom can be lower by 10 K than that at the high-elevation station.

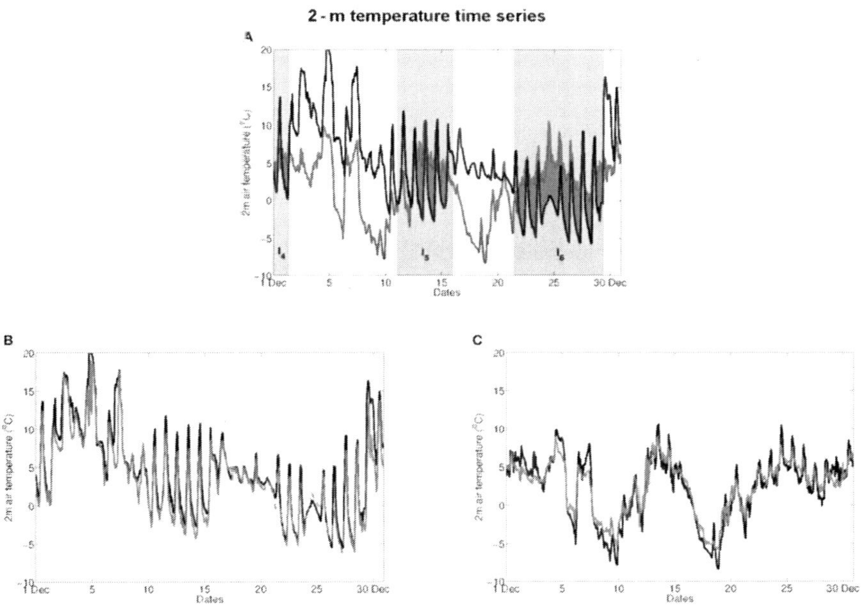

FIGURE 2. TIME SERIES OF 2-M AIR TEMPERATURE IN DECEMBER 2006 AT (A) THE LOW-ELEVATION STATION OF LV (BLACK) AND THE HIGH-ELEVATION STATION OF CR (RED). The gray-shaded zones refer to temperature inversions (the temperature in CR is higher than in LV); persistent inversions defined by criterion (1) are indicated by a yellow veil; **(B)** two low-elevation stations, LV (black) and POC (gray); **(C)** two high-elevation stations, CR (black) and LG (gray).

Inside the valley ABL, temperature records show that the temperature field is approximately horizontally homogeneous. Temperature time series at stations of similar elevations are indeed strongly correlated (r^2= 0.97), as can be seen for the two low-elevation stations of LV and POC (Figure 2B), or similarly for the two high-elevations stations of CR and LG (Figure 2C), for the whole month of December. On the opposite, high-elevation stations display a very distinct behavior than low-elevation stations (see Figure 2A) with a low correlation (r^2 = 0.36). It can be shown that the temperature difference between two stations belonging to the same category is two orders of magnitude smaller than the temperature difference between a high- and a low-elevation station, on average over the five considered months (November 2006 to March 2007). It follows that the temperature difference between the top and the bottom of the ABL is well estimated by the difference in temperature between any high-elevation station and any low-elevation station, and that the latter may be indicative of the ABL static stability. This is consistent with the results of Whiteman et al. (2004) who have shown that temperature vertical profiles can be well-approximated by pseudovertical profiles obtained from temperature stations on the sidewalls, in strongly stable and low synoptic wind conditions.

In valleys, inversion layers often fill the entire ABL up to the valley top (Whiteman, 1982). In the present case, the valley top is about 1500 m, which implies that the high-elevation stations are located just above the inversion. As shown below, this statement is consistent with results from numerical simulations of the Grenoble valleys.

A persistent inversion can therefore be detected by considering the temperature difference between a high- and a low-elevation station, divided by their elevation difference, $(\Delta T/\Delta z)_i$, where the index *i* refers to any couple of stations. We compute the average over 4 such couples of stations of the temperature difference across the ABL, which we refer as the *bulk temperature gradient* hereafter and denote $\Delta T/\Delta z$.

The detection of persistent inversions is then made using the following criterion:

$$\langle \Delta T/\Delta z \rangle_{24\,\mathrm{h}} > \langle \Delta T/\Delta z \rangle_{\mathrm{winter}} \text{ for more than 72 consecutive hours} \quad (1)$$

where $< >_{24\,\mathrm{h}}$ and $< >_{\mathrm{winter}}$ represent a 24-h-running average and the winter average (from November to March), respectively. For the winter of 2006–2007, $< \Delta T/\Delta z >_{\mathrm{winter}}$ is about −3 K.

The 24-h-running average bulk temperature gradient $< \Delta T/\Delta z >_{24\ h}$ is displayed over the whole winter in Figure 3, along with the constant value of −3 K. Using criterion (1), nine persistent inversions are detected, which are referred to as I_k, $1 \leq k \leq 9$. The December inversions are also indicated in Figure 2A. The characteristics of these nine inversions are displayed in Table 2, namely their period of occurrence with respect to criterion (1), the corresponding duration and an estimate of their strength as measured by the bulk temperature gradient averaged over their respective duration. The longest and strongest inversion is I_6, on which we shall mainly focus in the following.

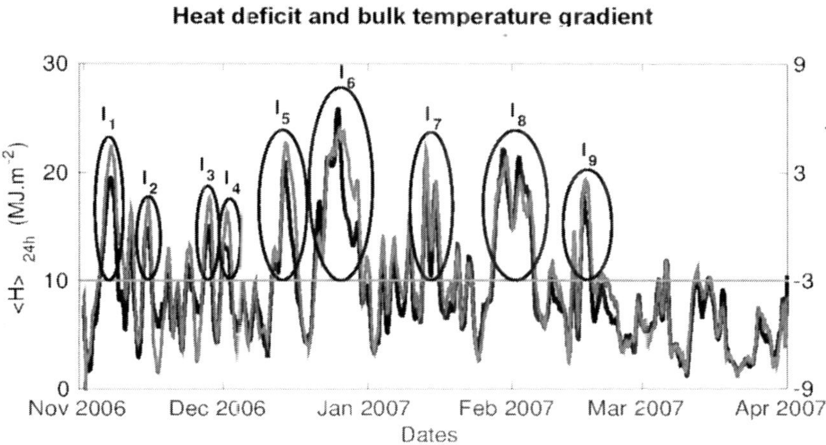

FIGURE 3. TEMPORAL EVOLUTION OF THE 24-H-RUNNING AVERAGE OF THE BULK TEMPERATURE GRADIENT $\Delta T/\Delta Z$ (RED) AND HEAT DEFICIT H (BLACK) DURING THE 2006–2007 WINTER. The 9 inversions I_k, $1 \leq k \leq 9$, detected by criterion (1) are circled in black. The persistence threshold $< \Delta T/\Delta z >_{winter}$ is indicated with a gray line. Extracted and adapted from Largeron and Staquet (2016).

Criterion (1) can be equivalently expressed in terms of the heat deficit H inside the inversion layer (Largeron and Staquet, 2016). The heat deficit is the amount of heat which should be provided to bring the temperature gradient to the adiabatic gradient, namely to fully mix the fluid column (Whiteman et al., 1999a,b). The 24-h-running average of the heat deficit is also displayed in Figure 3. As shown in this figure, the −3 K threshold for the winter-average bulk temperature gradient is equivalent to a threshold value of 10 MJ m^{-2} for the winter-average heat deficit.

TABLE 2. Persistent inversions detected using criterion (1).

Episode number	Persistent inversion period	Number of successive days	$(\Delta T/\Delta z)_{episode}$ (K km^{-1})
I_1	04 Nov 06 – 09 Nov 06	6.5	+0.79
I_2	13 Nov 06 – 15 Nov 06	3	−1.16
I_3	26 Nov 06 – 28 Nov 06	3.3	+0.36
I_4	30 Nov 06 – 02 Dec 06	3.2	−1.16
I_5	11 Dec 06 – 17 Dec 06	6.9	+1.29
I_6	20 Dec 06 – 30 Dec 06	11	+1.86
I_7	12 Jan 07 – 16 Jan 07	5.2	+0.29
I_8	27 Jan 07 – 05 Feb 07	10.5	+0.82
I_9	15 Feb 07 – 18 Feb 07	3.7	−1.41

The duration of each episode is indicated in columns 2 and 3 and a measure of the intensity of the episode, $(\Delta T/\Delta z)_{episode}$, is given in column 4.

It can also be shown (Largeron and Staquet, 2016) that (i) the nine detected episodes are strongly polluted, in the sense that the 24-h-running average concentration of PM_{10} exceeds the air-quality legal threshold of 50 µg m^{-3} for at least three consecutive days during each episode; (ii) all persistent inversion periods reported in Table 2 occur during and are triggered by anticyclonic conditions; (iii) all persistent inversions are generated by a mid-level warming and destroyed by a mid-level cooling.

3. THE NOCTURNAL WIND CIRCULATION IN THE GRENOBLE VALLEYS DURING PERSISTENT INVERSIONS

Figure 3 and Table 2 show that, among the nine persistent inversions of the considered winter, four of them (I_2, I_3, I_4, and I_9) last for a short duration (though greater than 3 days) and are of lower intensity. On the opposite, the five other persistent inversions last longer and are associated to stronger inversion strengths. These five episodes all involve a stagnation stage (see Largeron and Staquet, 2016), between the formation and the destruction of the inversion, during which the inversion displays similar features from one day to the next. The stagnation stage will also be referred to as the heart of the episode hereafter. In the following, focus is made on these five episodes for which numerical simulations are made, in order to investigate the main characteristics of the wind circulation in the Grenoble valleys during persistent inversions. We briefly describe the numerical model below, stressing the aspects relevant to the present study, before analyzing the wind circulation in the valleys. More details are provided in Largeron (2010).

3.1. Description of the Numerical Simulations

We performed high-resolution mesoscale simulations (according to the terminology defined by Cuxart, 2015) of the five episodes I_1, I_5, I_6, I_7, and I_8 using the Meso-NH numerical model developed by Meteo-France and the Laboratory of Aerology (Lafore et al., 1998).

The numerical model Meso-NH solves the Navier-Stokes equations in the anelastic approximation, thus filtering sound waves while allowing the density to vary in space and time. The model uses a terrain-following coordinate system along the vertical direction. The atmospheric model is coupled with the externalized land and ocean surface model SURFEX (Masson et al., 2013), to account for the surface fluxes and the evolution of four types of surfaces: nature, town, inland water and ocean with a large variety of ground covers. It can include a specific parameterization scheme for the urban energy budget (Masson, 2000). A full three-dimensional turbulence scheme that uses a prognostic equation for the turbulent kinetic energy is activated (Cuxart et al., 2000). The model may also include microphysical schemes for mixed-phase clouds and warm clouds and other specific parameterization for chemical species and aerosols, which we did not use in the present study.

As is now standard, a nesting of domains is used, allowing to capture the dynamical features of a wide range of scales and to focus on small scale processes in the innermost domain, which has the highest resolution (e.g., Zhou and Chow, 2013). The vertical resolution in each domain is the same, the size of the horizontal grid cells being decreased -in the present case by a factor of 3- from one outer domain to the inner one. For the present study, four nested domains are used, which are all centered on the Grenoble valleys. The largest domain D_1 has a horizontal resolution of 9 km and an extension of 675 × 675 km; and is forced at the boundary with 6-hourly fields from ECMWF operational analyses. The innermost domain is of size 30 × 30 km, with a horizontal grid size of 0.333 km. The databases Gtopo30 for the orography, with a resolution of approximately 1 km, and Ecoclimap 2 (Masson et al., 2003) for the ground cover are used, and are interpolated in the innermost domain to 1/3 km. Two configurations are considered, denoted C_1 and C_2 in Table 3 below, which mainly differ in the vertical resolution and duration of the computation. In configuration C_1, 60 levels are used in all domains, the vertical resolution at the ground being 15 m. In configuration C_2, a finer vertical resolution is used, of 4 m close to the ground, with 90 levels in total. The grid size is stretched along the vertical, the resolution being equal to 600 m at the top of the domain in each configuration. The duration of the computation is 10 h in C_1 and 53 h in C_2. The main features of the simulations are summarized in Table 3. The influence of these numerical choices (vertical resolutions and duration of the simulations) is investigated in the following Section 3.2 and in the Supplementary Material.

TABLE 3. Main parameters of the numerical simulations.

Numerical configuration	C1 (episodes I_1, I_5, I_6, I_7, I_8)	C2 (episode I_6)
Duration of the simulation	10 h	53 h
Simulation period	from 00:00 UTC to 10:00 UTC on 7 Nov for I_1, on 15 Dec for I_5, on 28 Dec for I_6, on 13 Jan for I_7, on 3 Feb for I_8	from 27 Dec, 18:00 UTC to 29 Dec, 23:00 UTC
Number of vertical levels (same in each domain)	60	90
Vertical resolution at the ground	$\delta z\vert_g = 15$ m	$\delta z\vert_g = 4$ m
Vertical resolution at the domain top	$\delta z\vert_t = 600$ m	
Vertical stretching	6% up to 2000 m ASL, 8% above	
Horizontal resolution	D1 : 9 km, D2 : 3 km, D3 : 1 km, D4 : 0.333 km	
Domain size	D1 : $(675km \times 675km)$ D2 : $(225km \times 225km)$ D3 : $(75km \times 75km)$ D4 : $(30km \times 30km)$	
Time step	D1 : 8s, D2 : 4s, D3 : 2s, D4 : 1s	
Lateral boundary conditions	D1 : ECMWF forcing D2 : 2-way grid-nesting D3, D4 : 1-way grid nesting	
Boundary conditions at the ground	SURFEX (with town-model TEB)	
Boundary conditions at the domain top	Rayleigh sponge	
Advection scheme of the scalar variables	Centered, positive definite 2th order in space and time	
Advection scheme of the vector variables	Centered, positive definite 4th order in space and time	
Pressure solver	Richardson (adaptative iteration)	
Turbulence model	1.5 order, TKE type, fully 3D	
Radiative scheme	ECMWF radiative scheme	
Time step of the radiative scheme	10 min	

In the third column, only changes relative to the second column are indicated; otherwise the line is left empty. The simulated periods are indicated at the beginning of the table.

3.2. Sensitivity Analysis of the Numerical Simulations

A first sensivity analysis has been made to investigate the numerical spin-up in the case of episode I_6 and configuration C_1. Starting the run at different times (0 h UTC on December 27, 18 h UTC on December 27 and 0 h UTC on December 28), results were compared at the end of the night (06:00 UTC) on December 28. This test, which is not presented here, leads to the conclusion that the numerical spin-up is about 4 h, namely, if comparison is made at the same time at least 4 h after the beginning of all the simulations, then the temperature and wind fields inside the valley are similar (see Largeron, 2010, for details). This result shows that the nocturnal quasi-steady circulation at valley-scale at the end of the night and in early morning (on which we focus in the following Section 3.3) can be analyzed by means of simulations starting at 0 h UTC in the C1 configuration. This choice appears facilitated by the presence of a persistent inversion which implies that an elevated inversion is already present and seen by the outer domain at the beginning of the simulations at 0 h UTC.

A second sensitivity analysis has been performed to investigate the effect of the vertical resolution, in the case of episode I_6, during the night from 18 h UTC on December 27 to 8 h UTC on December 28. This analysis is reported in the Supplementary Material. The main conclusion is that the vertical resolution close to the ground should be at least $\delta z|_g = 15$ m for the general features of the ABL dynamics and temperature profiles to have converged. Hence, a vertical resolution of $\delta z|_g = 15$ m can therefore be considered as sufficient to model these features.

However, comparisons with ground observations of 2-*m* air temperature data show that in a very shallow layer close to the ground, the temperature is only well reproduced with a fine resolution of $\delta z|_g = 4$ m in high-elevation stations, and still has a hot bias at the low-elevation stations.

For $\delta z|_g > 4$ m, simulations present a systematic hot bias at the bottom of the valley, and a systematic cold bias in altitude, which otherwise formulated, corresponds to a too weak temperature inversion compared to the observations. In this case ($\delta z|_g > 4$ m), surface fluxes suffer from a misrepresentation that in turn leads to a too weak cooling of the near-surface air in altitude and to an underestimated drainage of cold-air by katabatic winds toward the bottom of the valley.

In the following, consistently with these findings, the analysis made with the C_1 configuration ($\delta z|_g = 15$ m) then only deals with the general ABL dynamics for which a finest resolution leads to the same results. Fine-scale details of the near-surface ventilation and temperature gradient, reported in Sections 4 and 5, rely on the C_2 configuration ($\delta z|_g = 4$ m) instead.

3.3. Overall Nocturnal Circulation of the Five Episodes

Focus is made on one night of the five episodes I_1, I_5, I_6, I_7, and I_8, each simulation starting at 00:00 UTC and lasting for 10 h (see configuration C1 in Table 3). These nights are indicated in Table 3.

3.3.1. Synoptic Regime

As noted at the end of Section 2, each episode occurs under anticyclonic conditions, with low synoptic wind speed, of about 4 to 5 m s^{-1}. As for the wind direction, taken at the 600 hPa geopotential height (which is about 4200 m ASL in the present case of strong anticyclone), it hardly varies over the 10 h of the simulation but varies from one episode to the other: it flows either from the south–east (I_1, I_5), north–west (I_6), west (I_7) or from the east (I_8). The synoptic regime during each simulated period is very stable, the altitude of the 600-hPa geopotential height over the area varying by less than 15 m for I_1, I_6, I_7, I_8 and by 35 m for I_5.

These features of the synoptic wind computed by the model are consistent with those recorded from radio-soundings at the Lyon Saint Exupery airport, located about 80 km from the Grenoble site, which is a small distance compared to the synoptic scale. Wind recorded over 3000 m ASL there, the altitude of the highest summits of the Grenoble area, is representative of the synoptic wind over this site.

3.3.2. Valley Wind Circulation at the End of the Night

The key feature of the valley-scale circulation in the Grenoble valleys is that it consists in the superposition of several valley winds flowing from the different mountain ranges and valleys of the site. This circulation slightly varies from one episode to the other but common dominant features can be extracted, which are illustrated in Figure 4 by maps of the wind vector field at different altitudes between 350 and 3000 m ASL, for episode I_6. A sketch of this circulation is displayed in Figure 5 for each of the five episodes.

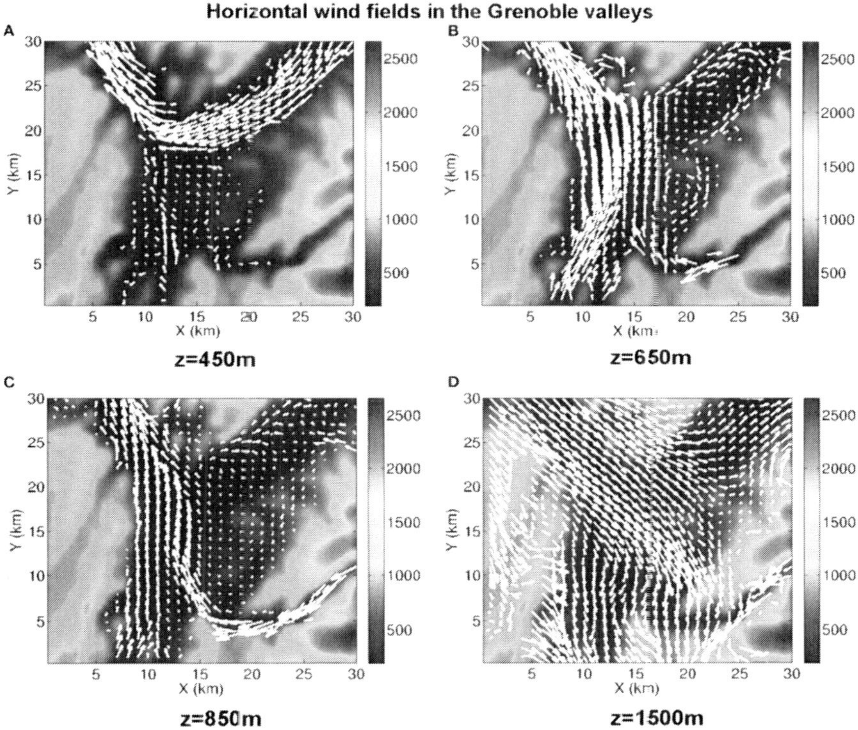

FIGURE 4. MAPS OF THE WIND AT A GIVEN ALTITUDE DURING EPISODE I_6 AT 06:00 UTC ON DECEMBER 28. (A) z = 450 m ASL; **(B)** z = 650 m ASL; **(C)** z = 850 m ASL; **(D)** z = 1500 m ASL.

During these five episodes, the circulation in the Grenoble valleys consists in four main winds which flow almost horizontally at different altitudes, resulting in a vertical layering in the ABL:

- *The Grésivaudan wind* flows along the bottom of the Grésivaudan valley from the north-east (the Grésivaudan valley has a very gentle slope descending toward Grenoble). This is a cold wind, which extends vertically over a layer of thickness 150–250 m, namely between the ground and about 500 m ASL (as shown in Section 4). It splits into two mass fluxes over the city, the major part flowing into the Voreppe valley toward the north-west, while a smaller part flows southward (Figure 4A and white colour current in Figure 5).

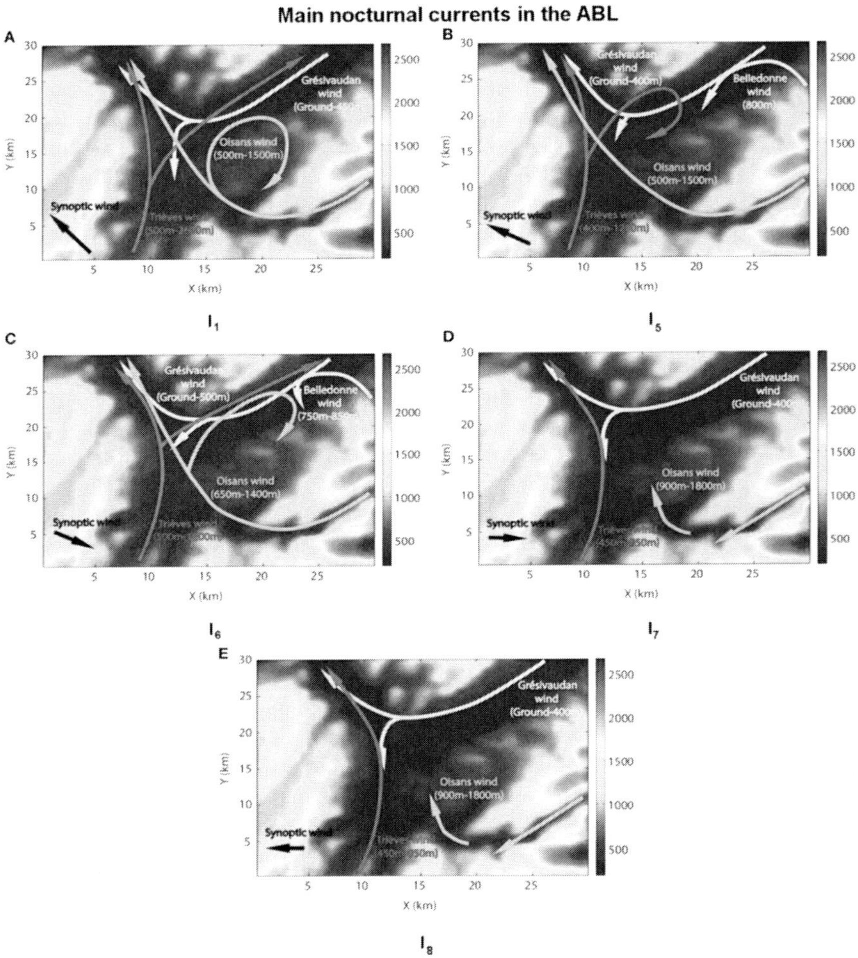

FIGURE 5. SKETCH OF THE MAIN NOCTURNAL CURRENTS DURING THE HEART OF EPISODE. (A) I_1(November 7); **(B)** I_5 (December 15); **(C)** I_6 (December 28); **(D)** I_7 (January 13); **(E)** I_8 (February 3).

- The second wind comes from the Trieves plateau located south of Grenoble (Figure 4B and red colour current in Figure 5). We refer to this wind as the *Trieves wind*. It flows between about 500 m ASL, therefore above the Grésivaudan wind, to an altitude of 1000 to 1200 m ASL depending upon the episode. This wind is then channelled into the Voreppe valley whatever the episode. A small part flows into the

Grésivaudan valley for episodes I_1 and I_6, in the direction opposite to that wind (and above it), while it recirculates as a large scale vortex over the city in I_5 (Figure 4B). A smaller vortex structure also forms along the flank of the Belledonne range, between 500 and 700 m ASL (as is visible in Figure 4B).

- The third wind flows from the Romanche valley and reaches the Grenoble basin from the south-east (Figure 4C and blue color current in Figure 5). This cold wind, referred to as the *Oisans wind*, meets the Trieves wind over the city between 600 and 1000 m ASL. A part of the mass flux is then channeled into the Voreppe valley and flows along with the Trieves wind (I_1, I_5, I_6). Another part recirculates as a large-scale vortex in the Grenoble basin and in the Grésivaudan valley (I_1, I_6). The thickness of the Oisans wind varies from one episode to the other, ranging from 500 and 1800 m ASL, which are roughly the lowest and highest altitude of the Romanche valley, respectively.
- Between 750 and 850 m ASL, a fourth wind, the *Belledonne wind* is sometimes observed (I_5 and I_6), which flows from the Belledonne range into the Grésivaudan valley while oscillating. Several large scale vortex structures are then visible in the latter valley.

From 1500 m ASL, which is the inversion altitude, up to 2500–3000 m ASL, the circulation appears to result from the channelling of the synoptic wind into the valleys. For episode I_6 (Figure 4D), this wind comes from the north-west in this range of altitudes. It is noticeable that this direction is opposed to, or at least very different from, that of the four winds just described, which flow below 1500 m ASL.

It appears from the five simulations that these local dynamics set in during a persistent inversion below 1500 m ASL, and are almost independent of the episode. The circulation inside the Grenoble valley can be decomposed into four winds, which are all associated with along-valley winds flowing almost horizontally at different altitudes, resulting in a vertical layering of the ABL. Large-scale vortices also form over the entrance of the Grésivaudan valley and along the flank of the Belledonne range.

3.3.3. Influence of the Synoptic Wind

Following Whiteman and Doran (1993), the direction of the wind in the Grésivaudan and Voreppe valleys is plotted vs. the direction of the synoptic wind for all five episodes in Figure 6. More precisely, for the Grésivaudan wind, an average of the wind direction from the ground to an altitude of 400 m ASL on all grid points of this valley is computed every hour of the simulated episode (see Table 3). The same average is applied to compute the Voreppe wind direction, the points considered being now in the

Voreppe valley, from the ground to an altitude of 600 m ASL. These grid points are indicated in Figure 7A.

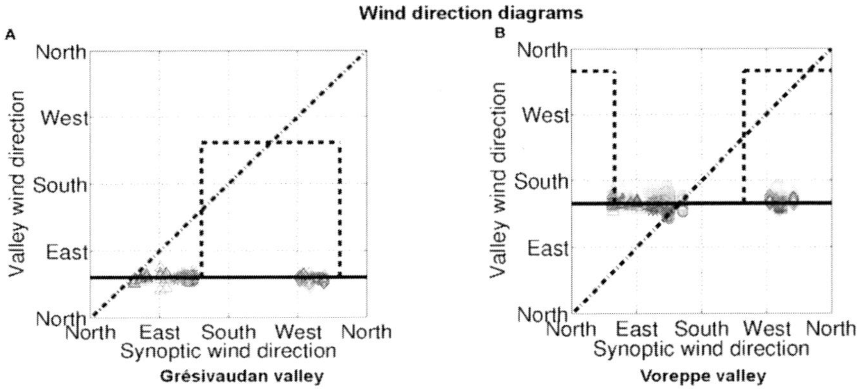

FIGURE 6. DIRECTION OF THE VALLEY-WIND VS. THAT OF THE SYNOPTIC WIND. The valley-wind direction is plotted every hour of the C1 configuration, each hour having a different color, and each episode is represented by a symbol: ∘: I_1, ⋆: I_5, □ : I_6, ◇: I_7, Δ: I_8. The horizontal black line is the direction of the valley close to the Grenoble basin. **(A)** Grésivaudan valley; **(B)** Voreppe valley. The valley-wind direction is averaged over the horizontal grid points displayed in Figure 7A, from the ground up to 400 m for the Grésivaudant valley and up to 600 m for the Voreppe valley. See text for details.

Figure 6 clearly shows that the wind keeps the same direction in either valley whatever the episode and, therefore, whatever the synoptic wind direction. For the Voreppe valley and episode I_6 for instance, the valley-wind flows toward the north–west, namely in the opposite direction to the synoptic wind. The same decoupling with the synoptic wind is observed for the Drac and Romanche valleys (not shown). This allows us to conclude that these winds are of thermal origin and result from the radiative cooling of the ground in a complex topography.

Note that the other processes leading to the formation of winds in the ABL, discussed by Whiteman and Doran (1993), are very unlikely and would have another signature. Indeed, a downward transport of horizontal momentum from above the valley, which would rather occur under unstable or neutrally stable conditions, would result in a linear dependency between the valley-wind direction and synoptic-wind direction (as indicated by the dashed dotted line in Figure 6). The channeling of the synoptic wind into the valleys (for a strong enough wind for instance) would lead to a step function (dashed line in Figure 6).

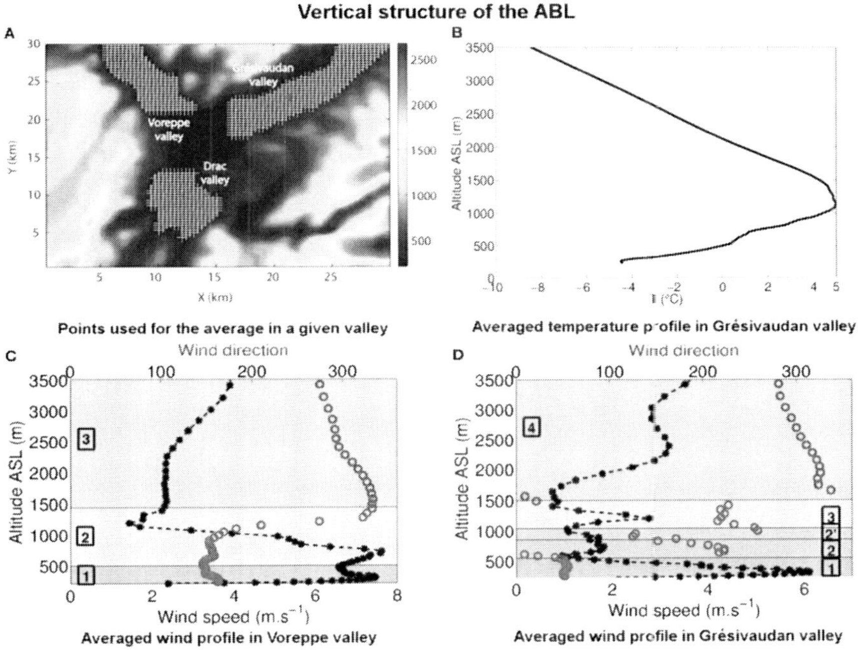

FIGURE 7. (A) The hatched zones are those over which the average for the Voreppe valley, the Grésivaudan valley and the Drac valley are performed. The point Gr, referred to in Section 5, is indicated with a red + symbol. **(B)** Vertical profile of temperature averaged over the Grésivaudan valley. **(C)** Vertical profile of wind speed (⋆) and wind direction (∘) averaged over the Voreppe valley. **(D)** Same as **(C)** in the Grésivaudan valley. The averages are performed over the horizontal grid points displayed in frame **(A)**. Profiles are shown at 06:00 UTC on December 28.

It follows that, during a wintertime persistent inversion, a unique organization of the valley-scale circulation within the Grenoble valleys sets in in the ABL, which does not depend upon the episode, because it is decoupled from the synoptic wind. This circulation is controlled by the local topography and the radiative cooling of the ground. Hence, a single episode may be considered as being representative of any persistent inversion of the winter when the circulation in the valley is considered.

4. SPATIAL STRUCTURE OF THE ATMOSPHERIC BOUNDARY LAYER

In the following focus is made on the heart of episode I_6, simulated with the finest resolution of the C_2numerical configuration and for a longer period of 2 days (see Table 3).

The choice of episode I_6 results from several considerations:

- I_6 is the strongest and the longest persistent inversion of the considered winter (see Table 2).
- The heart of episode I_6 is associated with a very low synoptic wind which flows from the north-west, namely in the direction opposite to most thermal winds of the Grenoble valleys, thereby waiving any doubt on the physical origin of the latter winds.
- The valley-scale circulation in the Grenoble valleys during episode I_6 is representative of the circulation during the other simulated episodes. This episode also clearly reveals the presence of the Belledonne wind, which is weaker in the other episodes.

4.1. Vertical Structure of the Nocturnal Winds

Vertical profiles of the wind speed and direction are displayed in Figure 7 for the Voreppe valley (Figure 7C) and for the Grésivaudan valley (Figure 7D). Results are shown at the end of the night (06:00 UTC) and are representative of the nocturnal vertical structure of the wind during the heart of the episode. The wind field is horizontally averaged over grid cells of the considered valley (displayed in Figure 7A) and the resulting mean speed and mean direction are computed as a function of altitude. The averaged vertical temperature for the Grésivaudan valley is also displayed in Figure 7B.

Figure 7D yields a refined view of the vertical layering described in the previous section for the Grésivaudan valley:

- The first layer (numbered as 1 in Figure 7D) lies above the ground (at about 220 m ASL), up to ≃500 m ASL, and is associated with the Grésivaudan wind. This wind blows from the north-east and is of rather large amplitude, up to 6 m s^{-1} reached at about 300 m ASL. Figure 7B shows that this layer is strongly stable, the average vertical potential temperature gradient $\partial\overline{\theta}/\partial z\partial\overline{\theta}/\partial z$ being equal to 22 K km^{-1} in the layer.

- The second layer (numbered as 2 in Figure 7D) surmounts layer 1 and is associated with a mean mass flux from the south-west, which is therefore up-valley and opposed in direction to the Grésivaudan valley. This layer is associated with the Trièves and Oisans winds which overlap in this range of altitudes. The layer is very inhomogeneous horizontally as it involves large scale vortices (of size close to half the valley width). The wind direction therefore strongly varies, implying a weak average wind speed. Like layer 1, layer 2 is strongly stable ($\partial\overline{\overline{\theta}}/\partial z \simeq \partial\overline{\theta}/\partial z \simeq 21$ K km^{-1}).

- An intermediate layer (denoted as 2') may be noticed between 800 and 950 m, which corresponds to the Belledonne wind. The mass flux flows here from the Belledonne to the Chartreuse range, with a direction comprised between 120 and 170° (namely from the south-east). This wind is weak, of less than 2 m s^{-1}.

- In layer 3, the mean mass flux is associated with the Trièves wind, blowing from the south-west. It has a clear jet type structure, with maximum value of about 2.7 m s^{-1} reached at 1250 m. This layer extends from about 1000 to 1500 m and is stable with a nearly uniform temperature (corresponding to $\partial\overline{\overline{\theta}}/\partial z \simeq \partial\overline{\theta}/\partial z \simeq 10$ K km^{-1}).

- Finally, the fourth layer (number 4 in Figure 7D) may be considered as a buffer zone between the top of the inversion layer and the bottom of the free troposphere. The synoptic wind is recognized here, being orographically channelled. It flows from the north-west with a slightly veering direction with altitude due to the shape of the mountain ranges. The stability of this layer is weaker, with $\partial\overline{\overline{\theta}}/\partial z \simeq \partial\overline{\theta}/\partial z \simeq 3.6$ K km^{-1}, which is that of the free troposphere above.

The complex wind structure in the Grésivaudan valley, where the signature of the different winds can still be clearly recognized, contrasts with that in the Voreppe valley displayed in Figure 7C. There, only three main layers can be distinguished, the lowest two layers being associated with thermal winds channelled into this valley (toward the north-west) after their coming out in the Grenoble basin. The first layer corresponds to the Grésivaudan wind, up to 500 m ASL. The second layer, from 500 m to about 1100 m, results from the overlapping of the Trieves and Oisans winds (locally a third maximum can be observed around 1000 m, associated with the Oisans wind). The maxima of these jets reach a higher value than in the Grésivaudan valley, of 7.5 m s^{-1}, due a strong pressure gradient between the Grenoble basin and the plain located northwest of the Voreppe valley, which accelerates the wind (not shown). The wind direction veers above 1000 m to align with that of the synoptic wind, marking the beginning of the buffer zone, which forms the third layer.

The Voreppe valley thus plays an important role in the ventilation of the Grenoble basin, in draining away the thermal winds. The air masses associated with these winds would otherwise partly accumulate in the Grenoble basin, possibly leading to an even higher pollution level during these wintertime persistent inversions.

The thermal winds and cooling of the ground at the bottom of the valley shape the temperature field, once anticyclonic conditions have set in. As a result, an inversion layer develops from the ground up to 1500 m, close to the altitude of the valley top as discussed in Section 2. Figure 7B thus shows that the mean temperature profile is nearly linear, the mean temperature at a height of 1000 m above the valley bottom being almost 10°C higher than at the bottom. The temporal evolution of this profile and associated structure of the inversion layer are analyzed further in Section 5.

4.2. Diurnal Circulation

On December 28 at the latitude of Grenoble, the astronomical sunrise is at 07:18 UTC, the sunset being at 16:02 UTC. Due to orographic shading, the local sunrise is delayed by 1 h in the major part of the valley bottom, and similarly, the local sunset is advanced by about 1 h in most of the Voreppe and Drac valley (but this is not the case in the Grésivaudan valley, which has an almost full sky view factor in the west direction, and is therefore not significantly shaded in the late afternoon by the Vercors range located westward). In the Grenoble basin, whereas the duration of the day is theoretically of 8h45, the duration of the period with a positive incoming insolation is then reduced to about 6h45. This implies that the diurnal cycle is dominated by its nocturnal component, which lasts almost 18 h.

Moreover, as will be shown in Section 5, due to both the low value of the wintertime insolation (further reduced by the orographic shading) and the strong stability over a significant depth of the inversion layer, daytime warming only affects a very shallow layer from the ground up to about 50 m above it. In this shallow layer, daytime convection occurs and triggers an up-valley wind in the main valleys of the site (not shown). The upper layers of the ABL, from 50 m above the ground up to the inversion top, are not affected by the daytime warming and the valley-scale circulation herein remains the same as during nighttime.

Consequently, on a diurnal cycle, the valley-scale circulation inside the ABL is dominated by the nighttime valley winds detailed in the previous section. This nighttime wind organization indeed prevails in all the ABL during the cooling stage (from about 15:00 UTC to 09:00 UTC), and above the shallow convective layer of thickness 50 m during the warming stage (from about 09:00 UTC to 15:00 UTC).

The nocturnal wind circulation described in Section 4.1 can therefore be considered as representative of the ABL kinematic structure all along the heart of a persistent inversion.

4.3. Recirculation and Stagnation Areas in the Grenoble Valleys

In this section, focus is made on the first 50 m above the ground in order to analyse the ventilation of the surface layer. The main motivation is to determine the areas where no ventilation occurs, referred to as stagnation areas, which have an increased impact on air quality and human health. For this purpose, we use the quantification method developed by Allwine and Whiteman (1994) for the Colorado plateau basin. This method relies on two indices, a stagnation index S and a recirculation index R, which are defined from the knowledge of the horizontal wind field components. The stagnation index, computed between two instants t_0 and t_1, represents the distance travelled by a fluid particle during this time interval. The recirculation index is also built with the wind field components, and with the stagnation index and takes values between 0 and 1: for $R = 0$, the fluid particle does not recirculate while for $R = 1$, the fluid particle exactly returns to its t_0-position at t_1. Allwine and Whiteman (1994) further introduce critical thresholds from which they define the state of the flow around a measurement station to be in a stagnation, recirculation or a ventilation state. The stagnation state is defined by $S \leq 130$ km, the recirculation state by $R \geq 0.6$ and the ventilation state by the two conditions $S \geq 250$ km and $R \leq 0.2$. Of course, the lower and upper bounds for S depend on the wind speed. This method has been used by numerous authors, both for *in situ* measurements (f.i., Venegas and Mazzeo, 1999) and numerical modeling (f.i., Levy et al., 2009).

The indices S and R have been computed for configuration C_2 in all points of the boundary layer. Averaging the indices over the first 50 m above the ground and over the 2 days of the simulation yields the maps of Figure 8, which displays the stagnation index in Figure 8A and the recirculation index in Figure 8B. Well-ventilated areas appear, which are the Voreppe valley, the Romanche valley and, to a lesser extent, the Grésivaudan valley. By contrast the Drac valley, the south-east of the Grenoble basin, and the flank of the east-face of the Chartreuse range, for which both $S \leq 130$ km and $R \geq 0.6$, appears as very critical areas for both stagnation and recirculation and, therefore, for the trapping of pollutants which would be emitted there. These results are consistent with the very weak wind field in the Drac valley shown on Figure 4A and with the persistent vortex structure which is present over the south-east of the Grenoble basin as shown on Figure 4B.

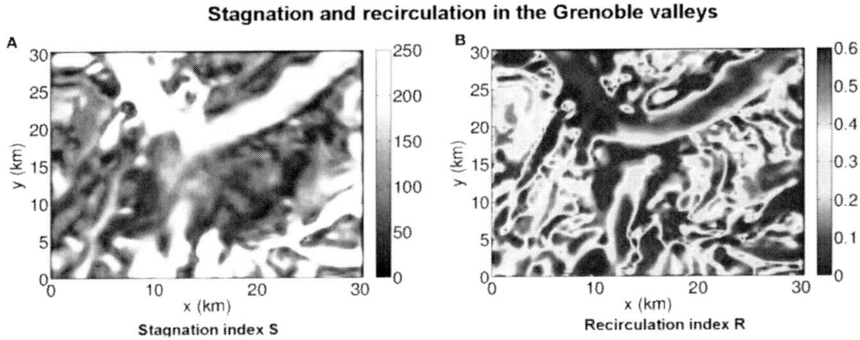

FIGURE 8. (A) Stagnation index S (in km) averaged over the first 50 m above the ground and over the 2 days of configuration C2. Stagnation areas appear in red and black. **(B)** Recirulation index R averaged as S. Recirculation areas appear in dark red. Ventilated areas ($S \geq 250$ km, $R \leq 0.2$) correspond to areas where S appears in white and R in blue.

5. ANALYSIS OF THE TEMPERATURE INVERSION DURING THE STAGNATION STAGE OF A PERSISTENT INVERSION EPISODE

5.1. Overall Evolution of a Persistent Episode

As discussed in the Introduction, the formation of the wintertime persistent inversion is controlled by the arrival of a warm air mass at mid-altitude (around 1500 m ASL), associated with very low wind speed; as shown in Section 3, the circulation of the air layer below becomes decoupled from the synoptic circulation and controlled by the thermal winds. The combined effect of cold air advected toward the bottom of the valley by the down-slope and valley winds and of radiative cooling leads to the progressive built-up of an inversion layer, the temperature gradient changing from adiabatic (and therefore negative) values to positive values in a day or two. As shown in Section 4, the inversion layer eventually fills up the whole valley, up to an altitude close to the top of the valley. This evolution is described in details in Largeron and Staquet (2016) and is similar to the one reported by Whiteman et al. (1999a) for the Colorado basin and by Reeves and Stensrud (2009) for valleys and basins in the Western United States.

The destruction of the inversion occurs when a cold front or trough, associated with strong winds and often with precipitation, reaches the region. The lower part of the atmospheric boundary layer becomes affected by the penetration of channelled winds which disturb and eventually destroy the inversion. This process is much more rapid than the formation of the inversion, as it involves mechanical and turbulent mixing processes, as opposed to the built-up of the inversion which results from radiative processes (Largeron and Staquet, 2016).

In the following, we focus on the thermal structure of the inversion during the stagnation stage of episode I_6. All results are shown at one point of the Grésivaudan valley, displayed in Figure 7A, and denoted as point Gr. This point, and the temperature profiles along the vertical line going through it, appear to be representative of the properties of the temperature field in the Grésivaudan valley.

5.2. Diurnal Forcing of the Atmospheric Boundary Layer

In numerical models, the balance of the surfaces fluxes is usually expressed as

$$R_n - G - LE - H = 0, \tag{2}$$

where R_n is the net incoming radiative flux, G the incoming heat flux into the ground, LE the latent heat flux and H the sensible heat flux (f.i., Stull, 1988).

In the SURFEX model, the surface temperature T_s obeys the equation (Noilhan and Planton, 1989)

$$\frac{\partial T_s}{\partial t} = C_T(R_n - LE - H), \tag{3}$$

where C_T is the heat capacity of the ground.

The four fluxes R_n, L_E, H and G are plotted vs. time in Figure 9A from 27 December 18:00 UTC to 29 December 20:00 UTC. The dashed lines correspond to the astronomical sunrise and sunset times (without taking into account orographic shading). In the context of the present study, no measurement of the surface fluxes are available so that the model performance can not be explicitly evaluated. Moreover, imbalances are frequently found in observations (e.g., Cuxart et al., 2015), pointing out some conceptual issues concerning the physical parameterizations of surface-atmosphere interactions currently used in numerical models (Foken, 2008). Nevertheless, simulated values suggest that:

- The sensible heat flux and the latent heat flux are much smaller than the radiative flux during both daytime and nighttime;
- The sensible heat flux decreases throughout the night and becomes more and more important as the night advances, whereas the latent heat flux remains negligible all night long;
- The surface flux balance is dominated by the radiative flux during the day and in the first part of the night; but in the second part of the night (after 23 h UTC), turbulence becomes a non-negligible component of the energy budget and partly compensates the long-wave outcoming radiative cooling.
- A radiative warming of the ground occurs for a short period of time, between about 09:00 UTC and 15:00 UTC, the radiative cooling therefore extending over the rest of the day for about 18 h.

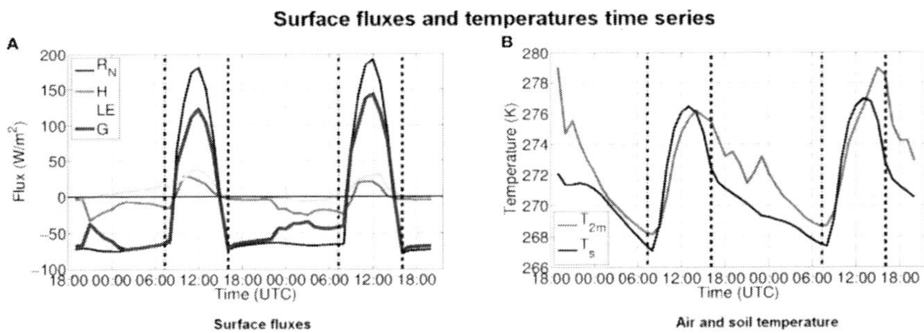

FIGURE 9. (A) Temporal evolution of the surface fluxes R_N, H, LE, $G = R_N - H - LE$ from 27 December, 18:00 UTC to 29 December, 20:00 UTC in the Grésivaudan valley at point Gr displayed in Figure 7A. The vertical dashed lines are the astronomical sunset and sunrise times for Grenoble. **(B)** Temporal evolution of the surface temperature (T_s) and the 2-*m* air temperature (T_{2m}).

The surface and 2-*m* air temperature responses are displayed in Figure 9B. The temperatures start to increase shortly after sunrise, up to about noon, consistent with the behavior of the radiative and sensible heat fluxes. A continuous decrease follows during the afternoon and night up to sunrise. These two stages will be referred to as the warming and the cooling stages hereafter.

5.3. Diurnal Evolution of the Inversion

Inversions layers are characterized by their height h_i and their strength $\Delta\theta$, where θ is the potential temperature, but also by their valley heat deficit (Whiteman et al., 1999a), already referred to in Section 2.2, which corresponds to the heat required to destroy the inversion. If a linear vertical profile of θ is assumed, and if the density ρ and the specific heat of air c_p are assumed to be constant, it can be shown that the heat deficit can be approximated by $0.5\rho c_p h_i \Delta\theta$ (Whiteman et al., 1999b; Largeron and Staquet, 2016).

In the present case, the height h_i is computed as the elevation of the highest point in the ABL where $\partial\theta/\partial z > |\gamma_{adiab}|$ (or equivalently $\partial T/\partial z > 0$), γ_{adiab} being the adiabatic temperature gradient. The strength $\Delta\theta$ is then the difference between θ at this height and θ at the ground level.

The structure of the inversion layer in the Grésivaudan valley from 27 December, 18:00 UTC to 29 December, 20:00 UTC is illustrated in Figure 10A by a (z, t) diagram of $\partial\theta/\partial z$ (above the point *Gr*) up to 2000 m ASL. Each point in the figure is plotted with a color which depends on the local stability quantified by the value of $\partial\theta/\partial z$: the points where $\partial\theta/\partial z > |\gamma_{adiab}|$, associated with an inversion, are in yellow; the points where $0 < \partial\theta/\partial z < |\gamma_{adiab}|$, where the layer is moderately stable with no inversion, are in orange; and the points where $\partial\theta/\partial z < 0$ where the layer is locally unstable (convective) are in black. Figure 10A shows that the ABL is entirely stable all along the diurnal cycle, except in a shallow layer close to the ground (of 50 m depth, namely about 5% of the inversion layer thickness) during the warming stage, and at a very few localized points around 700 m ASL at certain times, due to the presence of a strong shear. Also, in a very limited part of the ABL, static stability is moderate with no inversion (orange color) but these zones tend to disappear during the cooling stage. The major part of the ABL is otherwise filled with a strong and deep inversion (yellow color), as we now discuss it.

The top of the inversion layer is underlined with a thick black line in Figure 10A. The inversion layer height little varies over the period, between 920 and 1255 m. The lowest values are observed during the warming stage, the inversion being slightly eroded both from the bottom by the shallow convective layer and from the top by warm subsidence (this erosion was classified as type II by Whiteman, 1982). The inversion then thickens during the cooling stage. In early afternoon, when the surface air temperature is the warmest, the inversion height is still of 920 m implying that the inversion is far from being destroyed.

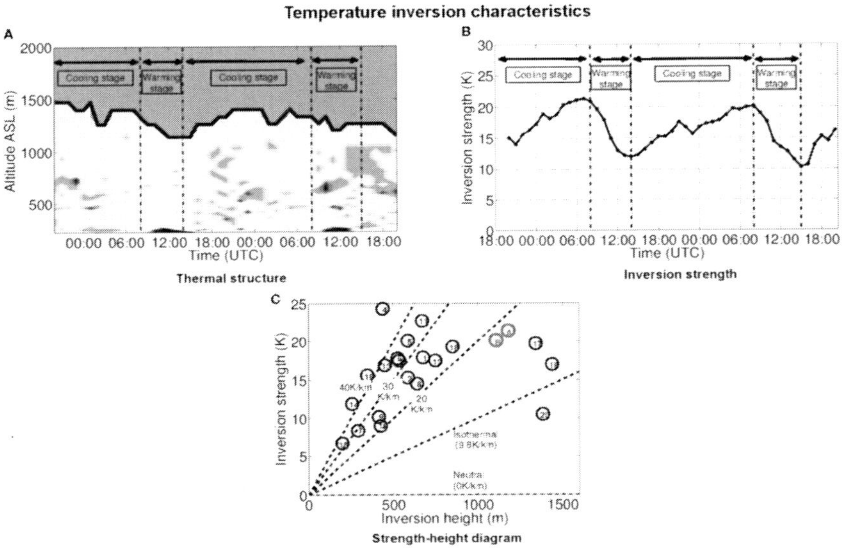

FIGURE 10. (A) Vertical profile of $\partial\theta\partial z\partial\theta\partial z$ above point Gr (see Figure 7A) from 27 December, 18:00 UTC to 29 December, 20:00 UTC. Black : $\partial\theta\partial z<0\partial\theta\partial z<0$. Orange : $0<\partial\theta\partial z<|\gamma adiab|0<\partial\theta\partial z<|\gamma adiab|$. Yellow : $\partial\theta\partial z>|\gamma adiab|\partial\theta\partial z>|\gamma adiab|$. **(B)** Inversion strength $\Delta\theta$ over the same time period. **(C)** Inversion strength vs. inversion height for 20 case studies of wintertime persistent inversions in the Rocky mountains (extracted and adapted from Whiteman et al., 1999b). Red symbols: parameters of the thermal inversion I_6 on 28 December 2006 (A) and 29 December (B).

The inversion strength $\Delta\theta$, displayed in Figure 10B, varies between 10.3 and 21.3 K over the period. The cooling stage is associated with an intensification of the inversion strength, with a maximum value of about 21.3 K, while the lowest values are reached during the warming stage (equal to 12 K on 28 December and 10.3 K on 29 December, at 14:00 UTC). Thus, the inversion strength decreases by nearly 50% over a day but its smallest value is still equal to 10 K, attesting that the inversion is persistent and strong.

Computation of the heat deficit leads to huge values at sunrise, of 23.4 MJ m^{-2} on 28 December and of 20.7 MJ m^{-2} on 29 December while the values remain strong at sunset, comprised between 11.7 and 12.8 MJ m^{-2} for these 2 days. These values are close to those inferred from the ground observations (see Figure 3). They are significantly higher than those estimated by Whiteman et al. (1999a) from inversion measurements in midsize Colorado valleys (1–8 MJ m^{-2} at sunrise) but of the same order as the values computed for the most extreme inversions in three valleys of size comparable to that of the Grenoble valleys.

Whiteman et al. (1999a) indeed compared the characteristics of inversions observed in 20 valleys of the Rocky mountains, which are reproduced in Figure 10C. Each valley is referred to by a number in a (h_i, $\Delta\theta$) diagram and the characteristics of the persistent inversion I_6 estimated on December 28–29 are superimposed. The heat deficit for a given valley is proportional to the area below the line joining the origin to the valley number, implying that the most intense inversions are found in the right upper corner of the diagram. These involve for instance the Bullfrog basin (number 17) located north-east of the Grand Canyon and inversion I_6. I_6 therefore ranges among the extreme inversions, which is consistent with this inversion to be the strongest of the winter of 2006–2007.

6. SUMMARY AND CONCLUSION

The purpose of this paper was to investigate the valley-scale circulation in the atmospheric boundary layer (ABL) of the Grenoble valleys, along with its thermodynamical structure, during persistent inversions of the 2006–2007 winter. Ground-based observations of temperature and high-resolution numerical simulations have been used for this analysis.

The first step was to detect persistent inversions from the 2-*m* air temperature data, a detailed analysis of which being reported in Largeron and Staquet (2016). A criterion was introduced for this purpose, based upon the temperature difference between the valley bottom and the valley top. This criterion was shown to be equivalently expressed in terms of the valley heat deficit inside the inversion layer, which represents the heat required to destroy the inversion layer. Similarly, Whiteman et al. (1999a,b) used such criterion to identify persistent cold-air pools in the Salt Lake valley. Nine episodes were detected from November to February, which all coincide with the nine strongest PM_{10} air-pollution episodes of that winter. These wintertime persistent inversions have a cumulated duration of one third of the total winter duration; they occur during anticyclonic conditions and are triggered by a mid-level advection of a warm air mass, as shown in Largeron and Staquet (2016). The simple method proposed here may serve as a reliable technique to detect persistent inversions from ground-based temperature data when no radio-soundings are available.

Numerical simulations were next performed with the MesoNH-SURFEX coupled model to analyse the persistent inversions, from formation to destruction. Only the stagnation stage, during which the inversion does not evolve from one day to the next, corresponding to the heart of an episode, is reported in the present paper. Several sensitivity and validation tests were performed to assess the validity of the computations, among which the sensitivity to the vertical resolution was investigated. It was found that a vertical resolution of 15 m at the ground allows the gross features of the atmospheric boundary-layer circulation to be reproduced, a resolution of at least 4 m at the ground being required for quantitative comparison with the 2-*m* air temperature to be obtained at the bottom of the valley. With a vertical resolution coarser than 15 m, the inversion layer is poorly represented, with a too weak strength due to a hot bias at the bottom of the valley and a cold bias at the valley top. This can be explained by the need to correctly represent the down-slope winds which drive and maintain the inversion layer in inducing cold-air drainage along the slopes, from the valley top down to the valley bottom.

Numerical simulations were thus performed in two steps. Firstly, among the nine persistent inversions, the five longest and strongest inversions, which exhibit a stagnation stage, were simulated as case studies, over a 10-h night with a vertical resolution of 15 m close to the ground. Focusing again upon one night in the heart of each episode, the main conclusions are that: (i) an inversion layer has formed up to an altitude of about 1500 m, whose absolute temperature gradient is nearly constant and of about 10 K km^{-1}. (ii) whatever the persistent inversion episode, the atmosphere can be decomposed into three layers : the free troposphere above 3000 m ASL, a buffer zone between 1500 and 3000 m ASL in which the large-scale synoptic wind is orographically channelled, and the inversion layer below 1500 m ASL. The inversion layer is decoupled from the atmosphere above whatever the episode, thereby recovering a common feature of wintertime persistent inversions. As a result, the atmospheric circulation in this inversion layer is controlled by thermal winds created by the nocturnal cooling of the ground; (iii) a local dynamics set in in the inversion layer, which is independent of the episode. These dynamics are characterized by a vertical layering resulting from the superposition of four main valley winds flowing from the tributary valleys of the Grenoble basin.

It follows that a detailed account of the heart of a persistent episode can be obtained by performing a high-resolution simulation of any of the five episodes. This is the second step of the numerical study. Two days during the heart of the longest and most extreme inversion were thus simulated with a 4-m vertical resolution close to the ground. The horizontal wind field appears to have a quasi-deterministic structure, with a veering shear with altitude, the wind speed and direction at a given altitude being controlled by a down-valley wind. Large-scale vortex structures appear at the entrance of the Grésivaudan valley and along the flank of the Belledonne range. A first guess of the impact of this circulation on air quality was addressed by computing the stagnation and recirculation indices introduced by Allwine and Whiteman (1994). Stagnation regions critical for air quality were thus identified, especially south to the agglomeration and along the flank of the Belledonne range where wind speeds remain low and vortices form.

A detailed account of the diurnal evolution of the thermal inversion was also obtained: the height of the inversion varies between about 900 and 1200 m, the lowest value occurring around 14:00 UTC when the inversion is eroded from below by a shallow convective layer of at most 50 m above the ground and from the top by a warm subsidence. The temperature difference between inversion top and bottom ranges between about 10 K (at 14:00 UTC) and 21 K (at 08:00 UTC). These features are those of a very strong and deep inversion and are actually similar to the most extreme inversions recorded by Whiteman et al. (1999a) in the Grand Canyon.

As already pointed out, no wintertime field campaign has ever been undertaken in the Grenoble valleys although this area is densely populated and strongly industrialized. Combined with the present results regarding stagnation of the atmospheric surface layer, this leads to severe human health issues. It would be very desirable to launch such a campaign, as the collected data would offer the possibility to get a better estimate of the validity of numerical simulations, regarding the quasi-deterministic pattern of the wind profiles or the thermal structure of the inversion layer we found. Field data would also allow us to get access to much finer scales than those analysed in this paper, such as surface fluxes, the knowledge of which is crucially needed to validate the ABL forcing of a surface-atmosphere coupled model or to improve the physical parameterizations. Obviously, complementary chemical and air quality studies would also benefit from acute knowledge of local fine-scale atmospheric dynamics.

AUTHOR CONTRIBUTIONS

YL: Author of the research presented in the manuscript and of the original manuscript. CS: Contribute to the manuscript.

CONFLICT OF INTEREST STATEMENT

The authors declare that the research was conducted in the absence of any commercial or financial relationships that could be construed as a potential conflict of interest.

The reviewer VM declared a shared affiliation, though no other collaboration, with one of the authors YL to the handling Editor, who ensured that the process nevertheless met the standards of a fair and objective review.

ACKNOWLEDGMENTS

The present work was supported by a funding from région Rhône-Alpes. Computations were carried out on the machines of the French computer center IDRIS.

SUPPLEMENTARY MATERIAL

The Supplementary Material for this article can be found online at: http://journal.frontiersin.org/article/10.3389/feart.2016.00070

REFERENCES

1. Air Rhône-Alpes (2011). Suivi des Polluants Réglementés Dans la Vallée de l'Arve. Technical report, Air Rhône-Alpes.
2. Allwine, K., and Whiteman, C. (1994). Single-station integral measures of atmospheric stagnation, recirculation and ventilation. Atmos. Environ. 28, 713–721.
3. Banta, R., Darby, L., Fast, J., Pinto, J., Whiteman, C., Shaw, W., et al. (2004). Nocturnal low-level jet in a mountain basin complex. Part I: evolution and effects on local flows. J. Appl. Meteorol. 43, 1348–1365.
4. Chemel, C., Arduini, G., Staquet, C., Largeron, Y., Legain, D., Tzanos, D., et al. (2016). Valley heat deficit as a bulk measure of wintertime particulate air pollution in the Arve River Valley. Atmos. Environ. 128, 208–215.
5. Couach, O., Balin, I., Jiménez, R., Ristori, P., Perego, S., Kirchner, F., et al. (2003). An investigation of ozone and planetary boundary layer dynamics over the complex topography of Grenoble combining measurements and modeling. Atmos. Chem. Phys. Discuss. 3, 797–825.
6. Cuxart, J. (2015). When can a high-resolution simulation over complex terrain be called LES? Front. Earth Sci. 3:87.
7. Cuxart, J., Bougeault, P., and Redelsperger, J. (2000). A turbulence scheme allowing for mesoscale and large-eddy simulations. Q. J. R. Meteorol. Soc. 126, 1–30.
8. Cuxart, J., Conangla, L., and Jiménez, M. (2015). Evaluation of the surface energy budget equation with experimental data and the ecmwf model in the ebro valley. J. Geophys. Res. Atmos 120, 1008–1022.
9. Cuxart, J., and Jiménez, M. (2007). Mixing processes in a nocturnal low-level jet: An les study. J. Atmos. Sci. 64, 1666–1679.
10. Cuxart, J., Jiménez, M., and Martínez, D. (2007). Nocturnal meso-beta basin and katabatic flows on a midlatitude island. Month. Weather Rev. 135, 918.

11. Drobinski, P., Steinacker, R., Richner, H., Baumann-Stanzer, K., Beffrey, G., Benech, B., et al. (2007). Föhn in the rhine valley during map: a review of its multiscale dynamics in complex valley geometry. Q. J. R. Meteorol. Soc. 133, 897–916.
12. Favez, O., El Haddad, I., Piot, C., Boréave, A., Abidi, E., Marchand, N., et al. (2010). Inter-comparison of source apportionment models for the estimation of wood burning aerosols during wintertime in an Alpine city (Grenoble, France). Atmos. Chem. Phys. Discuss 10, 559–613.
13. Fernando, H. (2010). Fluid dynamics of urban atmospheres in complex terrain. Annu. Rev. Fluid Mech. 42, 365–389.
14. Flamant, C., Drobinski, P., Nance, L., Banta, R., Darby, L., Dusek, J., et al. (2002). Gap flow in an alpine valley during a shallow south föhn event: Observations, numerical simulations and hydraulic analogue. Q. J. R. Meteorol. Soc. 128, 1173–1210.
15. Foken, T. (2008). The energy balance closure problem: an overview. Ecol. Appl. 18, 1351–1367.
16. Gohm, A., Harnisch, F., Vergeiner, J., Obleitner, F., Schnitzhofer, R., Hansel, A., et al. (2009). Air pollution transport in an Alpine valley: results from airborne and ground-based observations. Boundary-layer Meteorol. 131, 441–463.
17. Harnisch, F., Gohm, A., Fix, A., Schnitzhofer, R., Hansel, A., and Neininger, B. (2009). Spatial distribution of aerosols in the Inn Valley atmosphere during wintertime. Meteorol. Atmos. Phys. 103, 223–235.
18. Lafore, J., Stein, J., Asencio, N., Bougeault, P., Ducrocq, V., Duron, J., et al. (1998). The Meso-NH atmospheric simulation system. Part I: Adiabatic formulation and control simulations. Ann. Geophys. 16, 90–109.
19. Lareau, N., Crosman, E., Whiteman, C., Horel, J., Hoch, S., Brown, W., et al. (2013). The persistent cold-air pool study. Bull. Am. Meteorol. Soc. 94, 51–63.
20. Largeron, Y. (2010). Dynamics of the Stable Atmospheric Boundary Layer over Complex Terrain. Application to PM10 Pollution Episodes in Alpine Valleys. Ph.d. thesis, Université de Grenoble.
21. Largeron, Y., and Staquet, C. (2016). Persistent inversion dynamics and wintertime PM10 air pollution in Alpine valleys. Atmos. Environ. 135, 92–108.
22. Levy, I., Mahrer, Y., and Dayan, U. (2009). Coastal and synoptic recirculation affecting air pollutants dispersion: a numerical study. Atmos. Environ. 43, 1991–1999.
23. Lienhard, J., and van Atta, C. (1990). The decay of turbulence in thermally stratified flows. JFM 210, 57–112.
24. Mahrt, L. (2014). Stably stratified atmospheric boundary layers. Annu. Rev. Fluid Mech. 46, 23–45.
25. Martínez, D., Jiménez, M., Cuxart, J., and Mahrt, L. (2010). Heterogeneous nocturnal cooling in a large basin under very stable conditions. Boundary-layer Meteorol. 137, 97–113.
26. Masson, V. (2000). A physically-based scheme for the urban energy budget in atmospheric models. Bound. Layer Meteor. 94, 357–397.
27. Masson, V., Champeaux, J., Chauvin, F., Meriguet, C., and Lacaze, R. (2003). A global data base of land surface parameters at 1 km resolution in meteorological and climate model. J. Appl. Meteor. Climatol. 16, 1261–1282.
28. Masson, V., Le Moigne, P., Martin, E., Faroux, S., Alias, A., Alkama, R., et al. (2013). The surfexv7.2 land and ocean surface platform for coupled or offline simulation of earth surface variables and fluxes. Geosci. Model Dev. 6, 929–960.
29. Neff, W., and King, C. (1989). The accumulation and pooling of drainage flows in a large basin. J. Appl. Meteor. 28, 518–529.
30. Noilhan, J., and Planton, S. (1989). A simple parameterization of land surface processes for meteorological models. Month. Weather Rev. 117, 536.
31. Pope, C. A., Burnett, R. T., Thun, M. J., Calle, E. E., Krewski, D., Ito, K., et al. (2002). Lung cancer, cardiopulmonary mortality, and long-term exposure to fine particulate air pollution. JAMA 287, 1132–1141.

32. Price, J., Vosper, S., Brown, A., Ross, A., Clark, P., Davies, F., et al. (2011). COLPEX: field and numerical studies over a region of small hills. Bull. Am. Meteorol. Soc. 92, 1636–1650.
33. Reeves, H. D., and Stensrud, D. J. (2009). Synoptic-scale flow and valley cold pool evolution in the western united states. Weather Forecast. 24, 1625–1643.
34. Schnitzhofer, R., Norman, M., Dunkl, J., Wisthaler, A., Gohm, A., Obleitner, F., et al. (2007). Vertical distribution of air pollutants in the Inn Valley atmosphere in winter 2006. Geophys. Res. Abstr.
35. Silcox, G. D., Kelly, K. E., Crosman, E. T., Whiteman, C. D., and Allen, B. L. (2012). Wintertime PM 2.5 concentrations during persistent, multi-day cold-air pools in a mountain valley. Atmos. Environ. 46, 17–24.
36. Stull, R. (1988). An Introduction to Boundary Layer Meteorology. Kluwer.
37. Sun, J., Nappo, C., Mahrt, L., Belusic, D., Grisogono, B., Stauffer, D., et al. (2015). Review of wave-turbulence interactions in the stable atmospheric boundary layer. Rev. Geophys. 53, 956–993.
38. Venegas, L., and Mazzeo, N. (1999). Atmospheric stagnation, recirculation and ventilation potential of several sites in Argentina. Atmos. Res. 52, 43–57.
39. Vosper, S., Hugues, J., Lock, A., Sheridan, P., Ross, A., Jemmett-Smith, A., et al. (2014). Cold-pool formation in a narrow valley. Q. J. R. Meteorol. 140, 699–714.
40. Wei, L., Pu, Z., and Wang, S. (2013). Numerical simulation of the life cycle of a persistent wintertime inversion over salt lake city. Bound. Layer Meteor. 148, 399–418.
41. Whiteman, C. (1982). Breakup of temperature inversions in deep mountain valleys: Part I. observations. J. Appl. Meteorol. 21, 270–289.
42. Whiteman, C. (2000). Mountain Meteorology: Fundamentals and Applications. Oxford University Press, USA.
43. Whiteman, C., Bian, X., and Zhong, S. (1999a). Wintertime evolution of the temperature inversion in the Colorado Plateau Basin. J. Appl. Meteorol. 38, 1103–1117.
44. Whiteman, C., and Doran, J. (1993). The relationship between overlying synoptic-scale flows and winds within a valley. J. Appl. Meteorol. 32, 1669–1682.
45. Whiteman, C., Muschinski, A., Zhong, S., Fritts, D., Hoch, S., Hahnenberger, M., et al. (2008). Metcrax 2006. Bull. Am. Met. Soc. 89, 1665–1680.
46. Whiteman, C., Zhong, S., and Bian, X. (1999b). Wintertime boundary layer structure in the Grand Canyon. J. Appl. Meteorol. 38, 1084–1102.
47. Whiteman, C., Zhong, S., Shaw, W., Hubbe, J., Bian, X., and Mittelstadt, J. (2001). Cold pools in the Columbia Basin. Weather Forecast. 16, 432–447.
48. Whiteman, C. D., Eisenbach, S., Pospichal, B., and Steinacker, R. (2004). Comparison of vertical soundings and sidewall air temperature measurements in a small alpine basin. J. Appl. Meteorol. 43, 1635–1647.
49. Wolyn, P., and McKee, T. (1989). Deep stable layers in the intermountain western united states. Mon. Wea. Rev. 117, 461–472.
50. Zhong, S., Whiteman, C., Bian, X., Shaw, W., and Hubbe, J. (2001). Meteorological processes affecting the evolution of a wintertime cold air pool in the Columbia basin. Month. Weather Rev. 129, 2600–2613.
51. Zhou, B., and Chow, F. (2013). Nighttime turbulent events in a steep valley: a nested large-eddy simulation study. J. Atmos. Sci. 70, 3262–3276.

CHAPTER 2

Rainfall Variability and Trend Analysis of Annual Rainfall in North Africa

Zeineddine Nouaceur[1] and Ovidiu Murărescu[2]

[1]UMR CNRS 6228 IDEES, University of Rouen, Rouen, France
[2]Department of Geography, "Valahia" University, Târgovişte, Romania

ABSTRACT

The IPCC climate models predict, for the Maghreb countries, lower rainfall and increased aridity. Current observations in the three countries of central Maghreb (Morocco, Algeria, and Tunisia) are not consistent with these predictions. To demonstrate this new trend, a detailed regional analysis of rainfall evolution is conducted. This investigation is based on the calculation of the reduced centered index and the chronological graphical method of processing information (MGCTI) of "Bertin matrix" type. The results show extreme variability of this parameter and the severe past drought (more intense for Morocco, in which the drastic conditions from the seventies are observed). The results also show the beginning of a gradual return to wetter conditions since the early 2000s in Algeria and Tunisia and from 2008 for Morocco (this trend is confirmed by recent agricultural production data in 2011/2012 and 2012/2013).

1. INTRODUCTION

Climate change is today acknowledged by a large part of the scientific community. In its latest report, the IPCC [1] in 2013 evaluates the average trend of world temperature over the period 1880–2012 equal to 0.85°C with a degree of uncertainty ranging between 0.65°C and 1.06°C. The rise that occurred during the last decade (2003–2012) is +0.78°C (for a minimum of 0.72°C and a maximum of 0.85°C). The National Climatic Data Center, NCDC, considers the year 2014 as the warmest ever recorded, with an anomaly of +0.69°C (calculated for the period 1880–2014) [2]. The World Meteorological Organisation [3] considers the period 2011–2015 as the hottest on record, and the year 2015 as the hottest since modern observations began in the late 1800s. If, on a global scale, the increase in temperatures is unquestionable, the evolution of world pluviometry is much more contrasting as it is subject to a strong spatiotemporal variability. Despite this feature, and considering temperature rise, an increase in rainfall is likely to occur in certain region of the world favorable to such climate evolution. Indeed, the hydrological cycle acceleration under the influence of strong temperatures might lead to more rainfall and evaporation [4]. The results of several studies on rainfall evolution in many areas of the globe, as it happens in North Africa, show that climate change translates into wetter conditions [5, 6] as well as into a rainfall increase and repetition of extreme events (perceptible in the recent decades 1991–2010) [4, 7–9].

The Mediterranean area is known today as a "hot spot" of climate change [10–12]. This large inner sea stretching from the Strait of Gibraltar to the Lebanese coasts over an area of almost 4,000 km has a total surface of 2.51 million km. On the whole, this area is characterised by a well-defined summer drought and, depending on the subregions, by a maximum precipitation in autumn or winter [13]. During summer, rainfall is scarce. In this season, the Azores anticyclone moves north. When the anticyclone retreats south, it leaves free the passage to the ocean disturbances affecting North Africa. The rainy season begins in the fall and continues to the spring. The Mediterranean climate is subtropical (Csa and Csb, according to Koppen classification); however, cyclone activity presents large seasonal and spatial variability, with large differences from western to eastern Mediterranean and between cold and warm seasons [14]. The importance of this water mass on a regional scale explains the vast expanse in space of the Mediterranean climate domain. According to the results of various prediction of climatic models, both global and regional, a rise in temperatures and decrease in rainfall are expected [15–17]. Thus, by 2100, these regions should experience an average rise in temperature of 3-4°C, a decrease of rainfall, and increase of extreme events [12]. Philandras et al. [18] based on the future projections of regional climate model RACMO2/KNMI show that annual precipitation within the majority of Mediterranean is very likely to decrease on an average of 20%, during the period 2071–2100 compared to 1961–1990, under scenario "SRES" "A1B." The works of Giannakopoulos et al. [19] lead to the same result with a 20%

rainfall decline south of the Mediterranean Sea according to a climate warming scenario of +2°C. Tselioudis et al. [20], based on the results of first scenario, showed a marked decrease in winter precipitation value by the end of the 21st century. Dubrovský et al. [21] analyzed future climate conditions for the Mediterranean region based on 16 Global Climate Models. For temperature, the results show an increase for all parts of the Mediterranean. For precipitation, a decrease is noted except for the northernmost parts in winter. The intermodel agreement for the precipitation changes is lower than for temperature [21]. Giorgi and Lionello [11] estimate that significant Mediterranean rainfall decline is expected in connection with intensification of the anticyclonic circulation and a northward shift of the string of cyclonic depression areas. Alpert et al. [22] confirm a tendency of intense rainfall increase in Spain and Italy and also note an absence of trend in Israel and Cyprus. The combined effect of climate change and anthropic impact would entail a lack of water for almost 290 million people. At the same time, Algeria and Tunisia, which are part of the MENA region countries, have reached a state of absolute water scarcity.

In central Maghreb (Morocco, Algeria, and Tunisia), the rise in temperature is pursuant to the global situation. Thus, 2013 is one of the warmest years recorded in Tunisia since 1950 [23]. The rise of lowest temperatures () is however more evident in this part of North Africa [24–26]. Rainfall data of some terrestrial observations networks, on the south shore of the western Mediterranean basin, confirm this change for the last decade. However, the results show a contrary trend to climate models predictions [27, 28]. The first authors assign a tropical precipitation origin to the return of rains observed at the Moroccan stations of Safi and El-Jadida, while, within the whole of the central region of Morocco, drought conditions prevail. In the Middle Moroccan Atlas, a situation may be noted which differs from that described for the central Moroccan region [29]. Finally, Khomsi et al. [30] find no statistical significancy in the cumulative and extreme rainfall trends (summer and winter events) in 9 stations of this last country. In Tunisia, recent studies [31] show that annual precipitation trend is on the rise for 19 regions surveyed during 1977–2011. This analysis highlights a break of chronological sequences during the most characteristic periods of the great drought of the 1980s (1981–1992) and a return of rains in the early 2000s (2001–2009). This last trend is also confirmed by [32] for north-eastern Tunisia. The absence of a significant trend is however dealt with in other studies [33] over a long chronological sequence of Tunis, Manouba station. The same evolution may also be encountered in the intra-Tellian plains and basins [34]. Finally, Jebari et al. [35] have emphasized the tendency of rainfall decline and higher efficiency of solid transport and flows over 50 years' sequences (1960–2004).

This diversity of scientific opinions confirms the spatiotemporal variability of the Mediterranean pluviometry and particularly on the south shore of the basin [36, 37]. This peculiarity makes the identification of long-term significant trends difficult [38] and shows the need to study the precipitation trend according to a global regional approach which minimises the topoclimate effect and is informed of the spatiotemporal distribution of precipitation cycles.

2. MATERIALS AND METHODS

This article aims to analyze the precipitation trend for more than forty years of measurements on a wide northern strip in central Maghreb (Morocco, Algeria, and Tunisia). Thirty-five stations (14 in Morocco, 12 in Algeria, and 9 in Tunisia) which provide a common period of measurements were chosen for this study (1970–2013 for Algeria and Tunisia and 1970–2011 for Morocco) (Figure 1). The monthly precipitation data were provided by the Meteorological Services "The National Office of National Meteorology (ONM) for Algeria, the National Meteorological Institute (INM) for Tunisia, and the National Directorate of Meteorology (DNM) for Morocco". Quality and homogeneity of the data were tested by these services. The missing data were supplemented with information gathered from the NCDC "National Climatic Data Center" website (http:// www.ncdc.noaa.gov/).

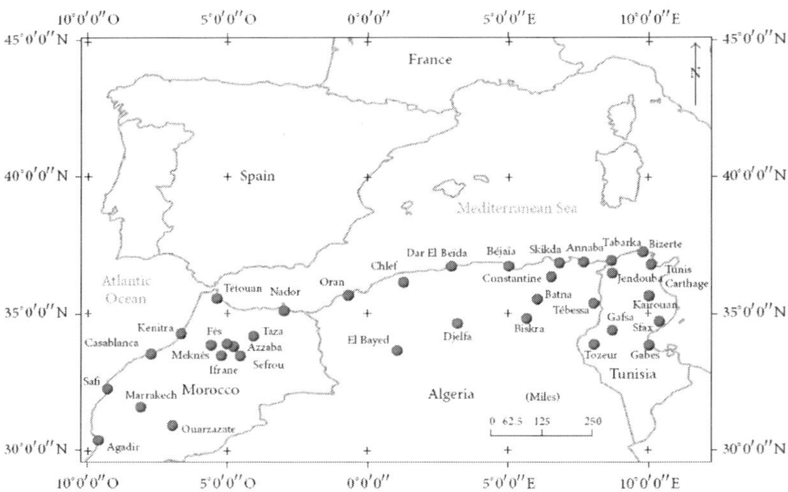

Figure 1: Spatial distribution of precipitation stations used for this study.

The method used for this study is the chronological graphic method of information processing (MGCTI) of "Bertin matrix" type. The MGCTI is an analytical method based on a statistical analysis and on a graphical representation of results. This method was used for the first time in 2013 [24]. The MGCTI is developed to facilitate the interpretation of the statistical results for the Mediterranean rainfall analysis, due to the high variability affecting this parameter.

The Bertin matrix was introduced to harmonize and consolidate information after the statistical treatment. The Bertin matrix is a manual and visual method of classification of information based on data. This matrix is used to group the data that have similarities according to all the criteria studied. It provides a simple and effective way to establish a multivariate typology based on observations of the user. This provides the concordant results (even rainfall character for all stations studied in the same year) but also identifies conflicting information (different characters between stations for the same year).

The MGCTI and its graphic representation allow a chronological reading and a spatial analysis of the phenomenon. This method has been successfully tested in many North African regions [28, 29, 32, 39] and in the Sahel [40]. A comparative study with the SPI (Standard Precipitation Index) method for detecting climate drought was conducted in 2015 [41]. This study showed the simplicity and clarity of the results obtained with the MGCTI method. In this paper, we test this method on a larger area consisting of different climatic regions of North Africa (more than 30 stations). The purpose of this article is to show the trend of rainfall over nearly half a century and to detect the date of changes of cycles.

The First Stage. An annual data precipitation (cumulative rainfall over a calendar year) hierarchy in terms of limit values (Q1, Q2, Median, Q3, and Q4) is done for all stations and for the entire series (Table 1).

Table 1: Pluviometry of central Maghreb.

Morocco	Agadir	Azzaba	Casablanca	Fes	Ifrane	Kenitra	Marrakech	Meknes	Nador	Ouarzazate	Safi	Sefrou	Taza	Tetouan
Min	81.43	147.00	107.72	200.00	517.10	252.46	95.20	181.50	154.55	25.08	147.06	191.73	114.10	308.40
Q1	127.60	248.00	273.98	339.90	639.30	357.38	142.00	380.30	250.91	72.74	241.18	401.70	418.50	454.06
Q2	190.00	302.00	331.14	446.10	869.20	501.64	198.61	478.50	285.45	93.71	318.49	485.82	509.70	588.32
Me	212.95	313.00	347.38	475.20	935.20	532.70	217.41	515.30	309.09	104.44	325.21	511.10	554.40	640.20
Q3	243.08	333.00	392.14	496.10	960.20	593.44	235.71	540.20	341.82	112.29	358.82	559.40	612.60	666.95
Q4	338.40	415.00	508.95	591.30	1099.90	695.00	306.40	679.40	392.73	154.27	466.39	673.00	790.00	801.40
Max	654.05	667.00	761.90	855.00	1742.10	927.87	463.76	1147.40	672.80	247.80	883.19	992.50	1005.80	1215.97
Algeria	Annaba	Batna	Bejaia	Biskra	Constantine	Dar El Beida	Djelfa	Chlef	El Bayed	Oran	Skikda	Tebessa		
Min	409.00	159.20	320.00	47.00	253.30	280.00	152.40	213.50	106.70	171.70	491.60	199.00		
Q1	531.10	234.00	631.00	69.07	426.00	475.40	249.00	298.00	219.10	239.50	615.90	282.30		
Q2	584.40	299.00	749.00	93.90	491.90	606.00	304.40	337.70	266.20	322.00	666.30	349.80		
Me	614.60	347.00	780.80	125.98	514.20	683.52	324.60	361.00	270.90	350.70	729.80	370.35		
Q3	662.44	407.30	833.00	133.61	570.30	717.50	342.30	419.10	304.60	404.20	764.90	406.80		
Q4	773.00	476.80	969.20	190.30	861.90	797.80	394.80	485.10	380.86	439.10	876.70	510.70		
Max	1126.00	362.19	1373.41	295.00	812.29	1169.00	512.70	577.11	546.57	609.00	1148.20	646.60		
Tunisia	Bizerte	Gabes	Gafsa	Jendouba	Kairouan	Sfax	Tabarka	Tozeur	Tunis					
Min	373.90	86.50	39.00	244.90	133.30	94.80	574.50	22.20	260.30					
Q1	467.40	121.10	106.90	334.20	221.20	142.90	790.50	46.40	353.10					
Q2	611.60	141.90	134.50	415.70	279.30	184.30	932.90	76.60	419.80					
ME	679.20	173.50	149.50	460.30	299.70	202.30	971.70	84.20	456.00					
Q3	724.30	205.90	169.60	490.30	324.00	232.40	1024.80	91.90	499.50					
Q4	785.60	311.30	216.40	572.60	367.40	306.70	1161.90	114.20	593.70					
Max	983.10	473.70	502.30	840.10	518.70	449.40	386.40	168.20	1011.60					

Depending on data position in relation to limit values, the years are considered as(i)very dry, below the first quintile;(ii)dry, between the first and the second quintile;(iii)normal with trends towards drought, between the second quantile and the third quintile;(iv)rainy, between the third and the fourth quintile;(v)very rainy, above the fourth quintile.

The Second Stage. A recoding of values is made by means of a range of colours (the colour varying in terms of the annual cumulative rainfall position in relation to limit values). This first processing is followed by a reordering procedure (permutations of columns) in order to get a ranking that allows the visualisation of a homogenous coloured structure (Bertin matrix) (Figure 2). This procedure allows for the visualisation of the climate parameter evolution in terms of two dimensions (time and space).

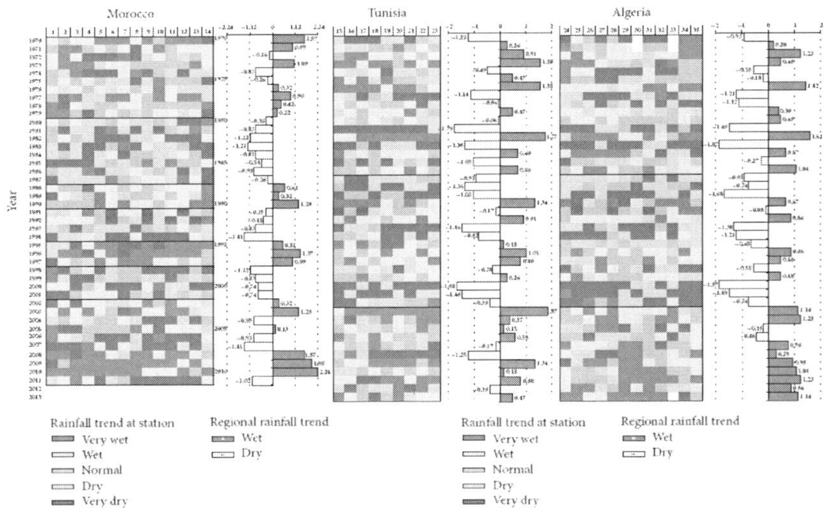

Number	Station	Longitude	Latitude	Elevation (m)
1	Taza	−4.02	34.21	521
2	Azzaba	−4.73	33.83	689
3	Sefrou	−4.49	33.49	801
4	Nador	−2.92	35.13	226
5	Tétouan	−5.33	35.58	10
6	Agadir	−9.55	30.38	11
7	Fès	−4.97	33.93	598
8	Casablanca	−7.67	33.56	62
9	Kénitra	−6.6	34.3	17
10	Meknès	−5.53	33.88	549
11	Ifrane	−5.17	33.5	1660
12	Safi	−9.23	32.28	45
13	Marrakech	−8.03	31.61	474
14	Ouarzazet	−6.9	30.93	1187
15	Gabes	10.06	33.53	4
16	Kairouan	10.06	35.4	60
17	Sfax	10.41	34.43	21
18	Tozeur	8.06	33.55	87

Number	Station	Longitude	Latitude	Elevation (m)
19	Tunis	10.14	36.5	4
20	Bizerte	9.48	37.15	5
21	Gafsa	8.49	34.25	313
22	Tabarka	8.54	36.58	66
23	Jendouba	8.48	36.29	143
24	Oran	−0.36	35.38	90
25	Dar El Beida	3.13	36.41	25
26	Annaba	7.49	36.5	8
27	Skikda	6.54	36.53	1.3
28	Béjaia	5.04	36.43	1.74
29	Chlef	1.2	36.13	143
30	Biskra	5.44	34.48	82
31	El Bayed	1	33.4	1341
32	Djelfa	3.23	34.2	1180
33	Tébessa	8.07	35.25	820.47
34	Constantine	6.37	36.13	685
35	Batna	6.19	35.45	822

Figure 2: MGCTI, classification of annual precipitation according to quintiles Q1, Q2, Q3, Q4, and Q5 (measurement period from 1970 to 2013).

The Third Stage. To determine the typical breaks and periods, a second procedure is conducted. It consists in assigning a number ranging from 1 (very dry year) to 5 (very wet year) according to the already determined features assigned to each year. The sum of numbers of all stations for each year is centered and reduced, thus getting a regional index (RI) varying from +1.80 for a very wet year to −1.80 for a very dry year. The "RI" is calculated as follows:where is year value, is the series average, and is standard deviation.

$$RI = \frac{(X_i - X)}{S}, \quad (1)$$

The projection of the result on a graph allows for the visualisation of the evolution of the phenomenon on a regional scale in a first stage and, in a second stage, for the determination of data on breaks and trend change.

3. RESULTS AND DISCUSSION

3.1. Towards the End of Climatic Droughts in Central Maghreb?

3.1.1. The Morocco Situation

The results of statistical and graphical processing of central Maghreb pluviometry show strong variability (typical of the Mediterranean climatic domain) and an organisation structured into three main climatic periods (Figure 2).

A period qualified as wet is visible on the matrix during 1970–1979. This decade contains 70% of the years with positive regional index. The highest values are noted at the beginning of the period (+1.57 for 1970 and +1.09 for 1973). Almost 50% of annual rainfall accumulations are considered as wet and very wet. The percentage of average year reaches 25.71%, while, during dry years, it represents 27.14%.

In 1980, a new drier climatic phase began and lasted until 2001. Only 30% of the years of this long period stretching over more than two decades show a positive regional index. The proportion of dry and very dry years reaches 49.01%, that of normal years 18.75%, and that of rainy and very rainy years 32.23%.

The regional graph allows us to distinguish three dry subperiods (negative regional index) with two rainy periods of three successive years in between:(i)From 1980 to 1987, an uninterrupted succession of years with negative regional indices can be noted. It is a period of severe drought which marks a large part of the Moroccan territory. The values are relatively high for 1982 (−1.12), 1983 (−1.21), and 1986 (−0.93). For all Moroccan stations studied, the proportion of dry and very dry years reaches almost 56%, while rainy and very rainy years account for a little more than a quarter, with 24%. This long period characterised by a lack of humidity is followed by three years (1988–1990) during which positive regional indices and a humid trend are recorded.(ii)From 1991 to 1994, rainfall accumulations become negative again.(iii)The regional index reaches, during this final year, one of the lowest values of the series, −1.41.(iv)From 1995 to 1997, a short wet period marks the Moroccan pluviometry, with a spectacular increase of pluviometry (the regional index reaches +1.37 in 1996, one of the highest values of the first three decades of this series).(v)The stage covering the 1998–2001 period is marked by the return of drastic conditions. A little more than 60% of rainfall accumulations recorded at various stations are considered as dry or very dry.(vi)The last period (2002–2011) of the precipitation series is considered as wet (60% of the years recorded a positive index). At the same time, the percentage of rainy and very rainy years at all stations reaches 52.38%. The last years of the series provide evidence of change of trend for this decade. Indeed, from 2008 to 2010, the percentage of rainy and very rainy years goes from 78.57% to 85.71% and 92.85%.

3.1.2. The Algeria Situation

The MGCTI graphical matrix (Figure 2) shows, for Algeria, an organisation of rainfall accumulations into three characteristic periods.

A first stage of strong variability in 1970–1986. No genuine climatic trend stands out in this period. All stations under study are characterised by an almost even distribution of extreme classes of annual rainfall, rainy and very rainy (40%), dry and very dry (39.50%), and normal (20.5%). The regional index is negative for more than 47% of the years as opposed to 52.94% for positive values. What is remarkable about this period is undoubtedly a powerful interannual oscillation between dry year and wet year. A sequence of three rainy years (1971–1973) can be noted at the beginning of the series, while, for the rest of the period, trend inversions (dry year, rainy year) are for two successive years at most.

A second stage starts in 1987 and finishes in 2002. It is marked by a drying trend because 55.60% of the years are considered to be dry and very dry, while rainy and very rainy years totalize no more than 26.58%. At the same time, the regional index is negative for almost 68% of this period as compared to 31.25% for positive values. The persistence of drought conditions never exceeds three consecutive years. Extreme negative values for this dry stage are recorded in 1989 (−1.68), 2000 (−1.87), and 2001 (−1.49).

The last period of this precipitation series (2003–2013), in which 82% of the years have a positive regional index, is described as rainy. On the whole, this last stage distinguishes itself through an uninterrupted sequence of 7 years (2007–2013) with positive regional indices. The percentage of rainy and very rainy years recorded at all stations (55.72%) is in net increase, while dry and very dry years, with 20.61%, are in net decrease as compared to the previous period.

3.1.3. The Tunisia Situation

Precipitation data recorded in 1970–2013 in Tunisia (Figure 2) are characterised by a significant variability previously emphasized during the study of Morocco and Algeria series. The MGTCI graphical matrix shows once more three major characteristic periods:(i)The first stage starts in 1970 and finishes in 1986 and is marked by a lack of trend. The regional index is positive for almost 53% of years and negative for the remaining 47%. The percentages calculated at all stations for wet and very wet classes show that almost 42% of years belong to this precipitation category. The dry and very dry classes represent just over 39% of years. This first period, just like that analyzed for Algeria, is characterised by an important variation of extreme years (two years, namely, 1981 with a negative index of −1.79% and 1982 with a positive index of +1.77, show this particularity). The beginning of the period tends less to this oscillation (a sequence of three rainy years is noticeable from 1971 to 1973).(ii)The second period is marked by a dry trend which lasts 16 years on this territory. The regional index shows that almost 63% of years record a negative value. The years belonging to the dry and very dry class totalize almost 55% for all stations. Rainy and very rainy years only account for 22.90%. Successive stages with the same trends exceed no more than three years (1987–1989 and 2000–2003 for negative indices and 1995–1997 for positive indices). Four years stand out through negative indices which exceed −1: 1988 (−1.36), 1989 (−1.03), 1993 (−1.46), 2000 (−1.68), and 2001 (−1.46).(iii)The last period (2003–2013) is considered wet. The regional index is positive for a period of 8 years. The return of drought conditions is noted in 2007, in 2008 (index above −0.5), and in 2012 (index equal to −1.25). Rainy and very rainy years are evaluated for this phase at almost 48%, while dry and very dry years totalize 31.31%. These numbers thus indicate an increase of years in the first class of +25% as compared to the previous period and a regress of the arid and very arid class of −23.64%.

3.2. A Net Increase of Rainy and Very Rainy Years for the Decade 2002–2011

Figure 3 was completed by starting from the decennial averages of the categories of rainy and very rainy, dry and very dry years, and normal averages, calculated for the 1982–2011 interval, confirming the previous arguments for the return of rainfall conditions in central Maghreb.(i)The analysis of the decade 1982–1991 (map shown in Figure 3(a)) shows the predominance of drought conditions in nineteen stations which record more than 50% dry and very dry years and only 7 which correspond to wet and wetter conditions for a little over a half of the duration of the period studied.(ii)The 1992–2001 period (map shown in Figure 3(b)) shows a more drastic precipitation situation with 21 stations that display rainfall accumulations from the dry to the very dry category of over 50%. At the same time, one may note the low values of rainy and very rainy years that reach 50% only at Skikda, Biskra, and Tétouan and are below this level at the remaining stations.(iii)The analysis of the map shown in Figure 3(c) for the 2002–2011 period of measurements clearly illustrates the predominance of rainy and very rainy years at most stations (25). Dry years reach 50% only at three stations (Marrakech, Sefrou, and Sfax). As for rainy years, they exceed 40% at all stations, except Dar el Beïda, Djelfa, and Gafsa.

The last map (Figure 4) displays the differences between the last two decades (1992–2001 and 2002–2011), between precipitation accumulations of rainy and very rainy years and dry and very dry years. This graphical representation allows us to see the evolution of extreme rainfall for the two already discussed periods. It reveals the importance of positive differences which indicate an increase of the former category in 83% of the stations. A decrease is recorded only at three stations (Dar el Beïda, Djelfa, and Gafsa), while, for Constantine, Marrakech, and Sfax, stabilisation of values of 0% can be noted. This map confirms a return of more favorable conditions throughout the entire territory, as the decrease of dry and very dry years is common at 29% of the stations (of which 14% show stabilisation of this category of humidity for the two decades).(i)In Tunisia, southern stations (Sfax and Tozeur) do not follow this trend and are characterised by a rise in the percentage of dry and very dry years of +20% and +10%, respectively.(ii)In Algeria, the Constantine, Annaba, and Skikda stations located in the east of the country record a rising trend of 10% for the first station and stabilisation of values for the last two.

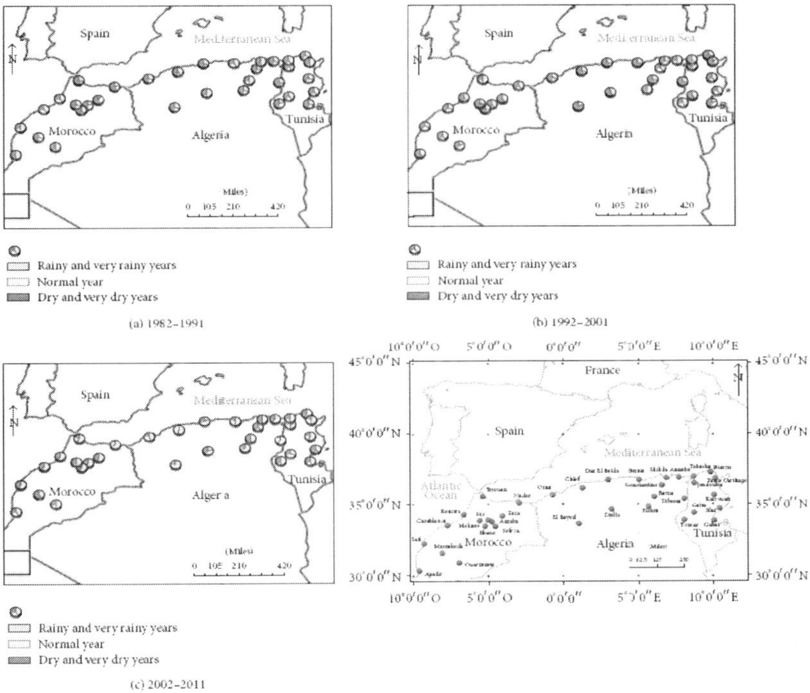

Figure 3: The evolution of precipitation accumulations of extreme years (1982–2011).

In Morocco, only one station (Marrakech) is not keeping up with the return of rainy conditions as it records a rise in dry and very dry years of 10%. On the Atlantic coast, Casablanca and Safi are the other two stations in this vast country that manifest a stable trend of extreme years regarding this last category.

The processing of the trend of extreme years from 1970 to 2013 (the percentage of stations which recorded a rainy and very rainy year, dry and very dry year or a normal one) for the three countries confirms previous results.

We can see in Figure 5(a) that, since the 2000s, the percentage developed for dry and very dry years has been clearly in constant decline in all three countries. At the same time, rainy and very rainy years show a rising trend confirmed starting with this last date (Figure 5(b)). Figure 5(c) concerning the class of normal years allows one to note that the percentage of stations, after displaying a trend of decrease from 1995 to 2000, increases starting with this last date for Algeria and Tunisia but decreases for Moroccan stations, which confirms a steady return of the wet stage in this country (considering the previously developed arguments for the other two classes).

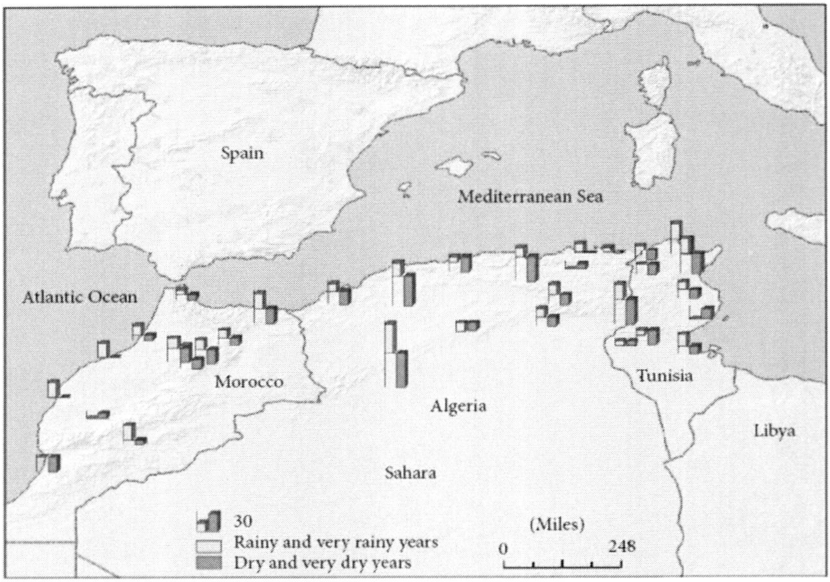

Figure 4: Differences between the last two decades (rainy and very rainy years and dry and very dry years).

The analysis of precipitation evolution in central Maghreb (Morocco, Algeria, and Tunisia) revealed a significant variability typical of the Mediterranean climate (which is stronger for the last two countries). A first wet period marks the precipitation series in Morocco (1970–1979) while, in the other two countries, from 1970 to 1986, no significant trend can be noted. This study also allowed us to highlight the durable and drastic nature of the climatic drought affecting Morocco for more than two decades (1980–2001). Indeed, precipitation deficits in these countries are intense and may persist for long periods of time (1980–1987, 1991–1994, and 1998–2001). Climatic drought recorded in Algeria and Tunisia is shorter (1987–2002) and the cycles of years of deficits not only are synchronised between the two countries but also are less extended in time and they never exceed three consecutive dry years (1987–1989, 1993-1994, and 2000–2002).

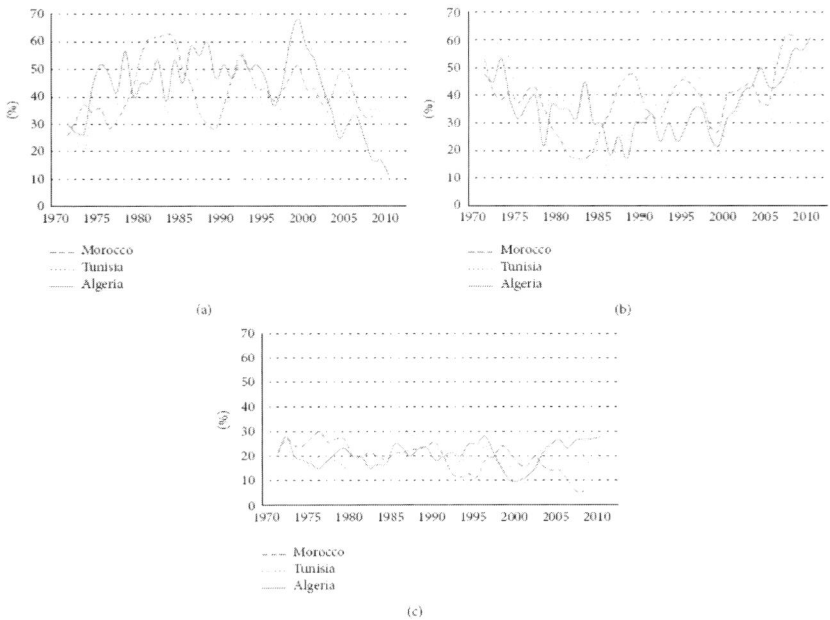

Figure 5: Mobile frequency averages of stations that recorded dry and very dry years (a), rainy and very rainy years (b), and normal years (c) (1970–2011 for Morocco and 1970–2013 for Algeria and Tunisia).

Finally, the rainfall return period is generalised in the Maghreb area under study. Starting with 2002 in Morocco and 2003 in Algeria and Tunisia, values confirm a return towards more rainy conditions, despite the persistence of extreme variability which is characterised by a brief return of dry and very dry years (2006 and 2008).

This new period marks a break from past droughty periods despite the return, once in a while, of deficient years (e.g., in 2005 and 2006 for Algeria and Tunisia and 2005, 2007, and 2011 for Morocco).

Figure 6, developed starting from the mobile average of all stations studied, sums up the main 3 stages that have affected Maghreb pluviometry during the last 44 years. The first humid stage is noticeable from 1970 to 1979, followed by the great climatic drought which lasts almost a half a century and, finally, the rainfall return which can be noted as of 2003.

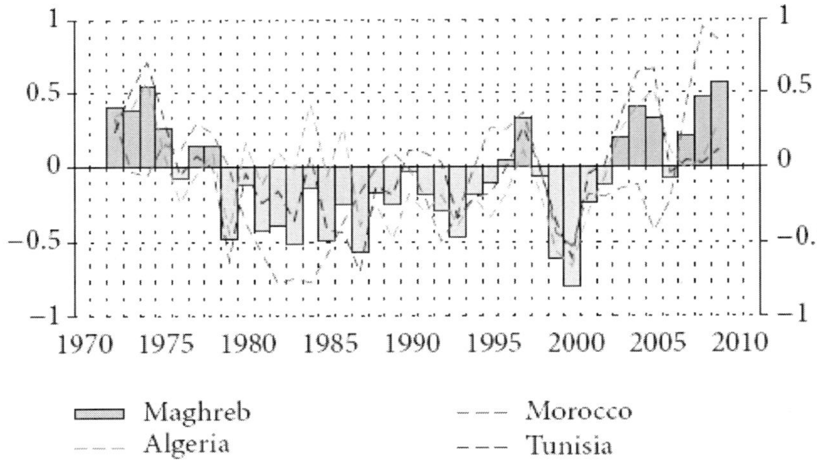

Figure 6: Reduced centered deviations of annual rainfall accumulations recorded in central Maghreb (mobile average calculated for 35 weather stations in 1970–2011).

Thus, south coast regions of western Mediterranean basin have been subjected, in the last years, to strong climate disturbances (these facts are not concordant with the results of climate patterns already cited for precipitations). The signs of change are extremely eloquent for the latest years in all this region of North Africa, as evidenced by the rainfall observed between September 2008 and September 2009 In Morocco [42]. The 2009-2010 rainfalls are also considered as exceptional and historic because they allowed for the filling of dams at their maximum capacity (the filling rate of the Al Wahda Dam, the largest of the kingdom, located in the Ouergha Wadi, was 98.5%). These rains ensured a record cereal production of 80 million quintals (for 2009/2010). This trend was confirmed for 2012/2013 because, with an average pluviometry of 450 mm, the excess rainfall on a national scale amounts to +20% as compared to a normal year [42]. At the same time, this agricultural campaign ensured a new record cereal production evaluated at 97 million quintals. For 2014, the filling of dams in operation reached nationwide a level of 72.38%, while westward, in areas that were affected by disturbances, an exceptional percentage of filling of 87% was noted. In the Moroccan South, observation data revealed a return towards more humid conditions in the Tafilet oasis [43–46] and a reinstatement of several "Foggaras" (a traditional hydraulic system affected by past climatic droughts). At times, this rainfall return occurs in extreme episodes as was the case of the recent floods (November 2014) which hit central and south-western Morocco (Agadir, Guelmin, and Marrakech areas). In the first area, rains recorded very high intensities (250 mm in 3 days from the 28th to the 30th of November 2014), which represents almost 90% of the observed average annual accumulation (280 mm).

As regards Algeria, the situation seems similar because the 2008/2009 agricultural campaign was described as highly satisfactory and the cereal production recorded the following year (2009/2010) is a record of 61.2 million quintals that has never been equalled. The hydrological situation is as satisfactory as in Morocco, judging by the Ministry of Agriculture publications: an unprecedented dam filling up which amounted to nearly 72% in 2010 and 81% in April 2013 for the 65 dams in operation and to 80.4% in March 2014, levels that had never been reached hitherto. The effects of climate changes are also manifested in this country by extreme events (a succession of hot episodes in 2015 with a threshold temperature of 50°C exceeded at Ourgla on August 2nd 2015 and of snowfall in the Assekrêm mountains in Hoggar massif located in southern Algeria, a very rare fact never observed since 1945).

In Tunisia, the situation is less obvious because three-quarters of the country have an arid and semiarid climate. This new trend is still noticeable through a rise in intense precipitation events and a net increase of the number of floods (Sfax city very well illustrates this; the city was flooded twice in 1969 and 1982 and, despite important improvement works carried out in 1984 and the past years, it was flooded again in 2009 and 2013) [47].

4. CONCLUSIONS

To conclude, we may say that the climate change observed during the last years is characterised by a rainfall return but with a far greater intensity. In Algeria, the severe disturbances (which lead to a pluviometry of over 30 mm/24 h) are on the rise in the last years (data, ONM). This again is not concordant with the IPCC conclusions which point out a decrease of disturbances over the Mediterranean space. At the same time, the temperature trend continues to be rising (despite a slight cooling of values recorded during the last years).

The variability of short-term global climate is generally associated with coupling phases of oceanic and atmospheric phenomena including El Niño Southern Oscillation (ENSO) and the North Atlantic Oscillation (NAO). While El Niño Southern Oscillation (ENSO) affects climate variability in the world, the North Atlantic Oscillation (NAO) is the climate model dominant in the North Atlantic region. The latter cyclic oscillation whose role is still under debate could explain the variability of rainfall in much of the Mediterranean area and support the hypothesis of a return of the rains marking the end of years of drought in central Maghreb

Further analysis of the response of rainfall to the North Atlantic Oscillation shows that drought seems more in line with the positive indicators of high intensity, demonstrating greater strengthening of the pressure in Azores and weakening of the Icelandic Low pressure. Under these conditions, the depressions are pushed towards northern latitudes, which promotes the establishment of a dry and mild weather on the periphery of the Mediterranean basin and North African regions.

When the signal fades or becomes negative, the pressure associated with the Azores high is lower compared to the normal value and, at the same time, the Icelandic Low is barely formed. According to this mode of traffic, the depressions corridor withdraws further south and thus affects the Mediterranean regions of the south shore which will be wetter.

This return of rainfall over central Maghreb, should it be confirmed, could mean the end of several decades of recurrent droughts and announce a durable return to "the normal." This assumption is supported by emphasizing the impact of different world climate oscillations (North Atlantic Oscillation; El Niño Southern Oscillation) for all the continents and particularly for the African continent [27, 48]. Do these facts announce a new climate phase which marks a break from past drastic conditions? Such question is difficult to answer today, but climate data and information concerning the hydraulics and agriculture of Maghreb confirm a completely unique situation in this vast area of North Africa.

Faced with such great changes that today affect the Maghreb region and given the complexity of spatial and temporal dimensions of the climatic signal, a more thorough research of causes and retroactions would allow for a better understanding of the mechanisms behind this new trend. These investigations should integrate and quantify the role of terrigenous aerosols in the dynamics of local climate and show the importance of sea surface temperatures in regulating rainfall.

ADDITIONAL POINTS

Check the following websites: WMO, World Meteorological Organisation: http:// www. wmo.int/pages/index_en.html ONM, Office National de la Météorologie (Algeria): http://www.meteo.dz/ Direction de la Météorologie Nationale (Morocco): http://www.marocmeteo.ma/ IMN, Institut de la Météorologie Nationale (Tunisia): http://www.meteo.tn

/default.html NCDC, National Climatic Data Center: http:// www.ncdc.noaa.gov/

COMPETING INTERESTS

The authors declare that they have no competing interests.

REFERENCES

1. IPCC (Intergovernmental Panel on Climate Change) GIEC, Changements climatiques en 2013, Les éléments scientifiques, résumé à l'intention des décideurs, service d'appui technique du groupe de travail I GTI, 2013, https://www.ipcc.ch/pdf/assessment-report/ar5/wg1/WG1AR5_SummaryVolume_FINAL_FRENCH.pdf
2. NOAA, National Climatic Data Center, State of the Climate, Global Analysis for Annual 2014, 2015, http://www.ncdc.noaa.gov/sotc/global/2014/13
3. WMO (Wolrd Meteorological Organization), "Hotter, drier, wetter. Face the future," Bulletin, vol. 65, no. 1, p. 64, 2016.
4. Wolrd Meteorological Organization (WMO), Statement on the Status of Global Climate in 2012, 2013.
5. M. H. I. Dore, "Climate change and changes in global precipitation patterns: what do we know?" Environment International, vol. 31, no. 8, pp. 1167–1181, 2005.
6. L. V. Alexander, X. Zhang, T. C. Peterson et al., "Global observed changes in daily climate extremes of temperature and precipitation," Journal of Geophysical Research Atmospheres, vol. 111, no. 5, Article ID D05109, pp. 1–22, 2006.
7. M. New, M. Todd, M. Hulme, and P. Jones, "Precipitation measurements and trends in the twentieth century," International Journal of Climatology, vol. 21, no. 15, pp. 1899–1922, 2001.
8. J. H. Christensen, B. Hewitson, A. Busuioc et al., "Regional climate projections," in Climate Change 2007: The Physical Sciences Basis. Contribution of Working Group I to the Fourth Assessment Report of the Intergovernmental Panel on Climate Change, S. Solomon, D. Qin, M. Manning et al., Eds., chapter 11, pp. 847–940, Cambridge University Press, Cambridge, UK, 2007, https://www.ipcc.ch/pdf/assessment-report/ar4/wg1/ar4-wg1-chapter11.pdf.
9. S. Planton, M. Déqué, H. Douville, and B. Spagnoli, "Impact du réchauffement climatique sur le cycle hydrologique," Comptes Rendus Geoscience, vol. 337, no. 1, pp. 193–202, 2005.
10. F. Giorgi, "Climate change hot-spots," Geophysical Research Letters, vol. 33, no. 8, 2006.
11. F. Giorgi and P. Lionello, "Climate change projections for the Mediterranean region," Global and Planetary Change, vol. 63, no. 2-3, pp. 90–104, 2008.
12. Blue Plan, Environment and development in Mediterranean, Climate change in the Mediterranean, no. 9, pp. 4, 2008
13. E. Xoplakie, Climate variability over the mediterranean [Ph.D. thesis], University of Bern, Bern, Switzerland, 2002.
14. P. Lionello, J. Bhend, A. Buzzi et al., "Cyclones in the Mediterranean region: climatology and effects on the environment," in Mediterranean Climate Variability, P. Lionello, P. Malanotte-Rizzoli, and R. Abd Boscolo, Eds., pp. 324–372, Elsevier, Amsterdam, The Netherlands, 2006.

15. F. Jin, A. Kitoh, and P. Alpert, "Water cycle changes over the Mediterranean: a comparison study of a super-high-resolution global model with CMIP3," Philosophical Transactions of the Royal Society A: Mathematical, Physical and Engineering Sciences, vol. 368, no. 1931, pp. 5137–5149, 2010.
16. C. C. Raible, B. Ziv, H. Saaroni, and M. Wild, "Winter synoptic-scale variability over the Mediterranean Basin under future climate conditions as simulated by the ECHAM5," Climate Dynamics, vol. 35, no. 2-3, pp. 473–488, 2010.
17. S. O. Krichak, J. S. Breitgand, R. Samuels, and P. Alpert, "A double-resolution transient RCM climate change simulation experiment for near-coastal eastern zone of the Eastern Mediterranean region," Theoretical and Applied Climatology, vol. 103, no. 1, pp. 167–195, 2011.
18. C. M. Philandras, P. T. Nastos, J. Kapsomenakis, K. C. Douvis, G. Tselioudis, and C. S. Zerefos, "Long term precipitation trends and variability within the Mediterranean region," Natural Hazards and Earth System Sciences, vol. 11, no. 12, pp. 3235–3250, 2011.
19. M. Giannakopoulos, M. Bindi, M. Moriondo, and T. Tin, "Climate change impacts in the Mediterranean resulting from a 2°C global temperature rise," A Report for WWF, http://www.fao.org/fileadmin/user_upload/rome2007/docs/Climate_Change_Ada ptation_Water_Sector_NENA.pdf
20. G. Tselioudis, C. Zerefos, P. Zanis, P. Repapis, and I. Kapespmenakis, "Future trends in Mediterranean precipitation and possible connections with the phase pf the North Atlantic Oscillation," in Proceedings of the 9th Conference of Meteorology, Climatology and Atmospheric Physics, pp. 513–520, Thessaloniki, Greece, May 2008.
21. M. Dubrovský, M. Hayes, P. Duce, M. Trnka, M. Svoboda, and P. Zara, "Multi-GCM projections of future drought and climate variability indicators for the Mediterranean region," Regional Environmental Change, vol. 14, no. 5, pp. 1907–1919, 2014.
22. P. Alpert, T. Ben-Gai, A. Baharad et al., "The paradoxical increase of Mediterranean extreme daily rainfall in spite of decrease in total values," Geophysical Research Letters, vol. 29, no. 11, pp. 1–31, 2002.
23. WMO (Wolrd Meteorological Organization), "Statement on the status of global climate in 2012," Tech. Rep. 1108, 2013.
24. Z. Nouaceur, B. Laignel, and I. Turki, "Changements climatiques au Maghreb: vers des conditions plus humides et plus chaudes sur le littoral algérien ?" Physio-Géo, vol. 7, pp. 307–323, 2013.
25. Z. Nouaceur, B. Laignel, and I. Turki, "Changement climatique en Afrique du Nord: vers des conditions plus chaudes et plus humides," in dans le Moyen Atlas Marocain et Ses Marges Actes du XXVII Colloque International de Climatologie, pp. 399–405, Dijon, France, 2014.
26. M. G. Donat, T. C. Peterson, M. Brunet et al., "Changes in extreme temperature and precipitation in the Arab region: long-term trends and variability related to ENSO and NAO," International Journal of Climatology, vol. 34, no. 3, pp. 581–592, 2014.
27. A. Sebbar, W. Badri, H. Fougrach, M. Hsain, and A. Saloui, "Étude de la variabilité du régime pluviométrique au Maroc septentrional (1935–2004)," Sécheresse, vol. 22, no. 3, pp. 139–148, 2011.
28. Z. Nouaceur, "Évaluation des changements climatiques au Maghreb, Étude du cas des régions du quart nord-est algérien," in Proceedings of the Actes du 23rd Colloque de l'Association Internationale de Climatologie, 'Risques et Changements Climatiques', pp. 463–468, Rennes, France, 2010.
29. M. Amyay, Z. Nouaceur, A. Tribak, Kh. Okba, and A. Taous, "Caractérisation des évènements pluviométriques extrêmes dans le Moyen Atlas marocain et ses marges," in Actes du XXV ème Colloque International de Climatologie, pp. 75–80, Grenoble, France, 2012.
30. K. Khomsi, G. Mahe, Y. Tramblay, M. Sinan, and M. Snoussi, "Trends in rainfall and temperature extremes in Morocco," Natural Hazards and Earth System Sciences Discussions, vol. 3, no. 2, pp. 1175–1201, 2015.

31. M. Kortli, Changement climatique, eau et santé en Tunisie [Mémoire de Master], Institut National Agronomique de Tunisie, INRGREF, Tunis, 2012.
32. B. Laignel, Z. Nouaceur, H. Jemai, H. Abida, M. Ellouze, and I. Turki, "Vers un retour des pluies dans le nord-est tunisien?" in Proceedings of the Actes du XXVIIe Colloque de l'Association Internationale de Climatologie-2 au 5 Juillet 2014, pp. 727–732, Dijon, France, 2014.
33. N. Fehri, "L'aggravation du risque d'inondation en Tunisie: éléments de réflexion," Physio-Géo, vol. 8, pp. 149–175, 2014.
34. L. Henia and Z. Hlaoui, "La pluviométrie dans les plaines et bassins intratelliens en Tunisie: évolution récente et projection dans le futur," in Climat Montagnard et Risques, Actes du XXIVème Colloque de l'AIC (Rovereto), M. Fazzini and G. Beltrando, Eds., pp. 303–308, AIC et Université de Ferrare, 2011.
35. S. Jebari, R. R. Berndtsson, F. Lebdi, and A. Akissa Bahri, "Sediment, discharge and precipitation variation in the wadi Mellegue catchment during the 50 past years, Actes des dixièmes Journées Scientifiques de l'INRGREF, Hammamet 21-22 novembre 2007 Exploitation des ressources en eau pentru une agriculture durable," Annales de l'INRGREF, vol. 11, pp. 116–122, 2008.
36. C. Norrant, Tendances pluviométriques indicatrices d'un changement climatique dans le bassin méditerranéen de 1950 à 2000. Étude diagnostique [Ph.D. thesis], Université Aix Marseille I, Université de Provence, Marseille, France, 2004.
37. C. Norrant-Romand and A. Douguédroit, "Significant rainfall decreases and variations of the atmospheric circulation in the Mediterranean (1950–2000)," Regional Environmental Change, vol. 14, no. 5, pp. 1725–1741, 2014.
38. E. Morin, "To know what we cannot know: global mapping of minimal detectable absolute trends in annual precipitation," Water Resources Research, vol. 47, no. 7, Article ID W07505, 2011.
39. Z. Nouaceur, B. Laignel, and I. Turki, "Changement climatique en Afrique du Nord : vers des conditions plus chaudes et plus humides, dans le Moyen Atlas Marocain et ses marges," in Proceedings of the Actes du XXVII Colloque International de Climatologie, pp. 399–405, Dijon, France, 2014.
40. Z. Nouaceur, B. Laignel, and I. Turki, "Changement climatique au Sahel: des conditions plus chaudes et plus humides en Mauritanie?" Sécheresse, vol. 24, pp. 85–95, 2013.
41. Z. Nouaceur and B. Laignel, "Caractérisation des évènements pluviométriques extrêmes sur la rive Sud du bassin méditerranéen: étude du cas du quart nord-est algérien," in Proceedings of the Actes du XXVIII Colloque International de Climatologie, pp. 573–578, Liège, Belgium, 2015.
42. Ministry of Energy, Mines, Water and Environment, Ministère de l'Energie, des Mines, de l'Eau et de l'Approvisionnement, 2008-2009, 2009-2010, 2010-2011, 2011-2012, 201-2013. Situation des ressources en eau et du remplissage des barrages durant l'année hydrologique, Direction de la Recherche et de la Planification de l'eau, 9 p (Maroc, Morocco), http://www.mem.gov.ma/
43. R. Thierry, Au fond de la khettara Lahloua, Le journal de l'IRD, IRD, Institut de recherche pour le développement, juin/juillet/août, no. 40, pp. 16, 2007
44. M. El Faiz and T. Ruf, "An introduction to the khettara in morocco: two contrasting cases," in Proceedings of the 1st WATARID International Conference on Water, Ecosystems and Sustainable Development in Arid and Semi-Arid Areas (WATARID '06), G. Schneier-Madanes and M.-F. Courel, Eds., pp. 151–163, Springer, Ürümqi, China, October 2006.
45. Institut de Recherche pour le Développement (IRD), "Les réseaux d'eau anciens ressuscitent en Méditerranée," Actualité Scientifique 370, 2011.
46. M. Mahdane, S. Lanau, Th. Ruf, and M.-J. Valony, Gouvernance d'une oasis dont l'eau provient quasi exclusivement de l'exploitation traditionnelle des galeries drainantes, les khettaras: le cas de SKOURA au Maroc, Atelier international à Tunis 4–6, Novembre 2010

47. A. Daoud, "Retour d'expérience sur les inondations dans l'agglomération de Sfax (Tunisie méridionale) de 1982 à 2009: de la prévention à la territorialisation du risque," Revue Géographique de l'Est, vol. 53, no. 1-2, pp. 1–14, 2013.
48. J. I. López-Moreno, S. M. Vicente-Serrano, E. Morán-Tejeda, J. Lorenzo-Lacruz, A. Kenawy, and M. Beniston, "Effects of the North Atlantic Oscillation (NAO) on combined temperature and precipitation winter modes in the Mediterranean mountains: observed relationships and projections for the 21st century," Global and Planetary Change, vol. 77, no. 1-2, pp. 62–76, 2011.

CHAPTER 3

An Analysis of the Spring-to-Summer Transition in the West Central Plains for Application to Long Range Forecasting

Rosalie G. Newberry[1], Anthony R. Lupo[2*], Andrew D. Jensen[3], R. Antonio Rodriges Zalipynis[4]

[1]KXAN-TV, Austin, TX/KOMU-TV, Columbia, MO, USA

[2]Department of Soil, Environmental, and Atmospheric Science, University of Missouri, Columbia, MO, USA

[3]Department of Mathematics and Meteorology, Northland College, Ashland, WI, USA

[4]Department of Software Engineering, Faculty of Computer Science, National

ABSTRACT

The spring-to-summer transition is of special importance in long range forecasting, as the general circulation transitions to a less energetic regime. This affects the Midwestern United States in a profound way, since agriculture is very sensitive to the variability of weather and climate. Beginning at the local scale, surface temperature observations are used from a representative station in the West Central Missouri Plains region in order to identify the shift from late spring to early summer. Using upper-air re-analyses as a supplement, the 500-mb height observations are examined to find a spring-to-summer transition date by tracking the location of a representative contour. Each of these is used to identify spring-to-summer transition date and then statistical analysis is performed on this long-term data set. Finally, teleconnections, specifically the influence of El Niño Southern Oscillation (ENSO) and Pacific Decadal Oscillation (PDO), and blocking are examined in order to quantify interannual variability. It was found that examining these criteria, developed in an earlier study that covered a much shorter time period, produced similar statistics to this 68-year study of spring-to-summer transitions. It was also found that the onset of La Niña was associated with hotter summers in the region, a result first found in the earlier study, but this association was much stronger here.

Keywords: Interannual Variability, Summer Season Transitions, El Nino, Pacific Decadal Oscillation

1. INTRODUCTION

The West Central Plains, including Missouri, is a region that exhibits strong seasonality, and interannual and inter decadal variability. The transition between the cold and warm seasons is often difficult to forecast, and it also is indicative of a change to a less energetic weather regime [1] [2] . Seasonal transition is recognized as a problem in dynamic meteorology, but research on the Northern Hemisphere spring to summer transition is noted in fewer than ten academic research papers ([2] , and references therein). It was noted, however, that more than 9000 articles appear in a search for "seasonal transitions" in the American Meteorological Society's online journals (http://www.journals.ametsoc.org). There are, however, other types of seasonal regime transitions that have been studied and are important from synoptic or storm-property point of view as well [3] - [5] .

Recently, a diagnostic method based on the stability of the upper air flow as measured using enstrophy diagnostics [6] was shown to successfully identify the transition between high and low amplitude hemispheric-wide flow regimes [1] using data from as few as two 12-h time periods. The study of [1] used a quantity called the Wave Amplitude Index in order to identify these transitions using time series. Then, [2] used the Wave Amplitude Index to show that the spring to summer transition could be more gradual or quite abrupt in the west-central plains region. It is important to extend our knowledge in identifying these transitions, in order to improve the capability of long-range forecasting, which will in turn impact different sectors of Missouri's culture and economy.

The detection of the transition between spring and summer can be accompanied by a change in daily temperature and precipitation patterns. Spring is known to be mild and it is the season with the greatest total precipitation in the Midwest region, with the peak in total precipitation occurring in most places during May. The onset of the summer pattern, which is sometimes abrupt, is associated with a period of consistently high temperature, high humidity and dry periods interjected with fewer episodes of convective precipitation [2] [7] . Also, the 500 hPa geopotential height contours can be used as a measure of the amount of potential or in the case of kinetic energy a surrogate for energy in the atmosphere (e.g., [1] [2] [8]). By examining a change in the 500-mb height field over an extended period of time, periods of atmospheric excitation and relative mean energy states can be used as seasonal markers [8] .

Two important teleconnections which have an influence on the seasonal conditions in the study region are El Niño Southern Oscillation (ENSO) and the Pacific Decadal Oscillation (PDO). Their phases and interactions have been shown to have unique impacts on regional temperature and precipitation regimes for the seasonal scale (e.g., [9] - [13]). Teleconnections, or the relationship of general circulation and climate anomalies at distances of thousands of kilometers, were first noted by Sir Gilbert Walker in the early twentieth century [14] . Recently, [15] looked at significant rainfall patterns in South America as a function of ENSO mode, noting that different ENSO phases were associated with a quasi-symmetric rainfall pattern. Also, [16] argued for the expansion of the idea of ENSO teleconnections into the Intra-Americas Seas, i.e. the Gulf of Mexico and the Caribbean Sea, citing that atmospheric ENSO patterns affect the two bodies of water in opposite ways. This is especially true for tropical cyclones (e.g. [17] [18]). Also, many researchers have focused on the impact ENSO has on the yearly monsoons, including Australian and Asian monsoons (e.g. [19] [20]).

Some applicable studies to this research, though, examined ENSO effects for North America. The primary mechanism for the influence of ENSO on the weather of North America is via its impact on the typical configuration of the Pacific—North American area jet stream. For example, [21] and [22] were among the first to examine the impact of ENSO over North America and the rest of the globe. Also, [23] examined land falling hurricanes in the US with respect to ENSO. Then, [24] utilized National Centers for Environmental Prediction/ National Center for Atmospheric Research (NCEP/NCAR) reanalysis data to demonstrate the connection between strong ENSO phases and upper-air data in North America, including 500-mb geopotential heights and 850-mb specific humidity. Additionally, [25] proposed that ENSO impacts may be bi-modal based on the primary ENSO genesis region. Specifically, [25] found that the ENSO impact on North

America differs for an event that is an eastern Pacific versus a central Pacific ENSO event. Also, other researchers have coupled the North American ENSO effects with the impacts of PDO. Also, [26] examined dry and wet conditions in the U.S. Great Plains, demonstrating that ENSO and PDO coupling can intensify ENSO phases. Then [12] also found this impact, but they concentrated specifically on the Midwest to outline the regional effects of ENSO/PDO coupling. Another [27] investigated PDO and El Nino interactions and their impact on the temperature and precipitation regimes in the Amur River basin of far southeast Russia, including summer season transitions similar to [2] .

The purpose of this study is based on the framework of the research published by [2] on the spring to summer transition in the West Central Plains and their interannual variability in the temperature regime. Their study, however, examined only 20 year period from 1981-2000 and focused on ENSO variability. This study [2] investigated daily temperature and precipitation patterns using data from Jefferson City, MO and also used nine cooperative National Weather Service sites as a quality control. They [2] also utilized 500 hPa heights information in two ways; first to identify a minimum height requirement for the spring-to-summer transition date, and second as a surrogate for energy in the atmosphere using the [1] wave amplitude index. This research will examine interannual variability the temperatures of the region and the spring-to-summer transition over a longer time period than [2] , and include an analysis of decadal variability with respect to the PDO and interactions with ENSO modes. While the relationship with ENSO seems to be relatively weak, a stronger connection is found in relation to the transition of ENSO phase [28] .

2. DATA AND METHODS

2.1. Data

The study of [2] examined spring-to-summer transitions using observational data over the period 1981-2000. This work expands the same data set, and the time period for this research is 1948 to 2015. The 500 hPa heights are archived at the National Centers for Environmental Prediction/National Center for Atmospheric Research (http://www.esrl.noaa.gov/psd/data/reanalysis) [29] . These are a netCDF-based data set that can be used to display assimilated data from 1948 to the present. Also, [2] identified twenty summer transition dates. The mean, median and mode all fell in June; and only one date was selected in the months of May and July each (21 May, 1991 and 01 July, 2000). Then [2] examined temperatures and precipitation from April through September, their result allows this work to focus on the May through July period.

The statistical tests used in this study are found in standard statistical textbooks (e.g., [30] [31]). In order to test long-term linear trends, the analysis of variance (ANOVA) technique was used, which involves the F-test. A standard t-test was also used to test correlations for significance. Mean temperatures were tested in order to analyze and compare ENSO, and PDO related variability. A two-tailed standardized test statistic (z*) was the technique used for the comparison of the sample means. Means for the total time series studied served as the expected frequencies of occurrence. All statistical tests assumed the null hypothesis, or that no a priori relationship was present among the variables tested. Confidence levels of 90%, 95%, and 99% were all tested.

2.2. Location

The previous study [2] chose Jefferson City, Missouri as the principal site for data collection, and used this as the representative site of the entire East-Central Missouri Ozarks (Figure 1). Utilizing principal component analysis from [32] - [34] , and [2] showed that focusing on Jefferson City is sufficient to make conclusions about interannual variability in the East-Central Missouri Ozarks region as a whole. Thus following those works, this study uses Jefferson City as the representative data site, but for a slightly different region. Here we use the NWS climate divisions including the Northwest Prairie, Northeast Prairie, West Central Plains, West Ozarks, East Ozarks and the Bootheel (Figure 2), a similar region to [2] . Additionally, it was originally recognized that these divisions were categorized into subgroups that possessed similar climatological and geographical characteristics (Climatological Data: Missouri, 2013. http://www.ncdc.noaa.gov). This study includes the Northeast Prairie, the Eastern Ozarks, and the part of the West Central Plains. Lastly, the Jefferson City reporting site moved three times during the study period. Nonetheless, every Jefferson City site is less than one degree of longitude and latitude from another. The Jefferson City Water Plant is the station current location and the location with the most longevity, encompassing the bulk of data used here. As in [12] there is no systematic impact from these moves.

2.3. ENSO and PDO

El Niño Southern Oscillation (ENSO) and Pacific Decadal Oscillation (PDO) are the two main teleconnections considered for this research [12] , as these have the strongest correlation with regional weather [35] . Other teleconnections such as the North Atlantic Oscillation and Arctic Oscillation did not correlate as strongly. The Japan Meteorological Agency (JMA) ENSO index is available through the Center for Ocean and Atmospheric Prediction Studies (COAPS) from 1868 to present (JMA sea surface temperature (SST) ENSO Index http://www.coaps.fsu.edu/jma). The JMA SST ENSO Index has been widely used in other published works (e.g., [23] , [36] , and [37] are some recent examples), and is also used here (Table 1). Also, [38] found that, however, while the JMA index is more sensitive to La Niña events than other definitions, it is less sensitive than other indices to El Niño events. The JMA classifies ENSO phases using SST within the bounded region of 4°S to 4°N, 150°W to 90°W. The JMA defines the inception of an ENSO year as 1 October, and its conclusion on 30 September of the next year.

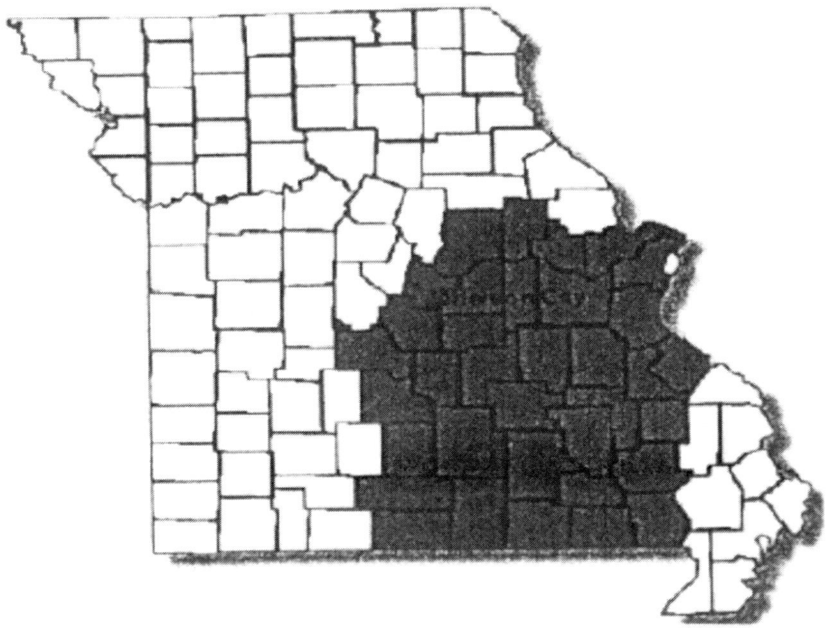

Figure 1. From Ratley et al. (2002), the location of the study region including Jefferson City, Missouri. This area is bounded by the edge of the East Ozarks to the west, the Mississippi River to the east, the Missouri River to the north and the Missouri/Arkansas border to the south.

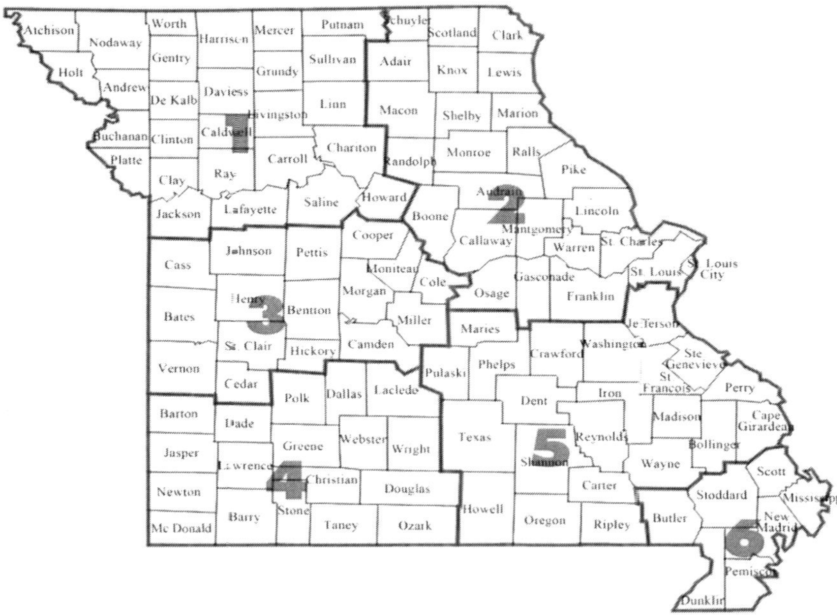

Figure 2. United States Department of Commerce Weather Bureau's Climatological Data: Missouri Section, Divisions, 1957 to Present, image courtesy of the Climate Prediction Center (CPC).

The Pacific Decadal Oscillation (PDO) positive and negative modes are catalogued also by the Center for Ocean-Atmospheric Prediction Studies (COAPS). The most important effect of PDO is how it interacts with ENSO during certain phases to create an enhanced effect on temperatures and precipitation variability (e.g., [23] [26]). The characteristics of these modes are less pronounced than those for ENSO due to the fifty- to seventy-year cycle of PDO (e.g., [39] [40]). COAPS describes the PDO phases (Table 2) as follows: the High (+) PDO phase is characterized by cold SSTs in the north central and western Pacific Ocean, warm SSTs off the western coast of North America, and a deep Aleutian low, while the Low (−) phase of the PDO is characterized by warm SSTs in the north central and western Pacific Ocean, cool SSTs off the western coast of North America, and no pronounced Aleutian low. The start date of the current PDO cycle from COAPS agrees with the independent analysis of [7] , based on the same SST patterns.

Table 1. Center for Ocean-Atmospheric Prediction Studies Japan Meteorological Agency El Niño Southern Oscillation Index, 1948 to present. Modes are El Niño (EL), La Niña (LA) and Neutral (NEU).

Year	Classification	Year	Classification	Year	Classification
1948	NEU	1970	LA	1992	NEU
1949	LA	1971	LA	1993	NEU
1950	NEU	1972	EL	1994	NEU
1951	EL	1973	LA	1995	NEU
1952	NEU	1974	LA	1996	NEU
1953	NEU	1975	LA	1997	EL
1954	LA	1976	EL	1998	LA
1955	LA	1977	NEU	1999	LA
1956	LA	1978	NEU	2000	NEU
1957	EL	1979	NEU	2001	NEU
1958	NEU	1980	NEU	2002	EL
1959	NEU	1981	NEU	2003	NEU
1960	NEU	1982	EL	2004	NEU
1961	NEU	1983	NEU	2005	NEU
1962	NEU	1984	NEU	2006	EL
1963	EL	1985	NEU	2007	LA
1964	LA	1986	EL	2008	NEU
1965	EL	1987	EL	2009	EL
1966	NEU	1988	LA	2010	LA
1967	LA	1989	NEU	2011	NEU
1968	NEU	1990	NEU	2012	NEU
1969	EL	1991	EL	2013	NEU

Table 2. Center for Ocean-Atmospheric Prediction Studies Pacific Decadal Oscillation Index, 1910 to present. Modes are high (positive) and low (negative).

Year Range	Mode
1910-1924	-PDO
1925-1946	+PDO
1947-1976	-PDO
1977-1998	+PDO
1999-2014	-PDO

2.4. Ratley et al. 2002 Temperature Criterion

We used the criterion of [2] , which is set in order to filter out synoptic-scale perturbations. The thresholds for surface temperatures were set at fifteen days. The criterion used by [2] for identifying the spring to summer transition date using 500 hPa heights was the start date for the first period of at least 10 consecutive days where the heights are in excess of 5820 m over the entire study region. The 5820 m contour is frequently used as an indicator of the difference between the baroclinicity of the mid-latitudes and the quasi-barotropic nature of the subtropics [2] . This height minimum is typically the trailing equatorward edge of tightly-packed mid-latitude height gradients, as also demonstrated in a derecho study by [41] , and in study of summer perturbations by [42] . Additionally, since the 5820 height contour is at the equatorward edge of the 500 hPa height gradients, they frequently coincide with gradients of potential temperatures near 350 K on the 2.0 Potential Vorticity Unit (PVU) (e.g, [43]) in identifying the general location of the jet stream. Potential vorticity is conserved on surfaces of potential temperature, but the above argument is not strictly true as potential temperature or potential vorticity will not be strictly conserved on a height surface (e.g., [44]). Nonetheless, many papers have demonstrated the general utility of potential vorticity quantities in defining features in the jetstream. Thus we are confident that this choice of contour has dynamic relevance to the study here.

The temperature criteria for a summer start date are as follows from [2] , and using a 15-day window for surface based data is sufficient for removing smaller-scale impacts [45] :

1) The first date of a fifteen-day consecutive period where the mean temperature exceeds 70.0°F (21.1°C), and at least ten of those fifteen days have a temperature at, or exceeding, 75.0°F (23.9°C).

2) The first date of a fifteen-day consecutive period where the maximum temperature exceeds 77.0°F (25.0°C) and at least ten of those fifteen days have a temperature at, or exceeding, 82°F (27.8°C).

Temperatures are reported in the National Climatic Data Center in degrees Fahrenheit, and then converted to Celsius.

2.5. Blocking Criterion and Integrated Regional Enstrophy (IRE)

In order to identify blocking events, we used the criterion defined in [28] and references therein. In this work, the characteristics of blocking upstream in the Pacific Region (140°E - 100°W) are examined in order to determine what role the occurrence of blocking might play in the configuration of the jet stream over the region, in a manner similar to [28] . Here we examine spring and summer blocking (as in [28]) in order to determine if the spring season provides precursory indication of the summer season's activity and thus the seasonal conditions in the study region. All the events used for this study can be found at the University of Missouri's blocking archive (http://weather.missouri.edu/gcc). Blocking events are available from 1969-2014. Studies going back to the early 1980's (e.g., [46] [47]) show the relevance of the conditions upstream, including the flow regime to the conditions in the area of interest. The character of the Pacific Region flow and blocking correlates strongly to weather conditions in the study region (e.g., [35]).

The study of [1] used the wave amplitude index to quantify the kinetic energy in the northern hemisphere large-scale flow regime. Also, [2] demonstrated that the spring-to-summer transition was associated with a decrease in the wave-amplitude index. Here we used the integrated regional entrophy (IRE) to represent the large- scale flow regime. The study of [38] proposed that the stability of the Northern Hemisphere flow was approximately equal to the enstropy integrated over the Northern Hemisphere in a barotropic and frictionless flow. Then [6] (and references therein) demonstrated the utility of the integrated regional estrophy (IRE) diagnostic (based on [48]) to study the relationship of flow regime stability to the onset and decay of blocking. In a geostrophic flow velocity and vorticity are related to the gradient and laplacian of the height field, respectively. For the purpose of this discussion, then kinetic energy and enstrophy are the square of these quantities, and these quantities should be smaller overall when the weaker, more zonal Northern Hemisphere summer season flow regime [49] becomes established.

3. CLIMATOLOGICAL ANALYSIS

3.1. Spring to Summer Transitions in the Study Region

The results of this study produced a transition date using each criterion, where data was available, and a final spring-to-summer transition date based on these three criterions for the 68 year period (1948-2015). There were a few years before 1980 in which Jefferson City temperature data were absent, including 1967. The 500 hPa height criterion provided a date for every year except 1967. In order to provide a possible date for that year, Columbia Regional airport data were used to provide continuity for some of the analyses, and the date identified was June 3 and 4 for each temperature criterion. It was also noted that the 500 hPa height criterion beginning with June 4 was only met for seven consecutive days, and at nine of the ten day period (only one time did not). During the summer as a whole, the longest consecutive period for meeting the criterion was eight days. Thus, a transition date of 4 June for 1967 seems quite plausible based on data used for this analysis.

An examination of Table 3 shows that the mean spring and summer temperatures and their standard deviations for the entire period, and these are close to the current 30 year means and standard deviations. The trend for summer temperatures would display a cooling trend using linear regression, but this trend was not statistically significant, a result identical to those of [13] . The spring season, however, Figure 3(a) showed a significant warming (at the 90% confidence level), and this resulted in a statistically significant decrease in the spring to summer temperature difference (at the 90%). This result is consistent with observations from other parts of the earth, and is consistent with the prevailing idea that, in a warmer world, the cold season will warm at a faster rate [50] .

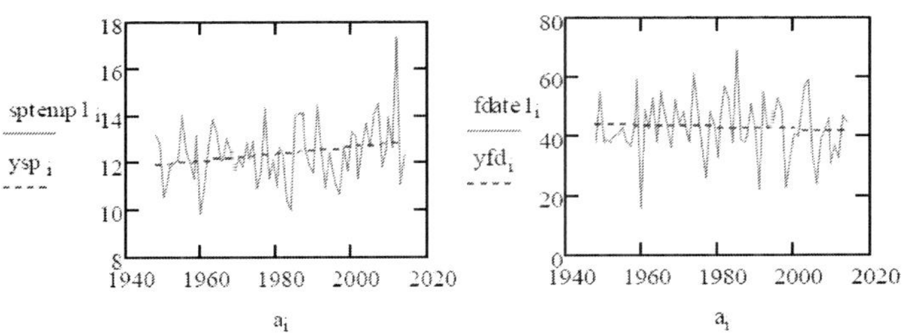

(a) (b)

Figure 3. (a) Mean spring season temperature (°C) versus time (years) and (b) the final date determined (days from 1 May) for each year. The blue dotted line is the linear regression line.

Table 3. Statistics for the three criterion used to identify the spring-summer transition date, and temperature data for the spring and summer seasons and their differences from 1948-2014.

Variable	Mean	Standard deviation
500 hPa criterion (days)	51-June 20	13.4
Mean temperature criterion (days)	42-June 11	9.2
Maximum temperature criterion (days)	38-June 7	14.1
Final date (days)	43-June 12	9.7
Summer temperature (°C)	24.3	1.1
Spring temperature (°C)	12.4	1.3
Difference (°C)	11.9	0.4

The average transition date for the entire data set, 1948-2015, is 12 June, not including 1967. The median transition date is also 12 June, and the median values for all variables in Table 3 were the same when the means were rounded. June was, by far, the most common month for final transition dates, accounting for 85.3% of all chosen values, and 78.0% of all transitions occurred within one standard deviation of the mean. Eight transitions occurred in May. These characteristics imply that the data set is close to a normal distribution, but one that is more peaked (kurtosis). Since the 1967 transition date was identified as 3 June, including it in the analysis does not change the overall statistics.

Concurrent with the decreased temperature differences between spring and summer, the trend for the final summer onset day is indicated to be occurring earlier (Figure 3), a trend that was significant at the 99% level for the 500 hPa data, and the mean temperature criterion data. The trend for the decreased temperature difference was significant at the at the 90% confidence level. Only the maximum criterion trended toward a later date, and this was significant at the 95% confidence level. These contradictory trends are possible since the summer minima trended higher, but also the dew point trend was significantly higher [13] . A higher dew point correlates with higher minima, and this is also consistent with the prevailing character of a warming climate [50] .

A further examination of the individual criterion demonstrates that the daily mean temperature and the final date chosen were similar statistically (Table 3) and the correlation between the two time series was 0.73, a value significant at the 99%. This was by far the strongest association in this data when cross correlating each time series. The 500 hPa and maximum temperature criterion were similar in their variation (standard deviation was 14 days), but the maximum temperature produced the earliest onset date and the 500 hPa the latest. While the 500 hPa criterion correlated with the final date at the 95% confidence level (0.28), the maximum criterion did not correlate with any other criterion at a statistically significant level. The 500 hPa height, the mean temperature criterion, and the final date chosen correlated positively with the summer mean temperature as might be expected, and these correlations were significant at the 99%, 90%, and 99% confidence level, respectively. There was only one other statistically significant relationship between these criterion and temperatures, and that was between the mean temperature criterion and the spring mean temperatures.

At the beginning of this work, it was proposed that the 500-mb height data would be the most useful criterion, as upper-air analyses are not subjected to surface observation errors and influences. However, there are errors inherent in the 500-mb data as well, and the years before the satellite era may possess larger error (e.g., [51]). Based on the statistical analysis above, an argument could be made that the mean temperature criterion would be the most useful for identifying the spring-summer transition date in the study region.

A comparison of this work and [2] (Table 4) shows a similar pattern to the statistics for the larger data set in that the 500 hPa criterion produced the latest date while the maximum temperature date was the earliest. Also, the mean temperature criterion and final date were within one day of each other in both studies. The study of [2] found that for 12 of the years (60%) of their 1981-2000 dataset, all three summer onset dates fell within a 10-day moving window. In the longer data set here, only 20 years (33%) were characterized in the same manner. In order to determine why this difference was so large we sampled the same 20 years as in [2] using our criterion. Here we found 11 years (55%) fit the same criterion, or a difference of only one year.

An examination of the two studies shows that a slightly different methodology was used to determine the onset date. The rules employed by [2] specified that the final date chosen must match one of the other criteria, which is more arbitrary. Here, we used similar rules except that the final date was the average date provided by the available criterion for that year, and thus, did not have to match any of the three criteria. The average of the two analyses gave final dates that were an average of 4 days apart. In this study of the period of [2] , five transition days (25%) fell within twenty-four hours of all criterion. Thirteen transition days (65%) fell within five days of one another, still within the time frame of the synoptic period. Seventeen days (85%) were within ten days, or planetary-scale time frame. The outcomes here chose an earlier transition date 45% of the time, and a later onset date 50% of the time, thus overall the change in assigning the final date did not result in large difference between the two studies.

Table 4. As in Table 3 except for relevant values from Ratley et al. (2002).

Variable	Mean	Standard deviation
500 hPa criterion (days)	48-June 17	12.0
Mean temperature criterion (days)	47-June 16	13.0
Maximum temperature criterion (days)	42-June 11	12.0
Final date (days)	43-June 15	10

3.2. Pacific Region Blocking

There has been speculation that frequency of Northern Hemisphere blocking has increased during the last two decades. Comparing our results to those of [52] (see Table 5 and Figure 4), indicate there has been a very sharp increases in the frequency and number of blocking events and blocking days in the Pacific region (140°E - 100°W) spring (significant at the 99% confidence level) and summer (significant at the 95% confidence level) seasons when the study is extended to the current time. There was, however, no significant change in the intensity of blocking in this region. The study of [52] covered the period 1970-2000, thus the sharp increase in blocking has occurred since this time. Also, [11] associated more and stronger Pacific Region summer blocking to cooler, and wetter summers in the region. Here we find a strong correlation between the occurrence of spring and summer blocking and regional summer temperatures (more blocking, cooler conditions for the study region), but not significant at the 90% and 95% confidence level as in [35] . This is especially true if only blocking east of dateline is included.

This and many other studies, even those going back to the 1980s (e.g., [53] [54]) show that the occurrence of Pacific Region blocking, especially in the East Pacific is associated with cooler winter temperatures in the region. This same correlation appears to be true for the summer season as well [11] . The occurrence of blocking, however, did not correlate with any of the onset criterion here for the final day chosen. This may be expected since the occurrence of blocking is quasi-random. However, the association of blocking with the character of seasonal temperature and precipitation characteristics is well-documented. Thus, it would seem that the influence of blocking may be more indirect and this issue will be explored versus interannual variations.

3.3. Synoptic Climatology and Dynamic Analysis

In order to examine the effectiveness of this criterion in identifying a date for the spring-summer transition, the 850 hPa and 500 hPa heights were used to examine the lower (Figure 5) and mid-tropospheric (Figure 6) air flow for the 30 day period before and after the nominal transition date in order to identify whether a difference existed. Five particular years were chosen in order to represent an early onset (2 June 2012) a late onset (28 June 2004), one close to the mean (11 June 2008), and two of particular interest (4 June, 1967—the date inferred above in section 3a for the year that lacked a criterion, and 17 June, 1971—Midwestern drought). Table 6 shows the summer mean temperature and total precipitation and their relative anomalies for each year relative to the 1981-2010 anomalies.

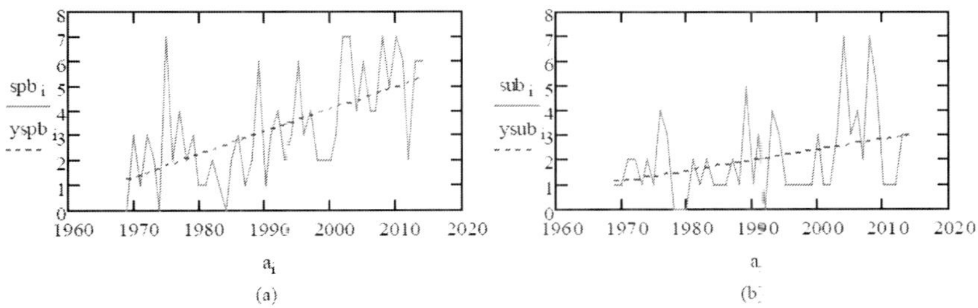

Figure 4. The frequency of (a) spring (FMAM) and (b) summer (JJA) blocking in the Pacific Region. The blue dotted line is a linear regression model.

Table 5. The mean spring and summer Pacific Region blocking (140°E to 100°W) from 1969-2015 as compared to Wiedenmann et al. (2002). The comparison with the earlier work is shown as a departure (current—Weidenmann et al., 2002).

	Number/Duration	Total days	Intensity
Spring	3.3/8.8 (+1.4/+1.6)	29.2 (−15.6)	3.33 (+0.73)
Summer	2.0/8.7 (+0.7/+1.7)	17.4 (+8.6)	2.15 (+0.04)

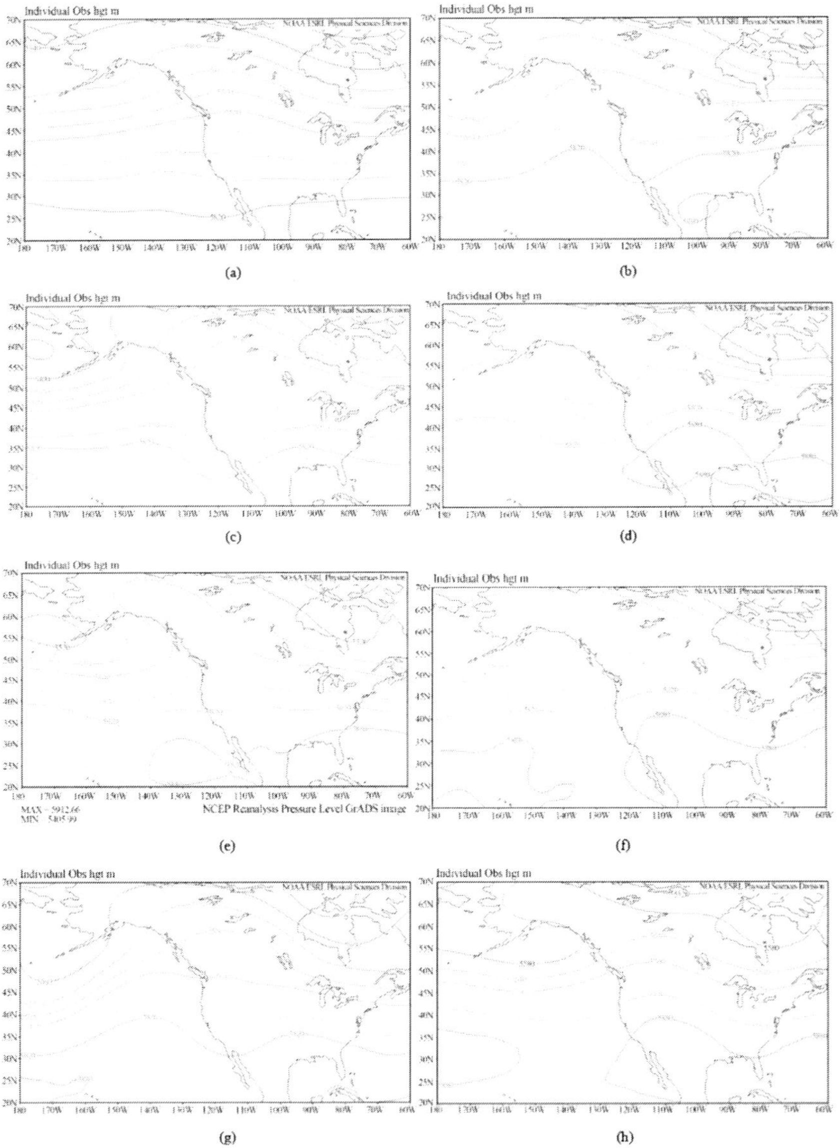
Individual Obs hgt m
NOAA ESRL Physical Sciences Division
MAX = 5912.66
MIN = 5405.99
NCEP Reanalysis Pressure Level GrADS image
(a)
(b)
(c)
(d)
(e)
(f)
(g)
(h)

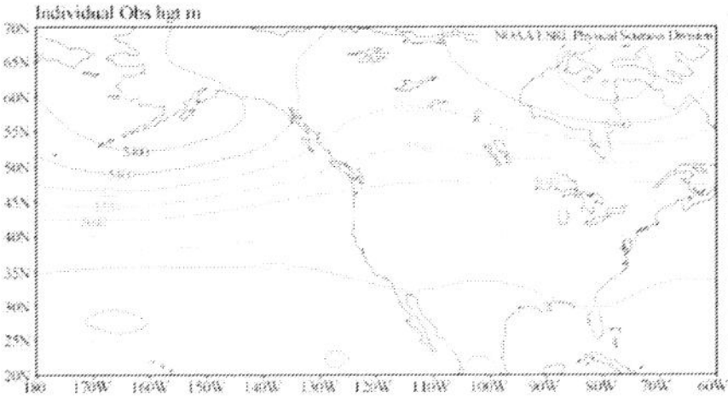

(i)

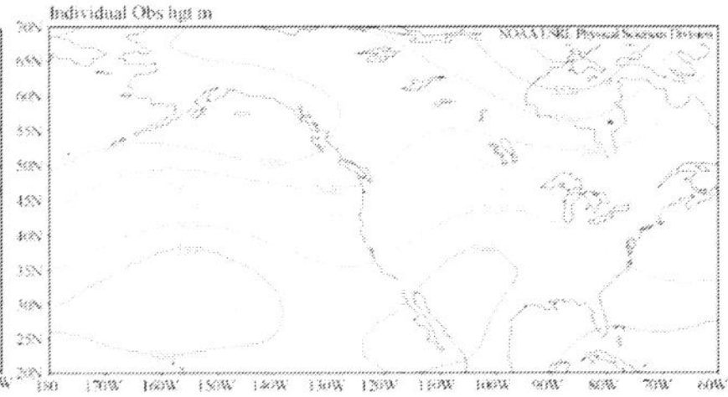

(j)

Figure 5. The mean 1200 UTC 500 hPa height map for the East Pacific and North America for the (a), (c), (e), (g), (i) 30 day period prior to, and (b), (d), (f), (h), (j) 30 day period following the identified onset date found in this study. The years used are (a), (b) 1967, (c), (d) 1971, (e), (f) 2004, (g), (h) 2008, and (i), (j) 2012. The identified dates for transition were, 4 June, 17 June, 28 June, 11 June, and 2 June, respectively. The contour interval is every 6 dam.

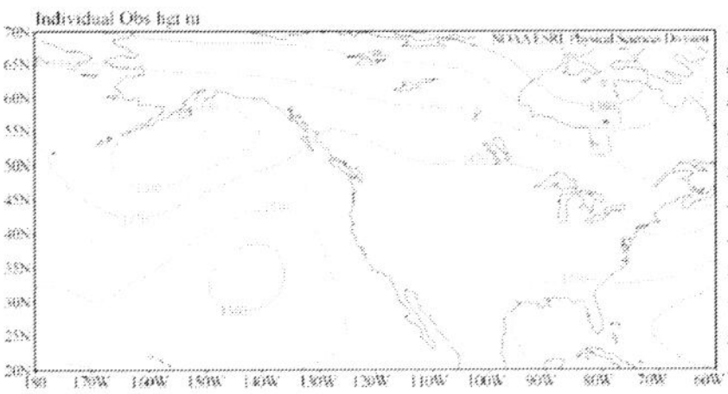

(a)

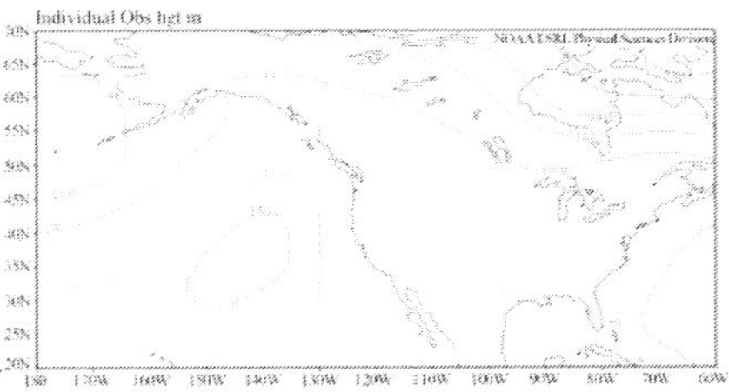

(b)

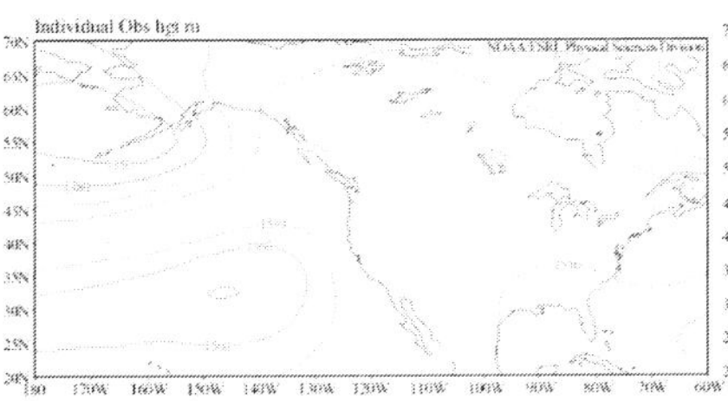

(c)

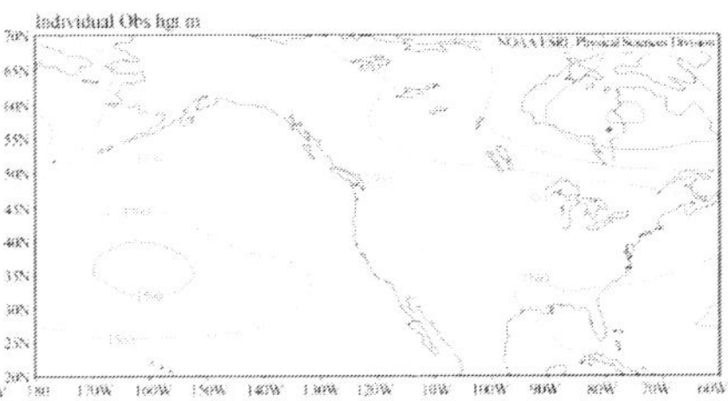

(d)

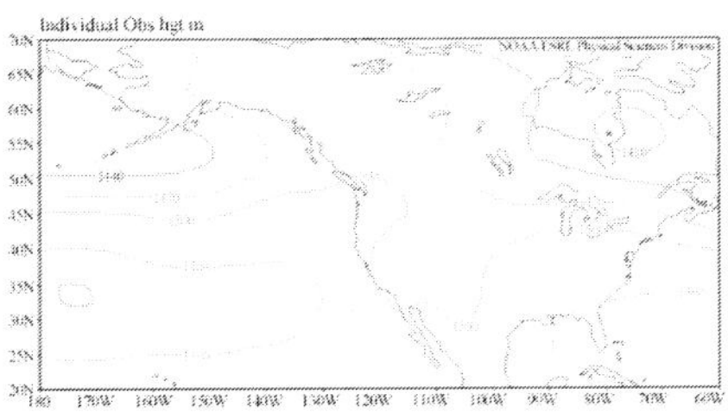

(e)

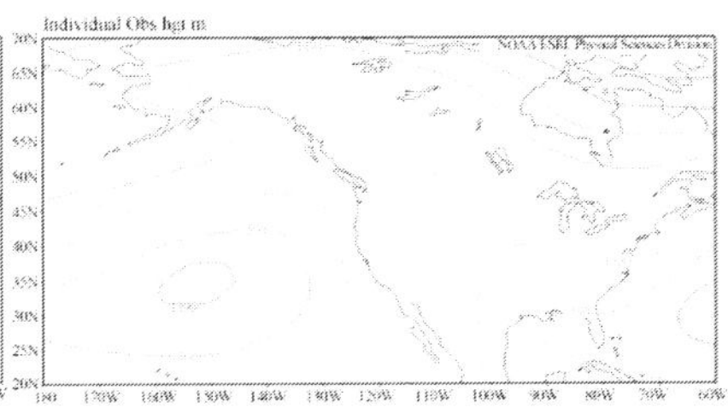

(f)

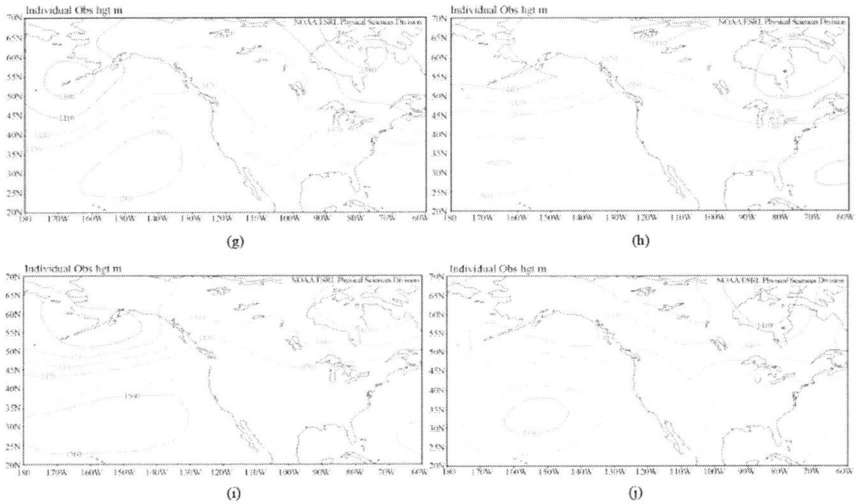

Figure 6. As in Figure 5, except at the 850 hPa level and a contour interval of 30 dam.

In all five cases at 500 hPa (Figure 5), the 5820 m contour was located within or equator ward of the study region in the 30 day period before the transition date. There was either split flow or ridging over the Rockies as is typical of the colder season. In the 2012 case, the flow was more zonal and the jet stream was clearly north of the region (but not the 582 m contour). In the 30 day period following the onset date, the 5820 m line and the jet stream was clearly poleward of the region, and summer season ridging was established in North America. Note that during the prior period for the year 2004 there was blocking in the Alaska Region. Similar to the case for the hot dry summer of 2012, it was noted that the years 1980 and 1954 also had earlier onset dates and a similar difference between the pre- and post-transition date (not shown). Thus, this criterion is rather effective for intensifying this change in the character of the 500 hPa flow over the region.

At the 850 hPa level (Figure 6), the post-onset dates for 1967, 1971, and 2012 show that the 850 hPa flow over the study region was more meridional and the stronger flow was to the west over Texas and Mexico, while the 2004 and 2008 flow regimes were more zonal and originally out of the gulf of Mexico. Table 6 shows that the former three years were dry over the study region, two of them extremely dry. This result for dry years is consistent with the study of [13] who looked at summer season dew points in the region. The latter two years were wetter than normal. During the pre-onset date period, the flow was more zonal over the study region for all the years except for the year 2012. Like the 500 hPa analysis, however, these were all different from the post- onset flow region.

Here we demonstrate using the integrated regional enstrophy (IRE) that the stability of the northern hemisphere flow (IRE) increases (decreases) in association with the time period identified for the spring-to-summer transition date. Figure 7 shows a time series of the IRE quantity from 1 May to 31 July for both 1971 and 2008. The transition date in 1971 was identified as 17 June. The IRE diagnostic for that year (Figure 7(a)) was higher for the period before 17 June, and considerably less following this date. However, the transition during May and June was more gradual. In the case of 2008, the transition date chosen was 11 June, and the IRE diagnostic increases rapidly (possibly indicating instability and flow regime transformation— [6]), and then decreases to a

Table 6. The mean temperatures (°C) and total precipitation (mm) and their anomalies for the years 1967, 1971, 2004, 2008, and 2012. The *represents years that were greater than one standard deviation from the normal (1.2°C and 13.1 mm).

Year	Temperature	Precipitation
1967	23.0	20.0
1971	24.2	18.2*
2004	21.9*	35.8
2008	23.7	47.1*
2012	26.8*	14.5*

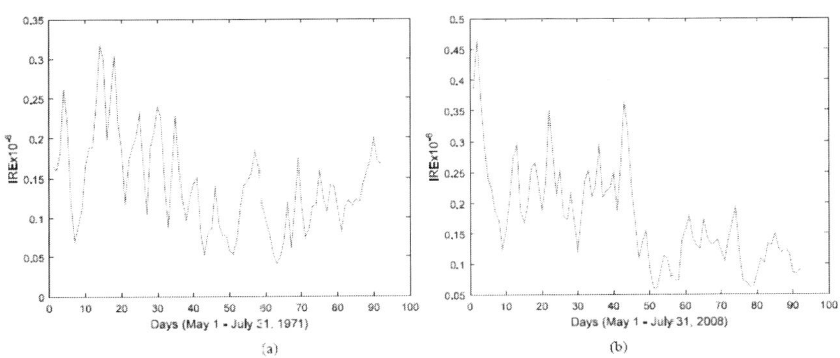

Figure 7. The Integrated Regional Enstrophy Diagnostic (IRE) for the time period 1 May to 31 July (in days from 1 May) for (a) 1971, and (b) 2008.

much lower level for the rest of the series. These examples demonstrate that the IRE diagnostic does indicate that the flow regime changes occur in the time frame identified by the criterion used here. Additionally, the IRE indicates that the summer season flow is more stable, and thus weather prediction is more favorable as shown statistically for the summer season in region [55] .

4. INTERANNUAL AND INTERDECADAL VARIABILITY

In order to identify the interannual and interdecadal variability a periodogram of the summer temperatures, the final date chosen, and Pacific Region blocking were constructed using the method of cycles (e.g., [12]) (Figure 8). All of these show statistically significant interannual variability, but there is little indication of interdecadal variability, as was found for regional precipitation events (e.g., [56] [57]), except for the blocking time series which shows significant variability beyond ten years. The interannual variability in Figure 8(a) for example can be inferred from the diagram which shows a significant peak at wave number 18, which in the 68 year data set represents variability with a period of three to four years. Interdecadal variability in the blocking occurrences (Figure 8(c)) is evidenced with a peak at wave number three and ten (46 years total) or at 15 and 5 years, respectively. Table 7 shows the number of early versus late onsets for each mode of ENSO and the PDO confirming that there is no strong interdecadal signal in the spring-to-summer onset. As may be expected for neutral years, there is no discernable tendency toward late or early onsets, and the same outcome can be inferred for years after El Niño onset, which [2] also found. However, for years after La Niña onsets, there is a stronger tendency for early onsets, especially during −PDO. Both La Niña years in Table 6 were early onsets. Additionally, [2] studied a period that contained few La Niña onsets, thus there is no basis for comparison with this study.

Table 8 shows the occurrence and of blocking within the Pacific region for each year shown in Figure 5, Figure 6. In section 3 we found a correlation between the more frequent occurrence of blocking and cooler, wetter summers. Both 2004 and 2008 were cooler, wet summers, while 2012 was the least blocked season after 2000. This season was very hot and dry. Additionally 1980 was tied for the least blocked summer season for the entire period 1969-2014, and 1971 was about average for the summer temperatures, but quite dry. The regular occurrence of blocking would allow for more variable summer weather and more frequent intrusions of cooler air in the region. Separating the occurrence of blocking by ENSO phase would demonstrate that more blocking occurred in years in which the transition was toward the El Nino (2.4 events) than those when the transition was the other way (1.75 events).

Table 7. The number of early or late onset dates for summer stratified by ENSO phase for +PDO/−PDO.

ENSO phase	Early onset	Late onset	On average date
El Nino (15)	3/4	2/6	0/0
Neutral (36)	4/9	8/11	3/1
La Nina (16)	0/10	1/4	0/1

Table 8. As in Table 5, except for the years in Figure 5. Figure 6 where blocking was examined.

	Number/Duration	Total days	Intensity
Spring 1971	1/14.0	14	2.25
2004	4/9.8	39	3.13
2008	7/9.9	69	2.50
2012	2/8.5	17	4.27
Summer 1971	2/5.5	11	2.29
2004	7/11.3	80	2.60
2008	3/9.7	29	1.75
2012	1/12	12	1.99

The study of [2] found no statistically significant differences between the summer temperatures as a function of ENSO in their 20 year study. Table 9 shows the mean seasonal temperature for each case in Table 7 using the larger data set. Similar to [56] or [57] , −PDO springs and summers show slight ENSO variation for spring and summer temperatures following El Niño onset averaging slightly warmer than summers following La Niña onset. This result matches the studies of [56] and similar studies from this group that there is weak ENSO variability in −PDO years. In +PDO yeas, which overlapped with most of the years in the [2] study, they found warmer summers following El Niño onset was statistically insignificant. Here we found that these summers were warmer at the 90% confidence level, but the spring-summer temperature difference is significantly larger at the 95% confidence level. As in Ratley et al. [2] , we can conclude that useful information may be found by relating the change in ENSO phase to summer temperatures, and [27] ([28]) found a similar phenomenon for a study of the summers of eastern (western) Russia, respectively.

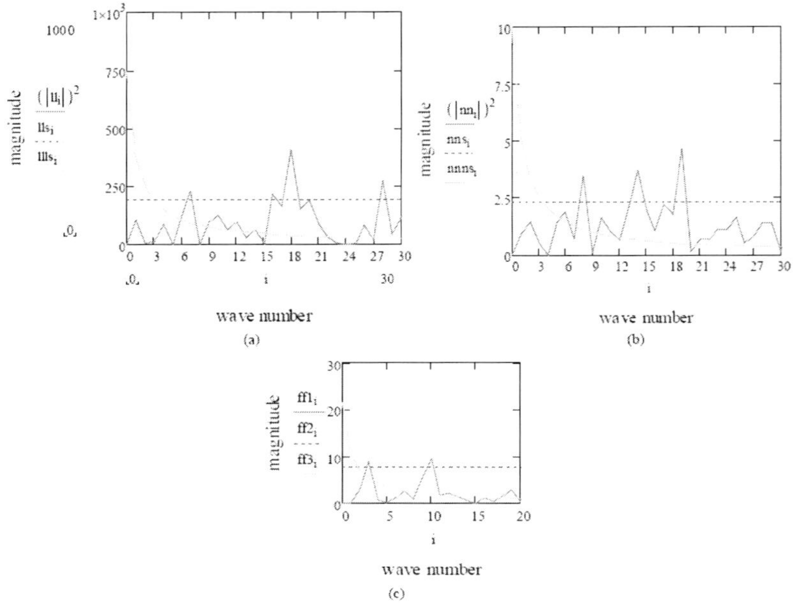

Figure 8. The power spectra versus wave number (per 67 years) for the (a) summer season temperatures, and (b) the final date time series from Figure 3(b). The blue (green) dotted line is the 95% confidence level for a white (red) noise spectrum. (see Wilks, 2006). For (c) as in (a) and (b) except per 46 years for the occurrence of Pacific region summer blocking (from Figure 4(b)).

In examining the changes in ENSO phase, defined as the difference between the new phase of 1 October and the phase of the current summer, [2] examined the tendencies for certain ENSO phase changes to be associated with early or late onsets. Here, we will define changes toward the La Niña (El Niño) direction as positive (negative), and no change in phase as a null case. Then, [2] found that positive (negative and null) transitions were associated with late (early) onsets. The results of our study indicate no such tendency in the much longer data set (Table 10). In fact it appears that all possible transitions are equally likely to produce early and late transitions overall.

Table 9. As in Table 7 except for mean spring and summer temperatures.

ENSO phase	Spring (°C)	Summer (°C)	Difference (°C)
El Nino (15)	12.5/12.8	24.5/24.6	12.0/11.8
Neutral (36)	12.0/12.6	24.2/24.5	12.2/11.9
La Nina (16)	11.9/12.2	23.1/24.1	11.2/11.9

Table 10. ENSO phase changes for early and late summer onset dates where a positive phase change is toward La Niña and negative is toward El Niño and null is no change in phase from one year to the next.

Direction of phase change	Early	Late
Positive		
EL-Neu	2	5
EL-LA	5	2
NEU-LA	2	1
Null		
EL-EL	1	0
NEU-NEU	12	10
LA-LA	3	3
Negative		
LA-NEU	5	1
LA-EN	3	1
NEU-EN	4	7

Finally, [2] found that summers associated with a transition toward La Niña or steady state La Niña conditions produced warmer, drier summers partly based on conjecture. Here, we tested all 67 seasonal transitions, and positive transitions were indeed warmest (24.6°C), followed by null transitions (24.2°C), and negative transitions (24.0°C). This result is not statistically significant, but does support the assertions of [12] and [13] who also found La Niña transitions to be warmer. Narrowing the subset to include only El Niño (−) and La Niña (+) transitions and their null cases (back-to-back years) for both demonstrates that La Niña transitions and steady state La Niña years produced a mean temperature of 24.8°C, versus 24.0°C for El Niño onsets and steady El Niño conditions. Including only La Niña or El Niño onsets and eliminating their null cases, the average summer temperature was 25.1°C versus 24.1°C. Thus, the conclusion of [2] that transitions toward La Niña were warmer is valid, or even stronger, in the much larger dataset.

5. DISCUSSION

The methodology of [2] for identifying the spring-to-summer transition was used for a longer dataset, this one covering a 68 year period from 1948-2015. They [2] used surface temperatures to derive a criterion based on the daily maximum and mean temperatures, and a 500 hPa height map criterion with the goal of identifying the spring-to-summer transition date for a 20 year period. They used each criterion in order to determine a date that matched one of the three dates found by the available individual criterion. This study modified the practice and identified the mean date of the three available criteria as the spring-to-summer transition date. It was found that:

- the statistics produced by this study were similar to those produced by the [2] study for the spring-to-summer transition date, in that the 500 hPa height mean transition date was the latest, and the maximum temperature mean transition date the earliest,
- the daily mean temperature criterion was the closest to the overall date (June 12) found by compiling the three criterion, it was only one day different in both this and the earlier study. This criterion and the final date were very strongly correlated, significant at the 99% confidence level.
- the mean spring temperatures trended upward at a significant rate, while the summer season temperatures were slightly cooler providing for a smaller annual cycle, and a significant trend (at the 99% confidence level) toward an earlier onset date,
- when comparing the results of this study to the smaller time period of the [2] work, it was found that our modification for identifying the final transition date had little impact on the overall statistics,
- Pacific region blocking showed a strong increase in occurrence and days, but not intensity, and more summer season blocking was associated with cooler summers and the transition toward El Niño,
- the 850 hPa and 500 hPa height fields show that the 30 day period before five chosen onset days was different from the 30 day period following onset. At the 500 hPa level, the 582 contour was located over or equatorward of the region during the spring and poleward during the summer. At 850 hPa, the summer season flow was more meridional and during the drier years the strongest flow was to the west of the study region,
- the [2] study found that summer seasons preceding La Niña onset were warmer and the transition date was later that the mean. Here we found that the association of La Niña onset summers being hotter summers was confirmed and the result was stronger than that of the [2] finding. However, this study did not find a tendency for early or late summer onsets in association with ENSO,

- the IRE diagnostic used for example in [6] indicated that the dates chosen for the spring-to-summer transition were close to decreases in the northern hemisphere enstrophy demonstrating that the criterion proposed here has utility in identifying the seasonal transition,
- and finally, this study showed only slight interdecadal variability in ENSO behavior, unlike other studies of different variables and seasons for this region, in particular it was noted that ENSO summers in the ?PDO phase were similar overall with little variation, while in the +PDO the summers following La Niña were cooler and associated with a smaller spring-to-summer temperature change.

6. CONCLUSION

In conclusion, these results would have applicability for seasonal or long-range forecasting in the region of study. The studies of [10] and [11] both demonstrate that SST anomalies and the occurrence of summer Pacific Region blocking correlate strongly to the summer weather and climate of the study region. These variables can then be used to predict the conditions of the summer season [11] . The information gained here provides another variable that can be used for long-range seasonal prediction, and this information should be on interest to the regional agriculture community.

ACKNOWLEDGEMENTS

The authors would like to thank Professor Neil I. Fox for his earlier comments on this work, as well as the anonymous reviewers who helped to make this a stronger contribution.

CITE THIS PAPER

Rosalie G. Newberry, Anthony R. Lupo, Andrew D. Jensen, R. Antonio Rodriges Zalipynis (2016) An Analysis of the Spring-to-Summer Transition in the West Central Plains for Application to Long Range Forecasting. *Atmospheric and Climate Sciences*,**06**,375-393.

REFERENCES

1. Hansen, A.R. (1986) Observational Characteristics of Atmospheric Planetary Waves with Bimodal Amplitude Distributions. Advances in Geophysics, 29, 101-134.
2. Ratley, C.W., Lupo, A.R. and Baxter, M.A. (2002) Determining the Spring to Summer Transition in the Missouri Ozarks Using Synoptic Scale Data. Transactions of the Missouri Academy of Science, 36, 55-62. [Citation Time(s):41]
3. Johns, R.H. (1993) Meteorological Conditions Associated with Bow Echo Development in Convective Storms. Weather and Forecasting, 8, 294-299.
4. Higgins, W.R., Yao, Y. and Wang X.L. (1997) Influence of the North American Monsoon System on the United States Summer Precipitation Regime. Journal of Climate, 10, 2600-2622.
5. Wang, S.Y. and Chen, T.C. (2009) The Late Spring Maximum of Rainfall over the U.S. Central Plains and the Role of the Low-Level Jet. Journal of Climate, 22, 4696-4709.
6. Jensen, A.D. and Lupo, A.R. (2013) Using Enstrophy Advection as a Diagnostic to Identify Blocking Regime Transition. Quarterly Journal of the Royal Meteorological Society, 139, 2-7.
7. Lupo, A.R., Kelsey, E.P., McCoy, E.A., Halcomb, C.E., Aldrich, E.M., Allen, S.A., Akyuz, F.A., Skellenger, S., Bieger, D.G., Wise, E., Schmidt, D. and Edwards, M. (2003) The Presentation of Temperature Information in Television Broadcasts: What Is Normal? National Weather Digest, 27, 53-58.
8. Spar, J. (1949) Energy Changes in the Mean Atmosphere. Journal of Meteorology, 6, 411-415.
9. Kung, E.C. and Chern, J.G. (1995) Prevailing Anomaly Patterns of the Global Sea Surface Temperatures and Tropospheric Responses. Atmósfera, 8, 99-114.
10. Lupo, A.R., Kelsey, E.P., Weitlich, D.K., Mokhov, I.I., Akyuz, F.A., Guinan, P.E. and Woolard, J.G. (2007) Interannual and Interdecadal Variability in the Predominant Pacific Region SST Anomaly Patterns and Their Impact on a Local Climate. Atmósfera, 20, 171-196.
11. Lupo, A.R., Kelsey, E.P., Weitlich, D.K., Davis, N.A. and Market, P.S. (2008) Using the Monthly Classification of Global SSTs and 500 hPa Height Anomalies to Predict Temperature and Precipitation Regimes One to Two Seasons in Advance for the Mid-Mississippi Region. National Weather Digest, 32, 11-33.
12. Birk, K., Lupo, A.R. Guinan, P.E. and Barbieri, C.E. (2010) The interannual variability of Midwestern temperatures and precipitation as related to the ENSO and PDO. Atmósfera, 23, 95-128.
13. Lupo, A.R., Smith, N.B., Guinan, P.E. and Chesser, M.D. (2012) The Climatology of Missouri Region Dew Points and the Relationship to ENSO. National Weather Digest, 36, 10-20.
14. Walker, G.T. and Bliss, E.W. (1932) World Weather V. Memoirs of the Royal Meteorological Society, 4, 53-84.
15. Grimm, A.M. and Tedeschi, R.G. (2009) ENSO and Extreme Rainfall Events in South America. Journal of Climate, 22, 1589-1609.
16. Muñoz, E., Wang, C. and Enfield, D. (2010) The Intra-Americas Springtime Sea Surface Temperature Anomaly Dipole as a Fingerprint of Remote Influences. Journal of Climate, 23, 43-56.
17. Zuki, Z.M. and Lupo, A.R. (2008) The Interannual Variability of Tropical Cyclone Activity in the Southern South China Sea. Journal of Geophysical Research, 113, D06106.
18. Lupo, A.R., Latham, T.K., Magill, T., Clark, J.V., Melick, C.J. and Market, P.S. (2008) The Interannual Variability of Hurricane Activity in the Atlantic and East Pacific Regions. National Weather Digest, 32, 119-135.
19. Fasullo, J. and Webster, P.J. (2002) Hydrological Signatures Relating the Asian Summer Monsoon and ENSO. Journal of Climate, 15, 3082-3095.

20. Wu, R. and Kirtman, B.P. (2007) Roles of the Indian Ocean in the Australian Summer Monsoon ENSO Relationship. Journal of Climate, 20, 4768-4788.
21. Ropelewski, C.F. and Halpert, M.S. (1987) Global and Regional Scale Precipitation Patterns Associated with the El Niño/Southern Oscillation. Monthly Weather Review, 115, 1606-1626.
22. Ropelewski, C.F. and Halpert, M.S. (1989) Precipitation Patterns Associated with the High Index Phase of the Southern Oscillation. Journal of Climate, 2, 268-284.
23. Bove, M.C., Elsner, J.B., Landsea, C.W., Niu, X. and O'Brien, J.J. (1998) Effects of El Niño on U.S. Landfalling Hurricanes, Revisited. Bulletin of the American Meteorological Society, 79, 2477-2482.
24. Shinker, J.J. and Bartlein, P.J. (2009) Visualizing the Large-Scale Patterns of ENSO-Related Climate Anomalies in North America. Earth Interactions, 13, 1-50.
25. Mo, K.C. (2010) Interdecadal Modulation of the Impact of ENSO on Precipitation and Temperature over the United States. Journal of Climate, 23, 3639-3656.
26. Hu, Z.-Z. and Huang, B. (2009) Interferential Impact of ENSO and PDO on Dry and Wet Conditions in the U.S. Great Plains. Journal of Climate, 19, 5500-5518.
27. Mokhov, I.I., Khon, V.C., Timazhev, A.V., Chernokulsky, A.V. and Semenov, V.A. (2014) Hydrological Anomalies and Trends in the Amur River Basin Due to Climate Changes. In: Extreme Floods in the Amur River Basin: Causes, Forecasts, and Recommendations, Roshydromet, Earth Climate Theory Studies, 81-121.
28. Lupo, A.R., Mokhov, I.I., Chendev, Y.G., Lebedeva, M.G., Akperov, M. and Hubbart, J.A. (2014) Studying Summer Season Drought in Western Russia. Advances in Meteorology, Special Issue: Large Scale Atmospheric Science, 2014, Article ID: 942027.
29. Kalnay, E., et al. (1996) The NCEP/NCAR 40-Year Reanalysis Project. Bulletin of the American Meteorological Society, 77, 437-471.
30. Neter, J., Wasserman, W. and Whitmore, G.A. (1988) Applied Statistics. 3rd Edition, Allyn and Bacon Press, Boston, 1006 p.
31. Wilks, D.S. (2006) Statistical Methods in the Atmospheric Sciences. 2nd Edition, Academic Press, San Diego, 627 p.
32. Park, C.-K. and Kung, E.C. (1988) Principal Components of the North American Summer Temperature Field and the Antecedent Oceanic and Atmospheric Condition. Journal of the Meteorological Society of Japan, 66, 677-690.
33. Lee, J.-W. and Kung, E.C. (2000) Seasonal-Range Forecasting of the Ozark Climate by a Principal Component Regression Scheme with Antecedent Seas Surface Temperatures and Upper Air Conditions. Atmósfera, 13, 223-244.
34. Palecki, M.A. and Leathers, D.J. (2000) Spatial Modes of Drought in the Central United States. Preprints of the 12th Conference on Applied Climatology, Asheville, 8-11 May 2000, 4 p.
35. Renken, J.S., Herman, J., Parker, D., Bradshaw, T. and Lupo, A.R. (2015) Using the Bering Sea and Typhoon Rules to Generate Long Range Forecasts. 39th NOAA Annual Climate Diagnostics and Prediction Workshop, St. Louis, 20-23 October 2014, 46-50. http //weather.missouri.edu/gcc/ 39cdpwdigest pdf
36. Pielke, R.A. and Landsea, C.N. (1999) La Niña, El Niño and Atlantic Hurricane Damages in the United States. Bulletin of the American Meteorological Society, 80, 2027-2033.
37. Huang, D., Ni, X., Tang, Q., Zhu, X. and Xu, D. (2012) Spatial and Temporal Variability of Sea Surface Temperature in the Yellow Sea and East China Sea over the Past 141 Years. In: Wang, S.-Y. and Gillies, R.R., Eds., Modern Climatology, Chap. 8, Intech Open, East Providence, 213-234.
38. Hanley, D.E., Bourassa, M.A., O'Brien, J.J., Smith, S.R. and Spade, E.R. (2003) A Quantitative Evaluation of ENSO Indices. Journal of Climate, 16, 1249-1258.
39. Mantua, N.J., Hare, S.R., Zhang, Y., Wallace, J.M. and Francis, R.C. (1997) A Pacific Interdecadal Climate Oscillation with Impacts on Salmon Production. Bulletin of the American Meteorological Society, 78, 1069-1079.

40. Minobe, S. (1997) A 50 - 70-Year Climatic Oscillation over the North Pacific and North America. Geophysical Research Letters, 24, 683-686.
41. Bentley, M.L. and Mote, T. (1998) A Climatology of Derecho-Producing Mesoscale Convective Systems in the Central and Eastern United States, 1986-1995. Part I: Temporal and Spatial Distribution. Bulletin of the American Meteorological Society, 79, 2527-2540.
42. Wang, S.-Y., Chen, T.-C. and Correia Jr., J. (2009) Climatology of Summer Midtropospheric Perturbations in the U.S. Northern Plains. Part I: Influence on Northwest Flow Severe Weather Outbreaks. Climate Dynamics, 31, 1221-1237.
43. Swanson, K.L. (2001) Upper-Tropospheric Potential Vorticity Fluctuations and the Dynamic Relevance of the Time Mean. Journal of the Atmospheric Sciences, 58, 1815-1826.
44. Lupo, A.R. and Bosart, L.F. (1999) An Analysis of a Relatively Rare Case of Continental Blocking. Quarterly Journal of the Royal Meteorological Society, 125, 107-138.
45. Yarnal, B. (1993) Synoptic Climatology in Environmental Analysis. CRC Press, Boca Raton, 195 p.
46. Namias, J. (1982) Anatomy of Great Plains Protracted Heat Waves (Especially the 1980 US Summer Drought). Monthly Weather Review, 110, 824-838.
47. Namias, J. (1983) Some Causes of United States Drought. Journal of Applied Meteorology and Climatology, 22, 30-39.
48. Dymnikov, V.P., Kazantsev, Y.V. and Kharin, V.V. (1992) Information Entropy and Local Lyapunov Exponents of Barotropic Atmospheric Circulation. Izvestiya, Atmospheric and Oceanic Physics, 28, 425-432.
49. Hurrell, J.W., van Loon, H. and Shea, D.J. (1998) Meteorology of the Southern Hemisphere. NCAR Tech. Memo. NCAR/CAS/95-08, 77 p.
50. IPCC (2013) Climate Change 2013: The Physical Scientific Basis, Contributions of Working Group I to the Fifth Assessment Report of the Intergovernmental Panel on Climate Change. http://www.ipcc.ch
51. Lupo, A.R., Smith, P.J. and Zwack, P. (1992) A Diagnosis of the Development of Two Extratropical Cyclones. Monthly Weather Review, 120, 1490-1523.
52. Wiedenmann, J.M., Lupo, A.R., Mokhov, I.I. and Tikhonova, E. (2002) The Climatology of Blocking Anticyclones for the Northern and Southern Hemisphere: Block Intensity as a Diagnostic. Journal of Climate, 15, 3459-3473.
53. Quiroz, R.S. (1984) The Climate of the 1983-1984 Winter. A Season of Strong Blocking and Extreme Cold in North America. Monthly Weather Review, 112, 1894-1912.
54. Jensen, A.D. (2015) A Dynamic Analysis of a Record Breaking Winter Season Blocking Event. Advances in Meteorology, Special Issue: Large-Scale Dynamics, Anomalous Flows, and Teleconnections, 2015, Article ID: 634896.
55. Lupo, A.R. and Market, P.S. (2002) The Verification of Weather Forecasts in Central Missouri and Seasonal Variations in Forecast Accuracy. Weather and Forecasting, 8, 891-897.
56. Berger, C.L., Lupo, A.R., Browning, P., Bodner, M., Rayburn, C.C. and Chambers, M.D. (2003) A Climatology of Northwest Missouri Snowfall Events: Long Term Trends and Interannual Variability. Physical Geography, 14, 427-448.
57. Lupo, A.R., Albert, D., Hearst, R., Market, P.S., Akyuz, F.A. and Allmeyer, C.L. (2005) Interannual Variability of Snowfall Events and Snowfall-to-Liquid Water Equivalents in Southwest Missouri. National Weather Digest, 29, 13-24.

CHAPTER 4

Ocean Global Warming Impacts on the South America Climate

Renato Ramos da Silva* **and Reinaldo Haas**

Climate and Meteorology Laboratory, Department of Physics, Federal University of Santa Catarina, Florianópolis, SC, Brazil

ABSTRACT

The global Ocean-Land-Atmosphere Model (OLAM) model was used to estimate the impacts of the global oceanic warming on the climate projections for the 21st Century focusing on the South America region. This new model is able to represent simultaneously the global and regional scales using a refining grid approach for the region of interest. First, the model was run for a 31-year control period consisting on the years 1960–1990 using the monthly Sea Surface Temperature (SST) from the Atmospheric Model Inter-comparison Project (AMIP) data as a driver for the ocean fluxes. Then, the model was run for the period 2010–2100 using the monthly projected SST from the Hadley Center model (HadCM3) as a driver for the oceanic changes. The model was set up with an icosahedral triangular global grid having about 250 km of grid spacing and with a refining grid resolution with the cells reaching about 32 km over the South America region. The results show an overall temperature increase mainly over the center of the Amazon basin caused by the increase of the greenhouse effect of the water vapor; a decrease on precipitation mainly over the northeast Brazil and an increase in the south and over the western Amazon region; and a major increase on the near surface wind speed. These results are similar to the global coupled models (GCMs); however, OLAM has a novel type of grid that can provide the interaction between the global and regional scales simultaneously.

Keywords: OLAM, regional climate modeling, climate change, South America, Brazil

INTRODUCTION

The major efforts to stablish future climate scenarios are based on global coupled models (GCMs). A series of ongoing numerical integrations of climate models are now taking place at several research centers. The latest Intergovernmental Panel on Climate Change (IPCC) report (IPCC, 2013) produced climate change projections for near-term (Kirtman et al., 2013) and long-term (Collins et al., 2013) based on the Coupled Model Intercomparison Project Phase 5 (CMIP5; Taylor et al., 2012).

For the South America region the majority of those GCMs show an overall increase of the near surface temperature, a decrease on the precipitation over the northeast region of Brazil, and an increase over the south of South America region (Collins et al., 2013; Kirtman et al., 2013).

The major climate variabilities over the South America are caused by the neighboring oceans (i.e., the Atlantic and the Pacific oceans). Near the equator gradients of temperature and pressure are the main forcing for the Hadley Cell formation (Lindzen and Nigam, 1987). Anomalies on the Atlantic SST and the trade winds can cause changes on the Hadley Cell strength (Servain et al., 2014), position (Wang, 2002; Xie and Carton, 2004) and advection of moisture and heat (Foltz et al., 2003) that produces variabilities on the moisture transport over the South America (Nobre and Shukla, 1996; Rao et al., 2006; Bombardi and Carvalho, 2011; Bombardi et al., 2014). Otherwise, during El Niño or La Ninã conditions on the Pacific the air-sea interactions can produce shifts on the position of the Intertropical Convergence Zone (ITCZ) and the Walker Cell causing extreme changes on the South America precipitation (Grimm, 2003; Kucharski et al., 2008); more specifically during El Niño climate analysis shows drier years over the northeast Brazil and higher precipitation over the south (Ropelewski and Halpert, 1987; Marengo and Hastenrath, 1993).

The majority of the global models have spatial resolution on the order of 0.5° to 4° and therefore do not resolve the regional meteorological features of South America forced by the complex surface heterogeneity and topography (Taylor et al., 2012).

To better resolve the regional climate the dynamic downscaling approach became an important resource. The dynamic downscaling method is characterized by the use of a higher spatial resolution regional climate model (RCM). A special project using RCMs is under development to provide dynamic downscaling of regional climate projections (Giorgi and Gutowski, 2015). While these models have higher spatial resolution for a given domain, they need atmosphere initial and lateral boundary conditions from a coarser global model. These RCMs have a one-way coupling with the GCMs and therefore the local changes do not affect the large scale.

A new approach was developed recently in order to provide a two-way coupling through the use of a global model with varying spatial resolution. One of these new type of models is the Ocean-Land-Atmosphere Model (OLAM) (Walko and Avissar, 2008a,b). This new model is able to represent simultaneously the global and the local atmospheric features through a grid refining approach.

Up to now, climate projections were conducted with global models that have low resolution, or with regional climate models that are dependent of a particular global model for boundary conditions. For the first time, we can use a new type of model such as OLAM that can do a global spatial coverage and resolve simultaneously the local meteorological systems using its refining grids. In this study we used OLAM to provide numerical simulations for the current climate and for the 21st century in order to evaluate the impacts of the global ocean warming on the climate projections for the South America region and try to answer the following: "What is the expected regional climate change caused by the projected oceanic global warming?; how these projected climate changes compares with results from other previous modeling estimates?; can OLAM provide a more refined regional climate projection as compared with the GCMs?"

DATA AND METHODS

In the next sub-sections we describe the OLAM model, the numerical experiments design and the data used as initial and surface boundary condition.

The OLAM Model

The OLAM model is an evolution of the Regional Atmospheric Modeling System (RAMS) model (Cotton et al., 2003) and uses a global icosahedral type of grid that can be triangular or hexagonal (Walko and Avissar, 2008a,b). The Navier-Stokes equations are solved on the global grid using the finite volume technique that maintains the conservation of mass, momentum and energy.

A series of recent studies showed that OLAM is able to simulate well the major weather systems (Ramos-da-Silva et al., 2014) and regional climate of the South America region (Medvigy et al., 2012, 2010). Furthermore, the model was used to understand the effects of the El Niño South Oscillation (ENSO) phenomena in the South America and the results showed that an spatial resolution over the Andes mountains have to be greater than 100 km to fully represent the effects of the ENSO on the climate of the Amazon region. Thus, since OLAM had shown a good skill for the South America local meteorological features, it can be used as a tool to evaluate the climate projections as impacted by expected oceanic warming.

The physical parametrizations options for OLAM are the same available at the RAMS model including the land-vegetation interaction (Walko et al., 2000a), the cloud microphysics (Walko et al., 1995, 2000b; Meyers et al., 1997), the radiation transfer (Chen and Cotton, 1987) and cumulus parametrization (Grell, 1993, 2002; Kain, 2004). The RAMS mesoscale model was used in several applications for the South America studies such as on the deforestation impacts (Gandu et al., 2004; Ramos da Silva et al., 2008); lightning formation (Gonçalves et al., 2012), convection transport (Freitas et al., 2000), and weather forecasting (Ramos da Silva et al., 2007; Longo et al., 2013). Therefore, the physical parameterizations have been very well-tested on several different systems for the South America region.

The Model Experiments Setup

In order to estimate the projected regional climate changes caused by the global oceanic warming, the OLAM model was set with a global icosahedral triangular mesh having about 250 km of grid spacing between cells. A nesting refinement approach was applied with focus over South America region with the innermost domain having a 32 km of grid spacing (Figure 1). This refining grid approach allows a two-way interaction between the large and the local scales. In the vertical direction the model was set with 45 atmospheric layers having about 200 m in the lowest layer and 2000 m in the upper stratospheric layers. For these spatial sizes the model had a time step of 30 s to maintain numerical stability.

The chosen physical parameterizations included the short and long-wave radiation transfer (Chen and Cotton, 1987), the cumulus parametrization (Grell, 1993), the soil-vegetation model (Walko et al., 2000a), the cloud microphysics (Meyers et al., 1997; Walko et al., 2000b, 1995), and the turbulent atmospheric fluxes parametrization (Smagorinsky, 1963).

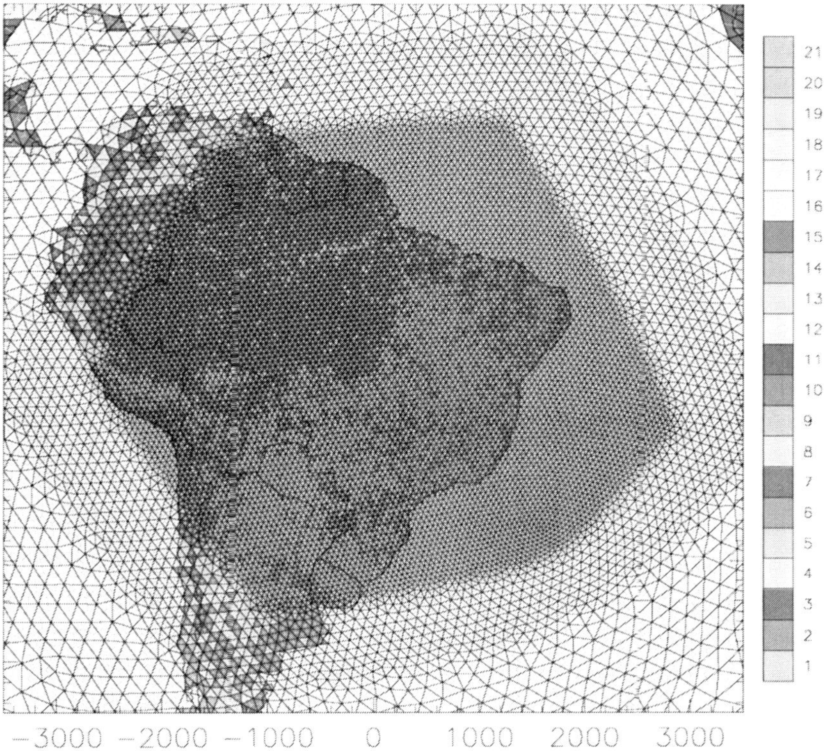

FIGURE 1. Model domain and spatial grid structure and surface heterogeneity.

Initially, the model was run for the period 1960–1990 to produce a climatological control run. The initial atmospheric conditions for geo-potential height, winds, air temperature, relative humidity, and soil moisture were obtained from the National Center for Environmental Prediction (NCEP) Reanalysis Project (Kanamitsu et al., 2002). This initial condition was set only for the onset of the integration because the model is global and the lateral nudging is not necessary. In this climatological simulation the oceanic conditions were updated based on the monthly observed SST data provided by the Atmospheric Model Intercomparison Project (AMIP) as described by (Taylor et al., 2000).

The Twenty-one century simulation was run for 2010–2100 having the same configuration of the control run but using monthly SST boundary projections from the Hadley Centre Coupled Model, version 3 (HadCM3; Gordon et al., 2000; Collins et al., 2001) from its A1B scenario.

This A1B projection assume that the anthropogenic emissions of CO2, CH4, N2O and sulfur-di-oxide based on a future world of very rapid economic growth and population that peaks by 2050 and declines thereafter based on more efficient technologies (IPCC, 2007).

The climate projections for the South America was produced by taking the difference between the mean field results for the periods 2011–2040, 2041–2070, and 2071–2100 and the results from the climatological run (i.e., 1960–1990).

Monthly precipitation data from the Global Precipitation Climatology Project (GPCP) was used to evaluate the model results for the 1960–1990 climatological run (Adler et al., 2003).

RESULTS

The GPCP observed accumulated precipitation data shows that while the Amazon region has the highest precipitation, the northeast region of Brazil has a very low accumulation (Figure 2A). Over the Amazon basin the model underestimated the precipitation as compared to the maps of GPCP observations (Figure 2B). For the climatological period (i.e., 1960–1990) there is only sparse pluviometric observed data over the region and the fields are obtained from extrapolations. Also, It seems that the differences between the model and the observations occurs over the Amazon River basin, thus although the model uses a 32 km of grid spacing, it may not be able to fully represent the squall lines that are the main precipitating meteorological system of this region (Cohen et al., 1995). Although, the model underestimates the total accumulation it represents the main spatial observed pattern (Figure 2B). Therefore, these results from the model can be used as a control run and the projected regional climate changes can be estimated through comparisons with the new runs for the 21st century.

Near Surface Temperature Changes

Figure 3 shows the results for the projected temperature change for 2011–2040. The results show increase over most of the north region, mainly over the Amazon region that can reach temperature increases higher than 4°C (around longitude 50 W); while a slight decrease is projected over the south of the domain. This warming over the Amazon was obtained also in other studies (Blázquez et al., 2012; Marengo et al., 2012) but here, the projected increase of temperature in this region is higher. For the period 2041–2070 the temperature increase reaches a larger domain with the highest increase occurring over the Amazon (Figure 4). For the period 2071–2100 the warming is spread for the entire continental domain reaching the highest

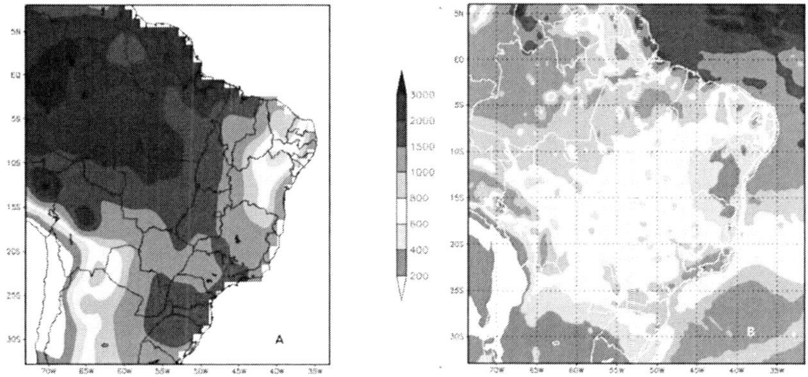

FIGURE 2. Accumulated precipitation (mm/yr) for the period 1960–1990 from (a) gpcp observations and (b) simulated by the olam model.

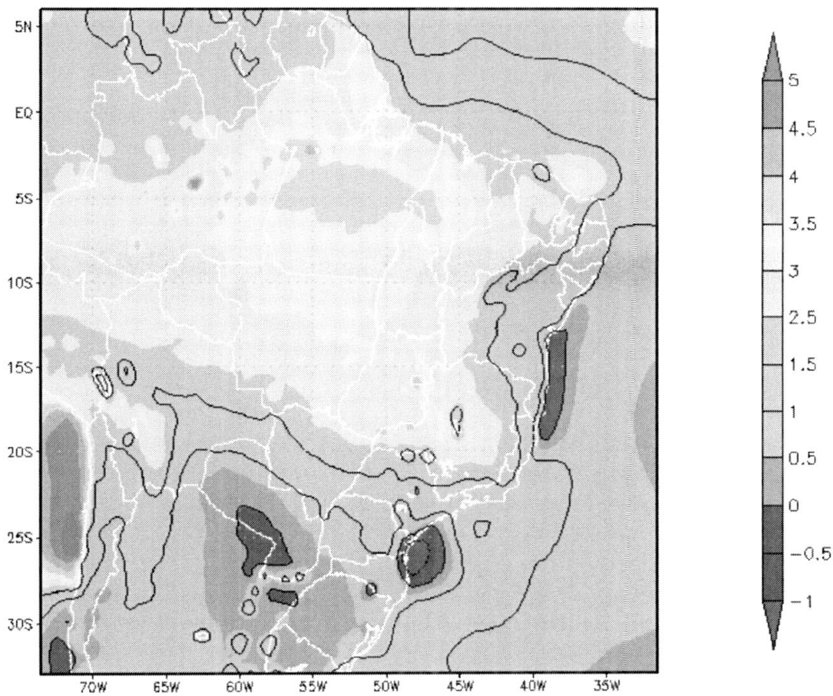

FIGURE 3. Projected surface air temperature changes (°c) for 2011–2040 relative to the climatological run (1960–1990).

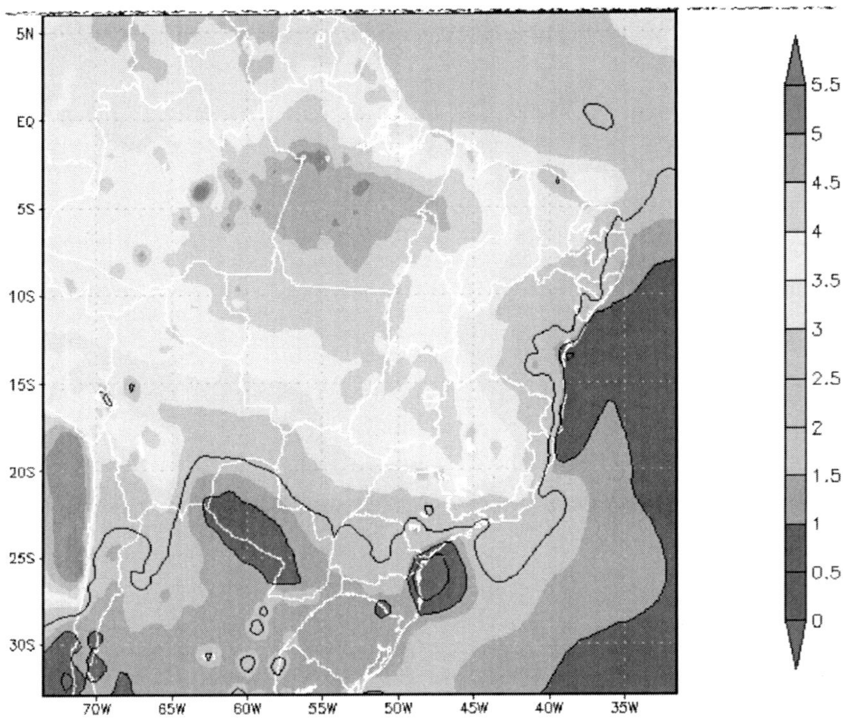

FIGURE 4. Same as figure **3**, but for temperature changes (°c) on the period between 2041 and 2070.

increase over the Amazon region in the order of 6°C (Figure 5). The major reason for this increase in temperature is due to the enhancement of the greenhouse effect of the water vapor. This greenhouse effect strength is shown by the increase in the downward long wave radiation at the surface that is higher over the tropical regions (Figure 6). The maximum increase of downward long wave radiation, greater than 60 Wm^{-2}, occurs at the same location. Thus, although the anthropogenic greenhouse gases trend were not considered on the OLAM regional climate projections, the greenhouse effect was enhanced by the feedback of the water vapor produced by the ocean warming.

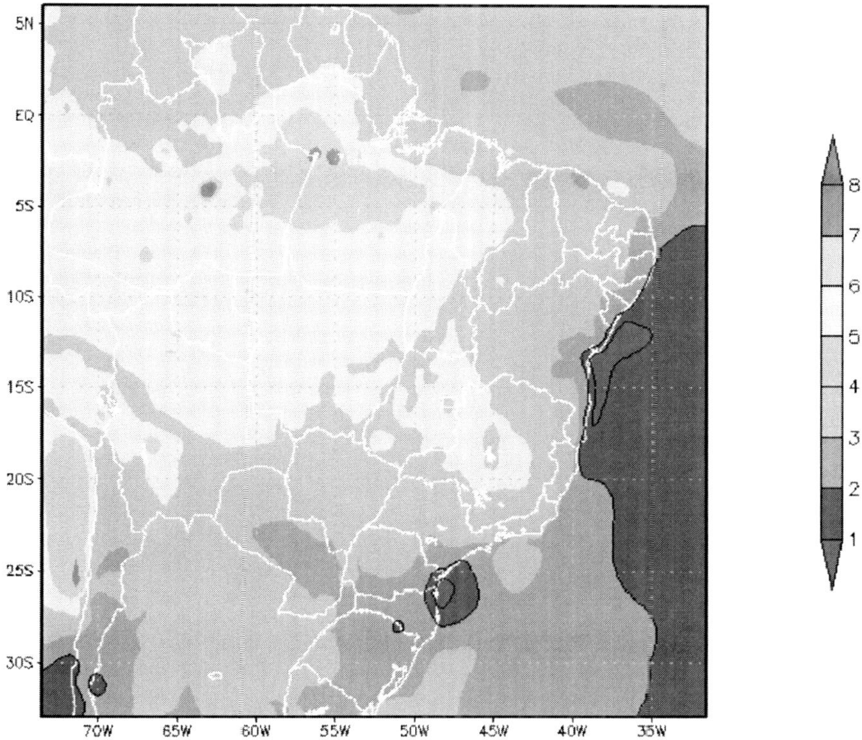

FIGURE 5. Same as figure **3**, but for temperature changes (°c) on the period between 2071 and 2100.

Precipitation Changes

Figures 7–9 show the projected precipitation changes for 2011–2040, 2041–2070, and 2071–2100, respectively. The results show a wider ITCZ that results in an increase on the precipitation over the north coast of South America. This broadening of the ITCZ causes a higher precipitation on the interface between the continent and the tropical Atlantic on the north and northeast of Brazil showing that OLAM is able to represent better the local regional changes in this location due to its better spatial resolution. The increase on the precipitation in this region is also associated with the wind results that show an increase of moisture convergence on the region (Figures 10–12).

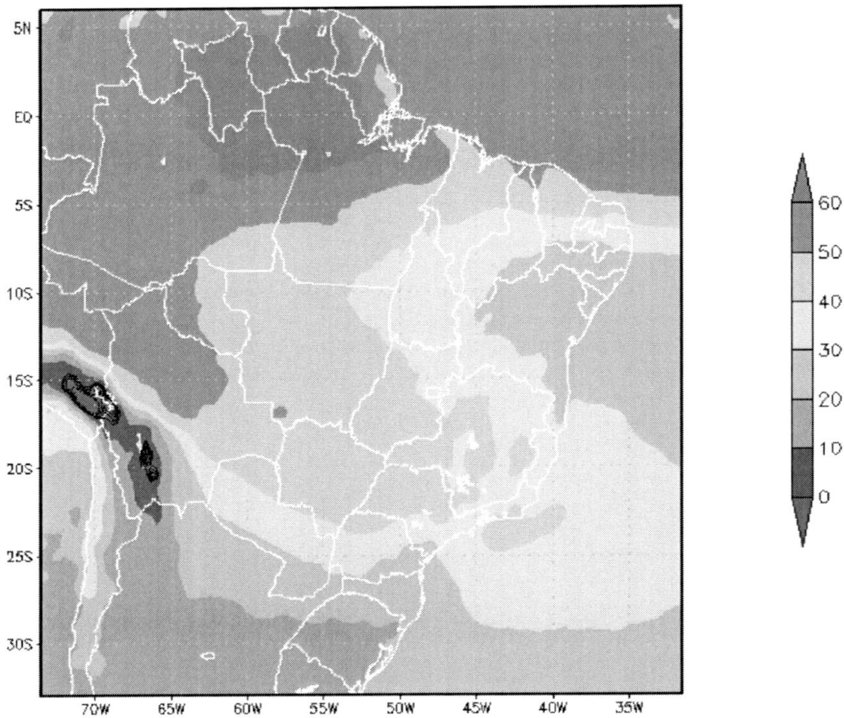

FIGURE 6. Downward long wave radiation at surface mean change (w m^{-2}) for the period 2071–2100 relative to the climatological run.

Further, analysis show a projected decrease on the precipitation mainly over the northeast region of Brazil and an increase over the west of the Amazon and over the south of the domain. However other regional changes are highlighted on some regions such as on the sea–land interface in the south Brazil where there is a stationary region of precipitation decrease that may be associated with the local ocean warming and with wind pattern changes (Figures 10–12). Recent studies show also regional changes associated with an expansion of the Hadley Cell in response to greenhouse forcing (Fu, 2015; Nguyen et al., 2015, 2013). In the west of the domain, over the Andes Mountains, there is a decrease on the precipitation that seems to be associated with the weakening on the moisture transport from the Amazon region (Figures 10–12). These projected precipitation changes in the South America are similar to the results of the GCMs used on the CMIP project, but can provide important regional features of the local climate that the GCMs can't provide.

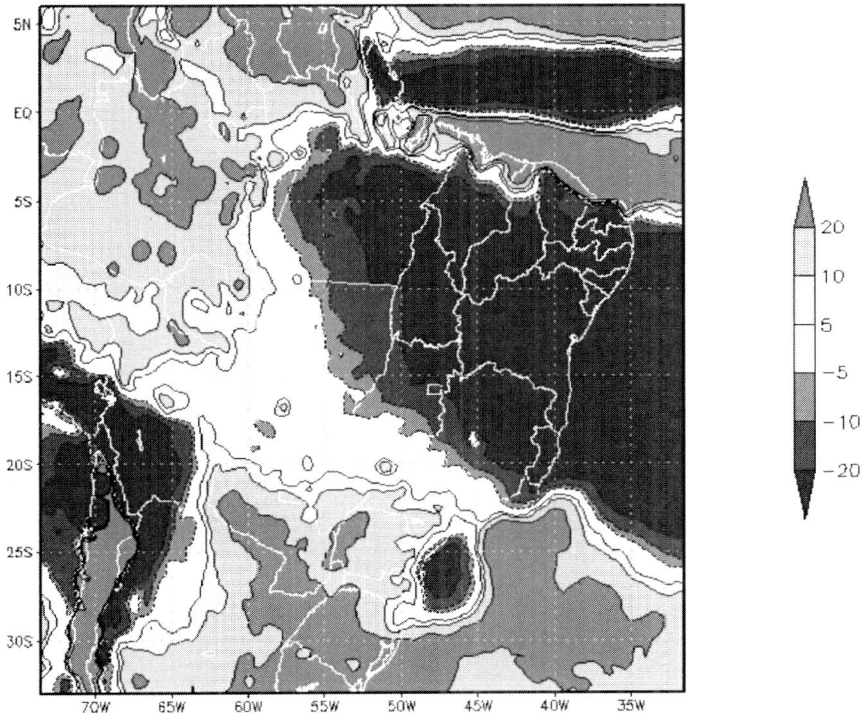

FIGURE 7. Projected change in precipitation (%) for 2011–2040 relative to the climatological run 1960–1990.

Wind Changes

Figures 10–12 show the projected wind changes for 2011–2040, 2041–2070, and 2071–2100, respectively. The results show an overall increase on the wind speed, mainly over the northeast region of Brazil where the winds are from the east. In some regions there is and increase greater than 1.5 m/s. In general the results show that regions of stronger winds can become even stronger. At the continental tropical areas the stronger winds are associated with the moisture convergence produced at the west region of the Amazon where there is an increase on the precipitation. This favors a stronger South Atlantic Convergence Zone (SACZ) that is aligned from the northwest to southeast where there is an increase on the precipitation that extends from the west Amazon to southeast Brazil (Figures 7–9).

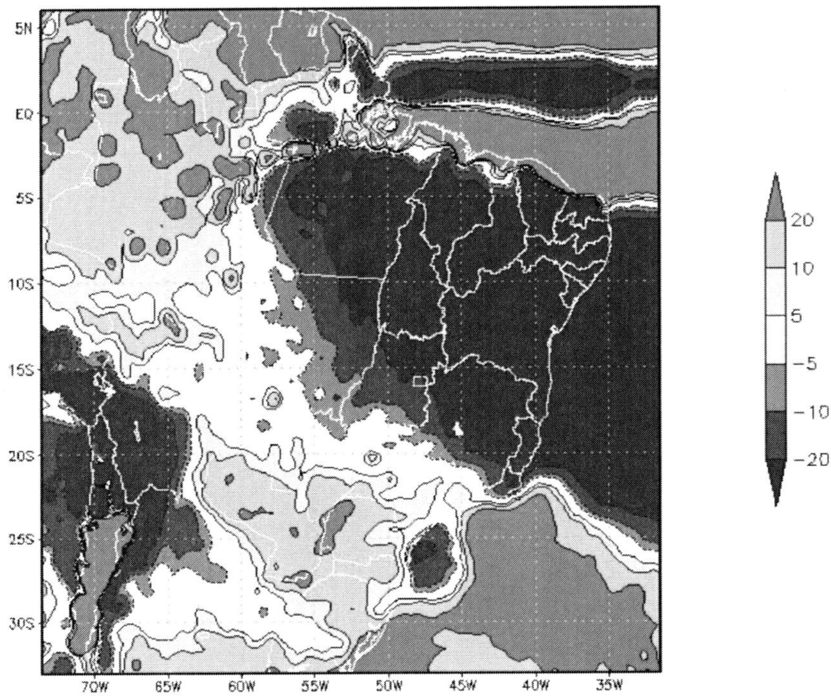

FIGURE 8. Same as figure **7**, but for precipitation changes (%) on the period 2041-2070.

DISCUSSION

Unless the regional climate limited area models, the new OLAM model has the capability to represent the large scale interaction with the regional local scale using a refining grid approach. Applied to study the impacts of global ocean warming on the South America climate, it shows a major increase on the temperature over the entire domain mainly over the Amazon region. Changes in the precipitation shows: an increase over the south Brazil and over western Amazon and a decrease over the northeast region of Brazil.

These projected changes in precipitation associated with changes on the convergence zones can impact the position and strength of the main tropical cells such as the Hadley Cell and the Walker Cell. While there is a broadening of the ITCZ associated with the Hadley Cell, the changes on the position of the Walker Cell are responsible for the drier climate over the northeast region.

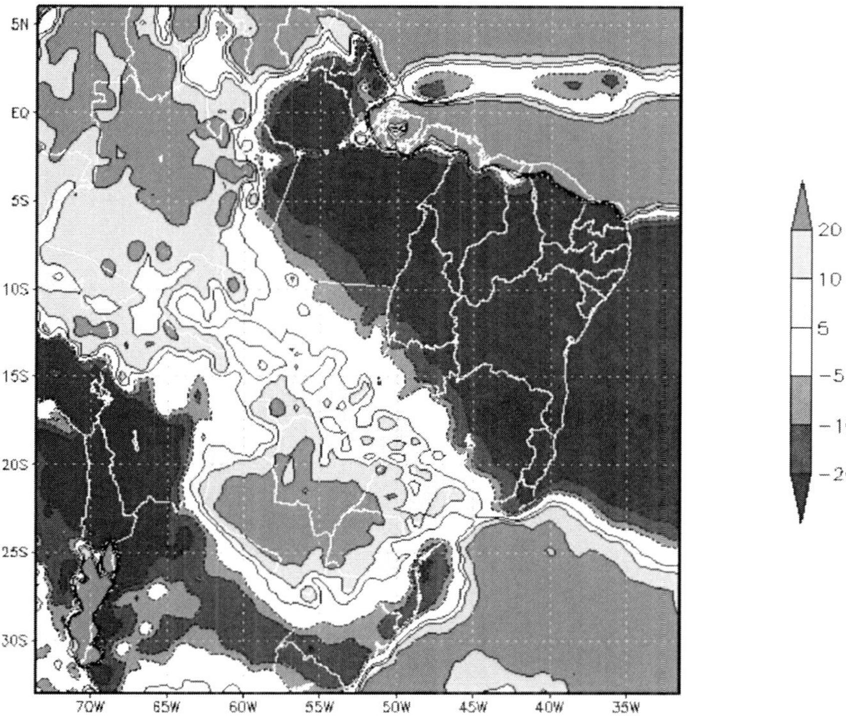

FIGURE 9. Same as figure **7**, but for precipitation changes (%) on the period 2071–2100.

Similar projections were obtained by the GCMs used on the CMIP project, but OLAM has a higher resolution and therefore it provides a detailed projection for the region. For instance, the model is able to resolve better the sea–land interfaces such as on the north region and the topography on the regions such as over the Andes Mountains. Similar projections were obtained by regional climate models applied for the South America region (Kitoh et al., 2011; Blázquez et al., 2012; Solman, 2013; Chou et al., 2014). However, those RCMs were limited area models and had to perform lateral nudging from other GCMs to produce their projections. OLAM has a new approach because it uses a global icosahedral grid along with a refining regional grid domain, and allows a two-way interaction between the global and local scales. In general, climate projections are obtained with the ensemble of various models. OLAM model adds an important new approach to this ensemble. Due to its capability to represent simultaneously phenomena of various scales, it can provide important improvements on the understanding of the physical processes involved on the regional climate change projections and its interaction with the global changes.

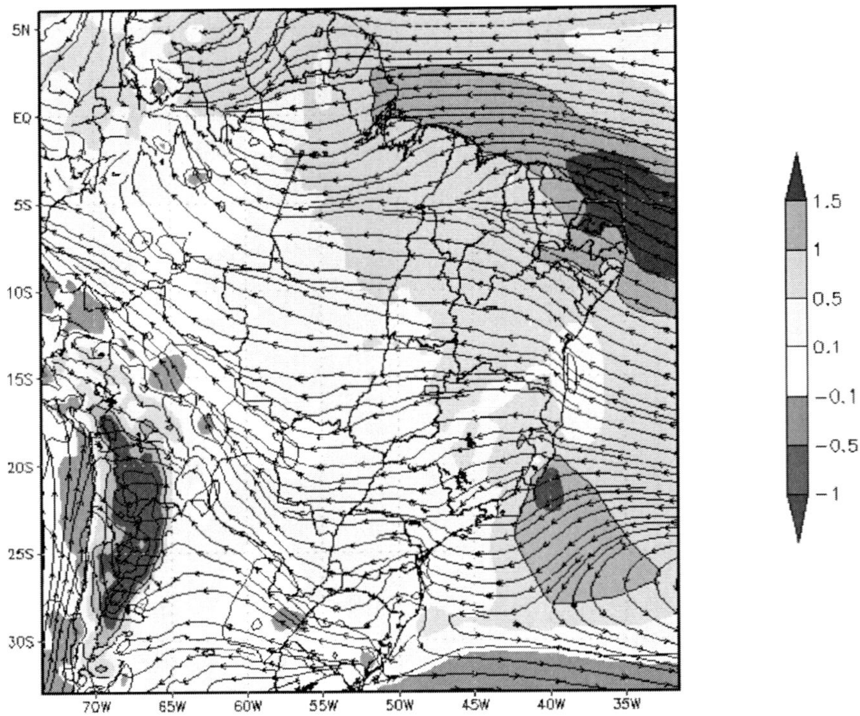

FIGURE 10. Wind changes (m/s) for the period 2011–2040 relative to the climatological run 1960–1990. The wind currents represent the mean wind field for the period.

The results obtained with OLAM did not included the greenhouse gases trends on its atmospheric model, however there was a noticeable enhancement on the greenhouse effect caused by the increase of the water vapor. The major results were a response from the ocean global warming. The South America region has a strong influence from the El Niño Southern Oscillation (ENSO) and La Niña events. Therefore, assuming that the Hadley Center model oceanic projections show a higher frequency of ENSO events (Collins, 2000), it is likely that those effects can have a crucial importance for the region's climate. The model results resemble the major findings from recent global and regional climate scenarios for the South America and therefore support the results for the climate projections obtained by the majority of those models for this region.

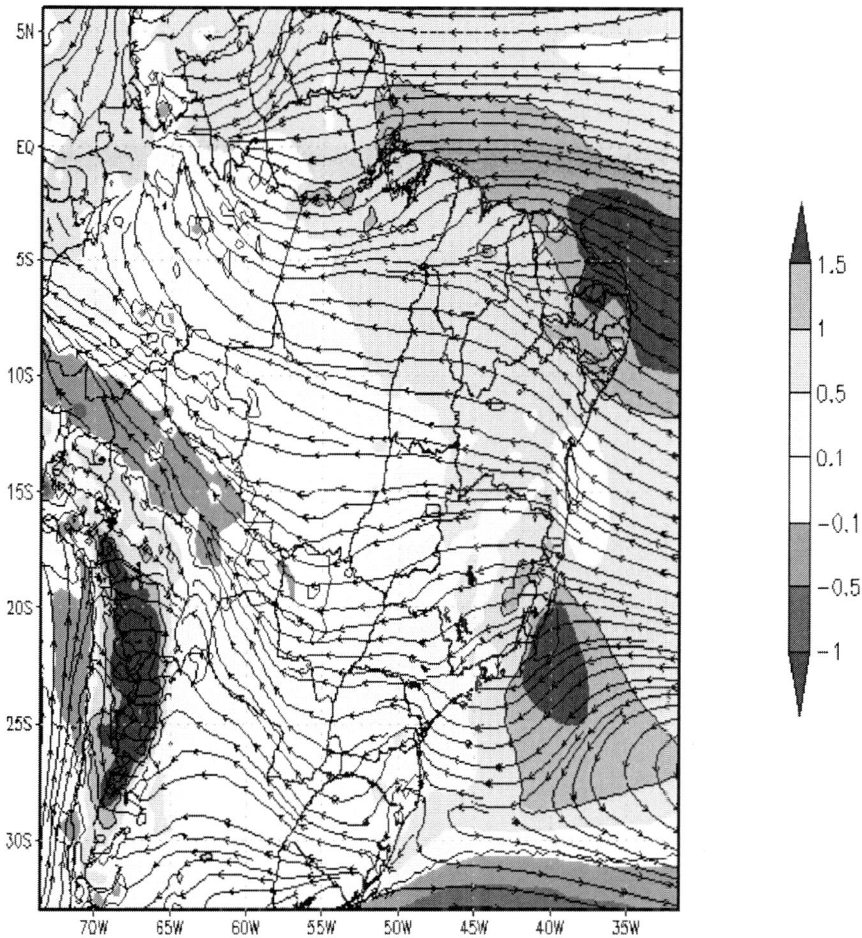

FIGURE 11. Same as figure **10** but for the wind changes (m/s) on the period 2041–2070.

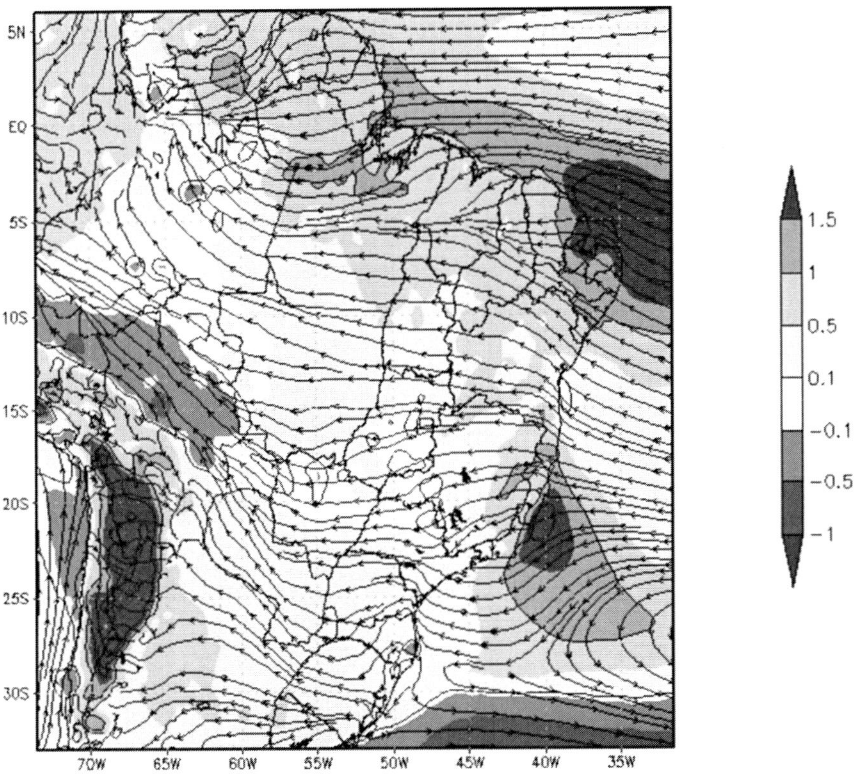

FIGURE 12. Same as figure **10**, but for the wind changes (m/s) on the period 2071–2100.

AUTHOR CONTRIBUTIONS

RD and RH worked together on the regional climate model setup, integration, and data analysis.

CONFLICT OF INTEREST STATEMENT

The authors declare that the research was conducted in the absence of any commercial or financial relationships that could be construed as a potential conflict of interest.

ACKNOWLEDGMENTS

We thanks to ANEEL and Tractebel Energy for their support on the development of this project.

REFERENCES

1. Adler, R. F., Huffman, G. J., Chang, A., Ferraro, R., Xie, P.-P., Janowiak, J., et al. (2003). The Version-2 global precipitation climatology project (GPCP) monthly precipitation analysis (1979–Present). *J. Hydrometeorol.* 4, 1147–1167.
2. Blázquez, J., Nuñez, M. N., and Kusunoki, S. (2012). Climate projections and uncertainties over south america from mri/jma global model experiments. *Atmos. Clim. Sci.* 2, 381–400.
3. Bombardi, R. J., and Carvalho, L. M. V. (2011). The South Atlantic dipole and variations in the characteristics of the South American Monsoon in the WCRP-CMIP3 multi-model simulations. *Clim. Dyn.* 36, 2091–2102.
4. Bombardi, R. J., Carvalho, L. M. V., Jones, C., and Reboita, M. S. (2014). Precipitation over eastern South America and the South Atlantic Sea surface temperature during neutral ENSO periods. *Clim. Dyn.* 42, 1553–1568.
5. Chen, C., and Cotton, W. R. (1987). The Physics of the marine stratocumulus-capped mixed layer. *J. Atmos. Sci.* 44, 2951–2977.
6. Chou, S. C., Lyra, A., Mourão, C., Dereczynski, C., Pilotto, I., Gomes, J., et al. (2014). Assessment of climate change over South America under RCP 4.5 and 8.5 downscaling scenarios. *Am. J. Clim. Chang.* 3, 512–525.
7. Cohen, J. C. P., Silva Dias, M. A. F., and Nobre, C. A. (1995). Environmental conditions associated with Amazonian squall lines. *A case study. Mon. Weather Rev.* 123, 3163–3174.
8. Collins, M., Tett, S. F. B., and Cooper, C. (2001). The internal climate variability of HadCM3, a version of the Hadley Centre coupled model without flux adjustments. *Clim. Dyn.* 17, 61–81.
9. Collins, M. (2000). The El Nino-Southern Oscillation in the second Hadley Centre coupled model and its response to greenhouse warming. *J. Clim.* 13, 1299–1312.
10. Collins, M., Knutti, R., Arblaster, J., Dufresne, J.-L., Fichefet, T., Friedlingstein, P., et al. (2013). *Climate Change 2013 The Physical Science Basis. Long-term Climate Change: Projections, Commitments and Irreversibility*, Working. Group. I Contribution to the Fifth Assessment. Report of the Intergovernmental Panel on Climate Change, 1029–1136.
11. Cotton, W. R., Pielke, R. A. Sr. Walko, R. L., Liston, G. E., Tremback, C. J., Jiang, H., et al. (2003). RAMS 2001: current status and future directions. *Meteorol. Atmos. Phys.* 82, 5–29.
12. Foltz, G. R., Grodsky, S. A., and Carton, J. A. (2003). Seasonal mixed layer heat budget of the tropical Atlantic Ocean. *J. Geophys. Res.* 108, 1–13.
13. Freitas, S. R., Silva Dias, M. A. F., Silva Dias, P. L., Longo, K. M., Artaxo, P., Andreae, M. O., et al. (2000). A convective kinematic trajectory technique for low resolution atmospheric models. *J. Geophys. Res.* 105, 24375–24386.
14. Fu, R. (2015). Global warming-accelerated drying in the tropics. *Proc. Natl. Acad. Sci. U.S.A.* 112, 3593–3594.
15. Gandu, A. W., Cohen, J. C. P., and de Souza, J. R. S. (2004). Simulation of deforestation in eastern Amazonia using a high-resolution model. *Theor. Appl. Climatol.* 78, 123–135.

16. Giorgi, F., and Gutowski, W. J. Jr. (2015). Regional dynamical downscaling and the CORDEX initiative. *Annu. Rev. Environ. Resour.* 40, 467–490.
17. Gonçalves, F. L. T., Martins, J. A., Albrecht, R. I., Morales, C. A., Silva Dias, M. A., and Morris, C. E. (2012). Effect of bacterial ice nuclei on the frequency and intensity of lightning activity inferred by the BRAMS model. *Atmos. Chem. Phys.* 12, 5677–5689.
18. Gordon, C., Cooper, C., Senior, C. A., Banks, H., Gregory, J. M., Johns, C., et al. (2000). The simulation of SST, sea ice extents and ocean heat transports in a version of the Hadley Centre coupled model without flux adjustments. *Clim. Dyn.* 16, 147–168.
19. Grell, G. A. (1993). Prognostic evaluation of assumptions used by cumulus parameterizations. *Mon. Weather Rev.* 121, 764–787.
20. Grell, G. A. (2002). A generalized approach to parameterizing convection combining ensemble and data assimilation techniques. *Geophys. Res. Lett.* 29, 1693.
21. Grimm, A. M. (2003). The El Nino impact on the summer monsoon in Brazil: regional processes versus remote influences. *J. Clim.* 16, 263–280.
22. IPCC (2007). *Climate Change 2007 - The Physical Science Basis.* Working Group I Contribution to the Fourth Assessment Report of the *IPCC.*
23. IPCC (2013). *Climate Change 2013: The Physical Science Basis.* Contribution of Working Group I to the Fifth Assessment Report of the Intergovernmental Panel on Climate Change.
24. Kain, J. S. (2004). The Kain–Fritsch convective parameterization: an update. *J. Appl. Meteorol.* 43, 170–181.
25. Kanamitsu, M., Ebisuzaki, W., Woollen, J., Yang, S.-K., Hnilo, J. J., Fiorino, M., et al. (2002). NCEP–DOE AMIP-II Reanalysis (R-2). *Bull. Am. Meteorol. Soc.* 83, 1631–1643.
26. Kirtman, B., Power, S. B., Adedoyin, J. A., Boer, G. J., Bojariu, R., and Camilloni, I. (2013). "Near-term climate change: projections and predictability," in *Climate Change 2013 - The Physical Science Basis, ed Intergovernmental Panel on Climate Change* (Cambridge: Cambridge University Press), 953–1028.
27. Kitoh, A., Kusunoki, S., and Nakaegawa, T. (2011). Climate change projections over South America in the late 21st century with the 20 and 60 km mesh Meteorological Research Institute atmospheric general circulation model (MRI-AGCM). *J. Geophys. Res.* 116, D06105.
28. Kucharski, F., Polzin, D., and Hastenrath, S. (2008). Teleconnection mechanisms of northeast Brazil droughts : modeling and empirical evidence. *Rev. Bras. Meteorol.* 23, 115–125.
29. Lindzen, R. S., and Nigam, S. (1987). On the role of sea-surface temperature-gradients in forcing low level winds and convergence in the tropics. *J. Atmos. Sci.* 44, 2418–2436.
30. Longo, K. M., Freitas, S. R., Pirre, M., Marécal, V., Rodrigues, L. F., Panetta, J., et al. (2013). The chemistry CATT BRAMS model (CCATT BRAMS 4.5): a regional atmospheric model system for integrated air quality and weather forecasting and research. *Geosci. Model Dev.* 6, 1389–1405.
31. Marengo, J. A., and Hastenrath, S. (1993). Case studies of extreme climatic events in the Amazon Basin. *J. Clim.* 6, 617–627.
32. Marengo, J. A., Chou, S. C., Kay, G., Alves, L. M., Pesquero, J. F., Soares, W. R., et al. (2012). Development of regional future climate change scenarios in South America using the Eta CPTEC/HadCM3 climate change projections: climatology and regional analyses for the Amazon, São Francisco and the Paraná River basins. *Clim. Dyn.* 38, 1829–1848.
33. Medvigy, D., Walko, R. L., Otte, M. J., and Avissar, R. (2010). The ocean–land–atmosphere model: optimization and evaluation of simulated radiative fluxes and precipitation. *Mon. Weather Rev.* 138, 1923–1939.
34. Medvigy, D., Walko, R. L., and Avissar, R. (2012). Simulated links between deforestation and extreme cold events in South America. *J. Clim.* 25, 3851–3866.
35. Meyers, M. P., Walko, R. L., Harrington, J. Y., and Cotton, W. R. (1997). New RAMS cloud microphysics parameterization. Part II: the two-moment scheme. *Atmos. Res.* 45, 3–39.

36. Nguyen, H., Evans, A., Lucas, C., Smith, I., and Timbal, B. (2013). The Hadley circulation in reanalyses: *climatology, variability, and change. J. Clim.* 26, 3357–3376.
37. Nguyen, H., Lucas, C., Evans, A., Timbal, B., and Hanson, L. (2015). Expansion of the Southern Hemisphere Hadley Cell in response to greenhouse gas forcing. *J. Clim.*, 28, 8067–8077.
38. Nobre, P., and Shukla, J. (1996). Variations of sea surface temperature, wind stress, and rainfall over the Tropical Atlantic and South America. *J. Clim.* 9, 2464–2479.
39. Ramos da Silva, R., Mota, M. A., Cohen, J. C. P., and Gandu, A. W. (2007). Progressos na Detecção e Previsão de Eventos Meteorológicos Extremos na Amazônia Oriental. *Boletim Soc. Brasil. Meteorol. SBMET* 31, 12–19.
40. Ramos da Silva, R., Werth, D., and Avissar, R. (2008). Regional impacts of future land-cover changes on the Amazon Basin wet-season climate. *J Clim.* 21, 1153–1170.
41. Ramos-da-Silva, R., Gandú, A. W., Cohen, J. C., Kuhn, P., and Mota, M. A. (2014). Weather forecasting for Eastern Amazon with OLAM model. *Rev. Bras. Meteorol.* 29, 11–22.
42. Rao, V. B., Giarolla, E., Kayano, M. T., and Franchito, S. H. (2006). Is the recent increasing trend of rainfall over northeast Brazil related to sub-Saharan drought? *J. Clim.* 19, 4448–4453
43. Ropelewski, C. F., and Halpert, M. S. (1987). Global and regional scale precipitation patterns associated with the El Niño/Southern oscillation. *Mon. Weather Rev.* 115, 1606–1626.
44. Servain, J., Caniaux, G., Kouadio, Y. K., McPhaden, M. J., and Araujo, M. (2014). Recent climatic trends in the tropical Atlantic. *Clim. Dyn.* 43, 3071–3089.
45. Smagorinsky, J. (1963). General circulation experiments wiht the primitive equations I. *The basic experiment. Mon. Weather Rev.* 91, 99–164.
46. Solman, S. A. (2013). Regional climate modeling over south america : a review. *Adv. Meteorol.* 2013:504357.
47. Taylor, K. E., Willianson, D., and Zwiers, F. (2000). *The Sea Surface Temperature and Sea-Ice Concentration Boundary Conditions For Amip II Simulations.* Livermore: University of California, Lawrence Livermore National Laboratory.
48. Taylor, K. E., Stouffer, R. J., and Meehl, G. A. (2012). An overview of CMIP5 and the experiment design. *Bull. Am. Meteorol. Soc.* 93, 485–498.
49. Walko, R. L., and Avissar, R. (2008a). The ocean–land–atmosphere model (olam). Part I: shallow-water tests. *Mon. Weather Rev.* 136, 4033–4044.
50. Walko, R. L., and Avissar, R. (2008b). The Ocean–Land–Atmosphere Model (OLAM). Part II: formulation and tests of the nonhydrostatic dynamic core. *Mon. Weather Rev.* 136, 4045–4062.
51. Walko, R. L. W. R, Cotton, M. P., Meyers, M. P., and and, Harrington, J. Y. (1995). New RAMS cloud microphysics parameterization Part I: the single-moment scheme. *Atmos. Res.* 38, 29–62.
52. Walko, R. L., Cotton, W. R., Feingold, G., and Stevens, B. (2000b). Efficient computation of vapor and heat diffusion between hydrometeors in a numerical model. *Atmos. Res.* 53, 171–183.
53. Walko, R. L., Band, L. E., Baron, J., Kittel, T. G. F., Lammers, R., Lee, T. J., et al. (2000a). Couple atmosphere-biophysics-hydrology models for environmental modeling. *J. Appl. Meteorol. Soc.* 39, 931–944.
54. Wang, C. Z. (2002). Atlantic climate variability and its associated atmospheric circulation cells. *J. Clim.* 15, 1516–1536.
55. Xie, S. P., and Carton, J. A. (2004). Tropical Atlantic variability: Patterns, mechanisms, and impacts. *Geophys. Monogr. Ser.* 147, 121–142.

CHAPTER 5

Conditions Supporting Funnel Cloud Development in Alaska

Stanley G. Edwin[1], Nicole Mölders[1,2*], Katja Friedrich[3], Sebastian Schmidt[3], Richard Thoman[4]

[1]*Department of Atmospheric Sciences, College of Natural Science and Mathematics, University of Alaska Fairbanks, Fairbanks, AK, USA*

[2]*Geophysical Institute, University of Alaska Fairbanks, Fairbanks, AK, USA*

[3]*Department of Atmospheric and Oceanic Sciences, University of Colorado at Boulder, Boulder, CO, USA*

[4]*National Weather Service Forecast Office, NOAA, Fairbanks, AK, USA*

ABSTRACT

The characteristics and climatology of funnel clouds in Alaska were examined using operational radiosondes, surface meteorological observations, and reanalysis data. Funnel clouds occurred under weak synoptic forcing between May and September between 11 am and 6 pm Alaska Daylight Time with a maximum occurrence in July. They occurred under Convective Available Potential Energy >500 $J \cdot kg^{-1}$ and strong low-level wind shear. Characteristic atmospheric profiles during funnel cloud events served to develop a retrieval algorithm based on similarity testing. Out of more than 129,000 soundings between 1971 and 2014, 2724, 442, and 744 profiles were similar to the profiles of observed funnel cloud events in the Interior, Alaska West Coast, and Anchorage regions. While the number of reported funnel clouds has increased since 2000, the frequency of synoptic situations favorable for such events has decreased.

Keywords: Alaska Funnel Clouds, Funnel Cloud Climatology, Similarity Retrieval Algorithm, Mesoscale Forcing for Funnel Cloud Formation in Alaska

1. INTRODUCTION

Over the last decade, Alaska funnel clouds received increasing attention from the media and public because of the increase of funnel cloud reports and interest in climate change [1] . Most reports come from the public in populated areas, weather forecasters or trained spotters, and passengers and pilots within flight corridors. Less than 2% of Alaska is developed (Figure 1), meaning that proba bility is high for events to remain unreported in unpopulated areas [2] . Furthermore, only eight operational Weather Surveillance Radar 88 Doppler (WDR-88D) radars exist in Alaska leaving large areas without radar coverage.

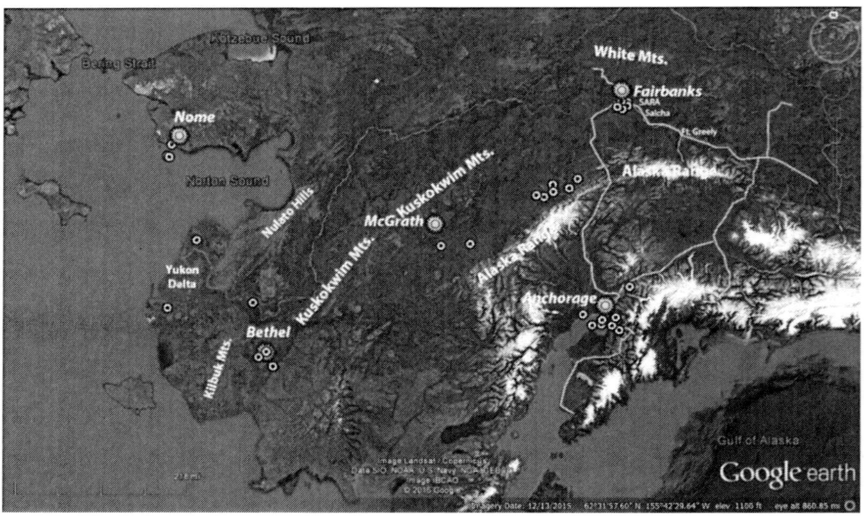

Figure 1. White circles denote the location of funnel cloud sightings in Alaska from 1955 to 2014. Yellow stars indicate radiosonde launch sites. Chartreuse lines are the paved roadnet. Names refer to locations mentioned in the text. SARS is the Small Arms Range Alaska. The Brooks Range is North of the map.

Alaskans have followed the reports of funnel clouds with great interest and concern as funnel clouds mean a threat to air traffic [1] . Due to Alaska's vast land and low population density (0.5 persons per square kilometer), many villages are off the road network (Figure 1). Using small aircrafts to reach one of the three major cities (Anchorage, population 300,950, Fairbanks 32,324, Juneau 32,660) for shopping, medical care, or visiting is as common to Alaskans like using subways, trams, taxis, and busses for metropolitan residents elsewhere in the United States. These aircrafts, which fly below 3 km above ground level, also deliver general supply, mail service, and medical transport to/from remote settlements [1] .

Three major mountain ranges, the Brooks Range, White Mountains, and Alaska Range, run from west to east. The Brooks Range is the northernmost, and the Alaska Range, the southernmost (Figure 1). Between the ranges are boreal forest-covered wide valleys.

This complex topography of high glacier-covered mountains, steep mountain passes, and wide, low populated valleys exposes aviation to several threats. In mountainous terrain, like along the coast, weather can change quickly [3] . Fog can close mountain passes and valleys, and thunderstorms may build in air corridors, and in-cloud icing may occur [4] . Funnel clouds add another potential flight hazard [1] .

No forecasting system exists for funnel clouds in Alaska [1] . The sighting and software detection protocols for tornadoes used by the National Weather Service (NWS) were developed for the Central United States. There, the majority of funnel clouds are associated with tornadoes and develop in high-reaching severe thunderstorms [5] [6] . As our analysis showed, none of the funnel cloud sightings in Alaska were associated with these types of mesoscale systems [1] . Consequently, the applied protocols may of limited or no use for Alaska.

Understanding the characteristics, forcing, formation and climatology of Alaska funnel clouds is vital towards forecasting them. The goal of this study was to establish a funnel cloud climatology based on radiosonde and reanalysis data to shed light on the characteristics and forming mechanisms. We hypothesized the following: a) Funnel clouds occur under distinct weak synoptic scale conditions that permit formation of topography induced mesoscale systems; b) The vorticity required to generate funnel clouds is due to interaction of the synoptic scale wind field and local wind systems. Consequently, similarity criteria can be determined to identify past and future potential funnel cloud events. While the latter is of interest for forecasting, the former permits assessment of frequency of funnel cloud occurrence to evaluate the increase in funnel cloud reports events since 2000.

2. DATASETS AND METHODS

2.1. Funnel Cloud Reports

The Alaska NWS has collected funnel cloud reports since 1955 [7] . This dataset excludes tornadoes except for persuasive purposes on radar detection capabilities. The dataset encompassed 43 funnel cloud reports between 1955 and 2014. Most reports came with broad location descriptions instead of geographic coordinates. We discarded two sightings due to missing date, time, and location data. We excluded water sprouts or funnel clouds spotted over ocean from our study.

All remaining 41 funnel cloud events occurred near populated areas, along air corridors, along the coasts and within major valleys (Figure 1). Under consideration of the Alaska climate divisions [3] [8] , we grouped the 41 reported events into three regions of similar atmospheric and landscape characteristics: a) The Interior with Fairbanks and McGrath characterized by the Tanana Valley with mountains to the North and South; b) the Alaska West Coast with Nome and Bethel influenced by the Beaufort Sea; and c) the Anchorage area governed by mountains and the northern coast of the Gulf of Alaska.

2.2. Soundings, Weather Radar, Reanalysis, and Surface Observations

In Alaska, operational twice daily (00 UTC, 12 UTC) radiosonde launches started in the 1940s. Archiving started 1948. The launch time changed in 1957. Since then, radiosonde data are available in winter at 2 pm Alaska Standard Time (AKST = UTC-8 h) the previous day and 2 am the same day, and in summer at 3 pm the previous day and 3 am Alaska Daylight Saving Time the same day (AKDT = UTC-9 h). The soundings provide pressure, height, temperature, wind speed and direction, and dew-point temperature.

Prior to 1971, most dew-point temperature data are missing. Due to technical issues, there are periods of missing dew-point temperature data or otherwise incomplete radiosonde data between 1989 and 1999. Out of the 41 funnel cloud reports, 32 events occurred within ±6 hours of the launch. Of these 32 sightings, 19 occurred within ±2 hours of a launch, and inside a 40-km radius of the site (Figure 1).

In Alaska, operational radar observations started in 1993, archiving in 1997. Between 1997 and 2014, ten funnel cloud events occurred within the 230-km radar range (four at Anchorage, two at Fairbanks, two at Bethel, two at Nome). Only four of them had latitude and longitude coordinates [7] for locating and analysis in radar data.

In Alaska, surface observations are sparse. Hourly observations of near-surface wind direction from the National Climate Data Center were available for some locations for limited time. We used the National Center for Environmental Protection/National Center for Atmospheric Research global reanalysis data [9] [10] at 300, 500 and 1000 hPa.

2.3. Data Processing

Progress in measurement techniques, digital possibilities and radiosonde-loca- tion retrieval increased the accuracy and data resolution over time [11] . Consequently, the radiosonde dataset is inhomogeneous with respect to data quality, accuracy, and quantity. The increased data quantity means an increase in vertical and temporal resolution during an ascent.

Investigations of climatology and changes in climatology require data of same or at least comparable resolution, accuracy, and quality [2] . We homogenized the radiosonde dataset by introducing mandatory pressure ranges (Table 1), in which we analyzed the observed funnel cloud events to determine profile characteristics and later to retrieve profiles like those observed during funnel cloud events.

To identify the synoptic scale situations under which funnel clouds occurred, we analyzed the reanalysis maps at 300, 500 and 1000 hPa for all 41 observed events. We made composite maps of synoptic situations of observed funnel cloud events for each region.

We also calculated composite maps using the reanalysis data of those days identified by the similarity retrieval algorithm (see Section 2.5) as potential funnel cloud events. Comparison of the composites of retrieved and observed funnel cloud events served to evaluate whether the retrieved events represented synoptic conditions in a mesoscale sense (weak synoptic forcing, strong wind shear in ABL, close to or saturated air in the lower troposphere with strong wind shear in the ABL) like on days with funnel cloud events. If so, the retrieval algorithm could identify synoptic situations suitable for funnel cloud events in numerical weather prediction (NWP) data.

2.4. Characteristics of Observed Funnel Cloud Events?Baseline Profiles

Theoretically (cf. [5] [12] [13]), the concentration of funnel cloud reports over the summer and between 11 am and 6 pm AKDT as well as the lack of thunderstorms near the reported location suggested that terrain- and ocean-induced circulations may be key in funnel cloud formation. During funnel cloud events, lifting condensation levels were between 1000 and 800 hPa. Close to or saturated conditions existed in a layer of 500 m to 2 km thickness (Figure 2).

Table 1. Baseline profile limits based on the sounding analysis: Mandatory pressure (p) ranges required to have valid measurements for being considered in the determination of the baseline profile limits of funnel cloud characteristics for the three regions for air temperature (T), dew-point temperature (T_d), and wind speed (v). The closest radiosonde site in a region was used except for events that fell between two sites. Baseline profiles and limits were then determined for the same event for both sites. In the columns named "events between Fairbanks and McGrath" and "events between Bethel and Nome," the first and second columns refer to the first and second mentioned sites. For further details see text.

Regions	Interior				Alaska West Coast			Anchorage area
Sounding site	Fairbanks	McGrath	Fairbanks and McGrath		Bethel	Bethel and Nome		Anchorage
p (hPa)				T (°C)				
250 - 150	−60.8 - −43	−71.2 - −54	−63.9 - −50.5	−66.2 - −58.6	−71.4 - −52.8	−70 - −57.4	−71.7 - −56.4	−78.0 - −63.0
350 - 250	−52.2 - −44	−55.4 - −48	−56.6 - −40	−53 - −49.1	−58.2 - −45.3	−57.7 - −49.6	−58.7 - −47.3	−77.5 - −42.0
350 - 250	−52.2 - −44	−47.8 - −26	−50.8 - −28.6	−36.6 - −31.2	−53.9 - −32.2	−41.7 - −32.9	−43 - −33.7	−50.0 - −38.3
550 - 450	−24.1 - −16	−27.9 - −21	−53.1 - −14.9	−47.9 - −10.3	−48.1 - −18.9	−31.1 - −19.7	−40.4 - −17.4	−42 - 25.7
738 - 650	−6.4 - −1	−11.6 - −5	−14.5 - −2.1	−4.4 - −7.5	−14.5 - −5.5	−12.6 - −4.6	−18.2 - .5	−36.0 - −1.2
888 - 813	4.4 - 11	−1.6 - 7	−4.6 - 3.4	−0.5 - 1.7	−0.5 - 1.5	−1.1 - 3.6	−0.8 - 5.7	−12.0 - 6.4
963 - 888	9.1 - 12	6 - 11	2 - 6.8	−0.5 - 6.7	1.1 - 9.2	1.6 - 7.8	2.1 - 9.8	−9.0 - 4.2
>963	10.8 - 21	14 - 21.3	5.3 - 9.9	5.1 - 9.7	6.5 - 12.3	5.6 - 9.6	2.7 - 12.9	−0.8 - 9.4
p (hPa)				T_d (°C)				
250 - 150	−77.6 - −53	−57.5 - −43	−56.7 - −51.4	−57.3 - −50.3	−57.2 - −45.1	−49 - −45	−56.7 - −42	−53.8 - −42.0
350 - 250	−61 - −50	−45.3 - −45	−47.6 - −45.2	−47 - −45.1	−48.3 - −38.4	−51.9 - −42.6	−48.3 - −45	−53.5 - −45.3
350 - 250	−61 - −50	−30.3 - −20	−32.3 - −29	−32.2 - −29.1	−33.4 - −25.7	−37.6 - −29.6	−34.9 - −30.7	−44.0 - −31.4
550 - 450	−27.8 - −21	−20 - −16	−23.2 - −16.3	−24.4 - −16.2	−22.6 - −13.7	−26 - −17.5	−23.8 - −18.4	−32.0 - −19.6
738 - 650	−9.5 - −1	−2.3 - 0	−8.9 − -2.2	−9.9 - −3.2	−6.6 - −0.6	−9.2 - −3.7	−7.3 - −2	−13.0 - −4.0
888 - 813	0.2 - 7	8.2 - 12	2.6 - 8.3	−0.4 - 5.6	1.7 - 8.8	−0.4 - 6.4	2.2 - 7.2	−2.4 - 5.4
963 - 888	4.4 - 12	14.2 - 16	6 - 13.9	3.9 - 11.2	4.4 - 13.9	3.7 - 10.6	7.2 - 12.6	−0.2 - 13.3
>963	9.6 - 15	6.8 - 29	6 - 17.5	5.9 - 13.5	9 - 22.5	8.19 - 18.7	9.7 - 20.8	9.5 - 15.3
p (hPa)				v ($m \cdot s^{-1}$)				

250 - 150	−77.6 - −53	−57.5 - −43	−56.7 - −51.4	−57.3 - −50.3	−57.2 - −45.1	−49 - −45	−56.7 - −42	−53.8 - −42.0
350 - 250	−61 - −50	−45.3 - −45	−47.6 - −45.2	−47 - −45.1	−48.3 - −38.4	−51.9 - −42.6	−48.3 - −45	−53.5 - −45.3
350 - 250	−61 - −50	−30.3 - −20	−32.3 - −29	−32.2 - −29.1	−33.4 - −25.7	−37.6 - −29.6	−34.9 - −30.7	−44.0 - −31.4
550 - 450	−27.8 - −21	−20 - −16	−23.2 - −16.3	−24.4 - −16.2	−22.6 - −13.7	−26 - −17.5	−23.8 - −18.4	−32.0 - −19.6
738 - 650	−9.5 - −1	−2.3 - 0	−8.9 − −2.2	−9.9 - −3.2	−6.6 - −0.6	−9.2 - −3.7	−7.3 - −2	−13.0 - −4.0
888 - 813	0.2 - 7	8.2 - 12	2.6 - 8.3	−0.4 - 5.6	1.7 - 8.8	−0.4 - 6.4	2.2 - 7.2	−2.4 - 5.4
963 - 888	4.4 - 12	14.2 - 16	6 - 13.9	3.9 - 11.2	4.4 - 13.9	3.7 - 10.6	7.2 - 12.6	−0.2 - 13.3
>963	9.6 - 15	6.8 - 29	6 - 17.5	5.9 - 13.5	9 - 22.5	8.19 - 18.7	9.7 - 20.8	9.5 - 15.3
p (hPa)				v (m·s^{-1})				
250 - 150	0.8 - 12	0.9 - 11	1.3 - 13.5	6.6 - 25.9	7.8 - 42	0.6 - 9	0 - 18.3	2.2 - 23.4
350 - 250	5.2 - 6	5.6 - 6	0 - 13.5	5.2 - 31.7	9.4 - 42.4	2.6 - 20.8	0 - 26	26.9 - 31.7
350 - 250	5.2 - 6	1 - 11	4 - 11.4	4.1 - 26	8.2 - 27.5	3.4 - 14.5	0.7 - 21	5.0 - 20.2
550 - 450	0 - 8	0 - 12	1 - 15.7	3.2 - 24.5	4.6 - 21.2	2 - 12.5	2.4 - 14.6	1.8 - 15.2
738 - 650	0 - 8	0.9 - 5	0 - 21.9	2.4 - 14.3	1.6 - 13.5	3.5 - 10.4	1.5 - 14.6	1.6 - 8.0
888 - 813	2.4 - 4	0.4 - 2 .	0 - 16.3	0 - 15.4	2.7 - 6.3	2 - 12.2	2.4 - 7.1	0.0 - 11.0
963 - 888	0.0 - 1.7	0 - 8	0 - 14.8	2 - 16.8	1.1 - 3.5	2.3 - 12.4	0.9 - 6.6	0.0 - 18.4
>963	1.6 - 4	0 - 3	0 - 4.6	0 - 9.5	0.7 - 3.1	0.8 - 7.3	2.3 - 5.5	2.0 - 6.4

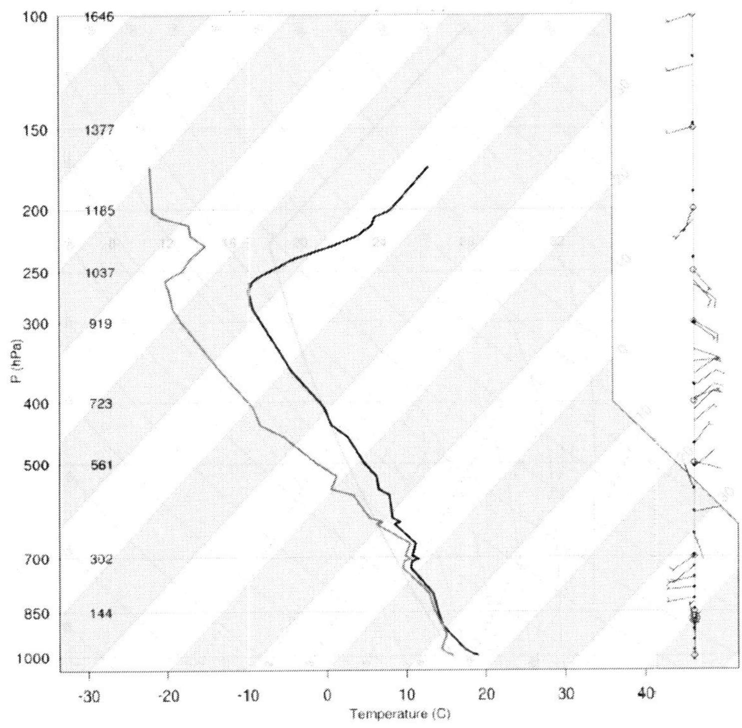

(a)

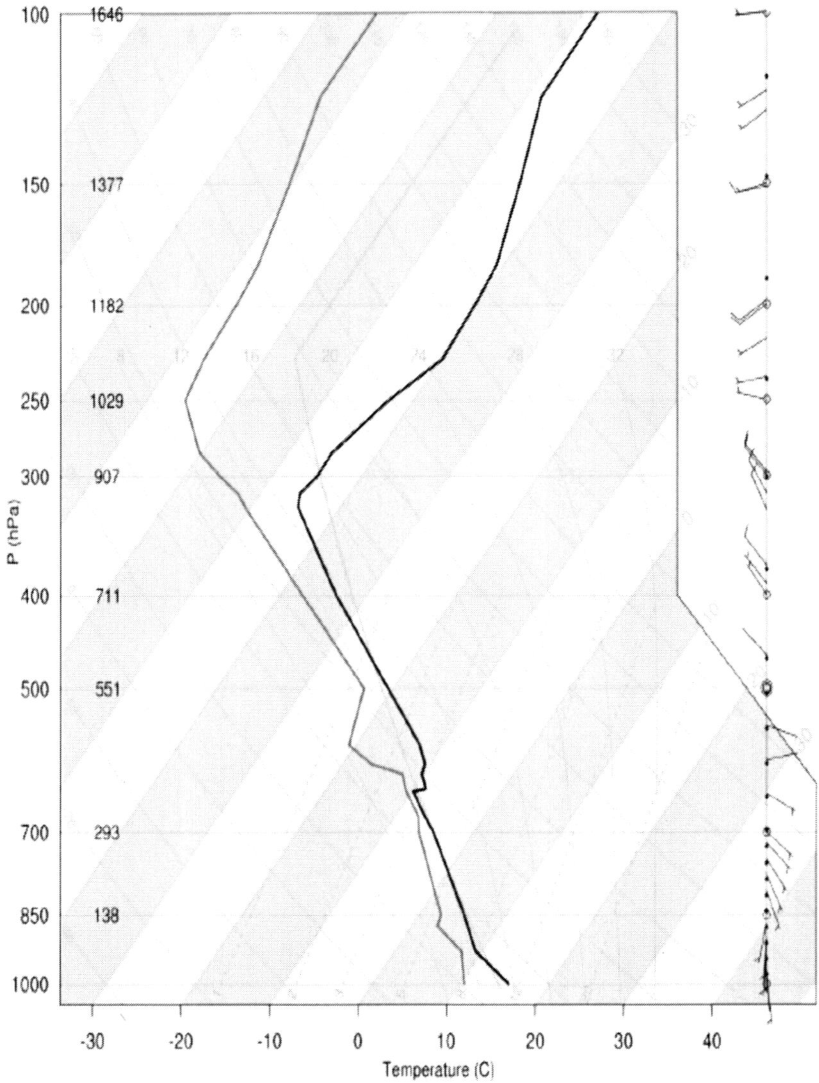
100
1646
150
1377
200
1182
250
1029
300
907
400
711
500
551
700
293
850
138
1000
P (hPa)
-30
-20
-10
0
10
20
30
40
Temperature (C)

(b)

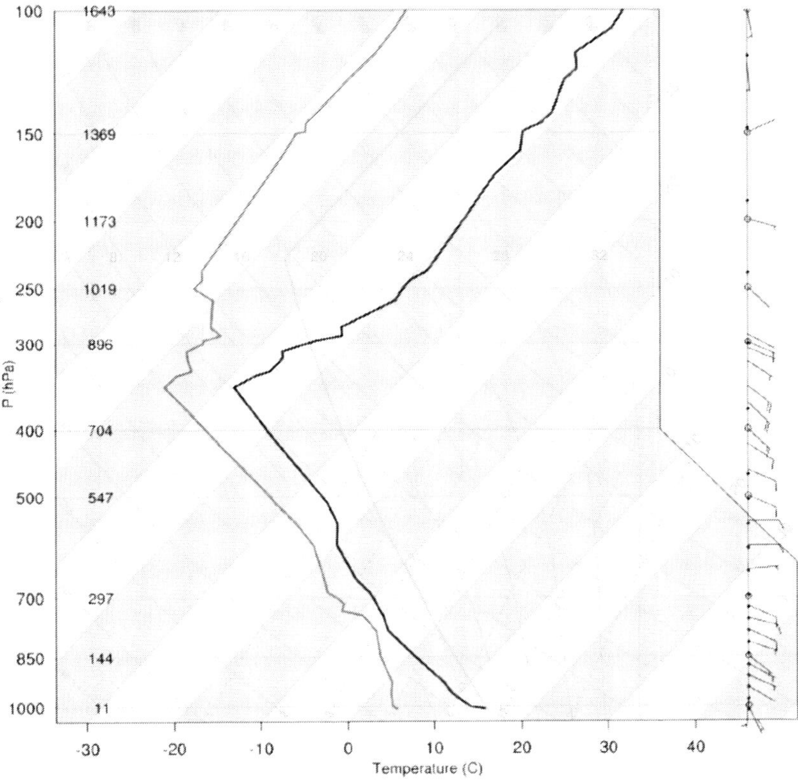

(c)

Figure 2. Examples of radiosonde profiles on days with funnel cloud sightings (a) at Fairbanks in the Interior on July 22, 2010 12 UTC; (b) at Bethel in the Alaska West Coast region on July 29, 2004 00 UTC and (c) close to Anchorage on July 1, 2012 00 UTC. At the time of the radio soundings, convective available potential energy (CAPE) values were 1403 $J{\cdot}kg^{-1}$, 1234 $J{\cdot}kg^{-1}$, and 939 $J{\cdot}kg^{-1}$, lifting condensation levels were at 950 hPa, 928 hPa, and 868 hPa and cloud precipitable water was 3 cm, 2 cm, and 1 cm at Fairbanks, Bethel, and Anchorage, respectively. The actual funnel clouds occurred in Fairbanks on 2010-07-22 9:47 am, and between Bethel and Nome on 2004-07-28 at 1:25 pm AKDT. No exact time was reported for the funnel cloud in the Anchorage region.

To understand the interaction of forcing at the synoptic scale and smaller mesoscale, we used the soundings closest to the observed funnel cloud event to assess the typical atmospheric conditions. Since dew-point temperatures were widely unavailable in the dataset prior to 1971, the number of funnel cloud events reduced to a total of 26 events with 8 for the Anchorage, 13 for the Interior, and 5 for the Alaska West Coast area.

In these regions, air and ground temperature conditions vary widely from May to September [3] . Given the small sample sizes for each region, means and standard deviations failed to represent the range of funnel cloud conditions and to create a representative baseline profile.

Therefore, for any pressure level within the limits of one of the pressure ranges, the associated height, air and dew-point temperature were stored for the funnel cloud events for each radiosonde site. In case of funnel cloud sightings between two radiosonde stations (Fairbanks and McGrath, Bethel and Nome), we stored baseline profiles for both sites.

In a next step, the maximum and minimum limits of the stored variables were determined from all data available for that pressure range and radiosonde site. Table 1 lists the resulting minimum and maximum profiles that give the range of baseline profiles of funnel cloud events in the respective regions.

Since the events occurred in different places (Figure 1) with different topography (e.g. valley height, slope direction, steepness), wind direction differed largely among events in the atmospheric boundary layer (ABL). Therefore, no baseline profile ranges were determined for wind direction.

2.5. The Similarity Retrieval Algorithm

Once the baseline profile limits were derived, we applied the following retrieval algorithm for similarity testing. We searched the entire sounding database between May and September between 1971 and 2014 for other days with profiles that fulfill the baseline profile limits given in Table 1.

For profiles fulfilling these conditions, CAPE was calculated. Since reported funnel cloud events that were close in time or space to the launch time and site had CAPE about 500 $J \cdot kg^{-1}$ or higher (Figure 3(a)), all those profiles falling within the baseline profile limits (Table 1) with CAPE > 500 $J \cdot kg^{-1}$ were considered as having atmospheric conditions that could potentially lead to a funnel cloud event. Note that the atmospheric conditions during the funnel cloud event, which occurs close to the site and close to the launch time, are represented better by the observed profile than the atmospheric conditions of events farther away or with a large difference between the time of the sounding and event.

In the case of funnel clouds between two sites, the similarity retrieval algorithm screened for potential events using the data of both radiosonde sites and the baseline profile limits determined for funnel clouds between

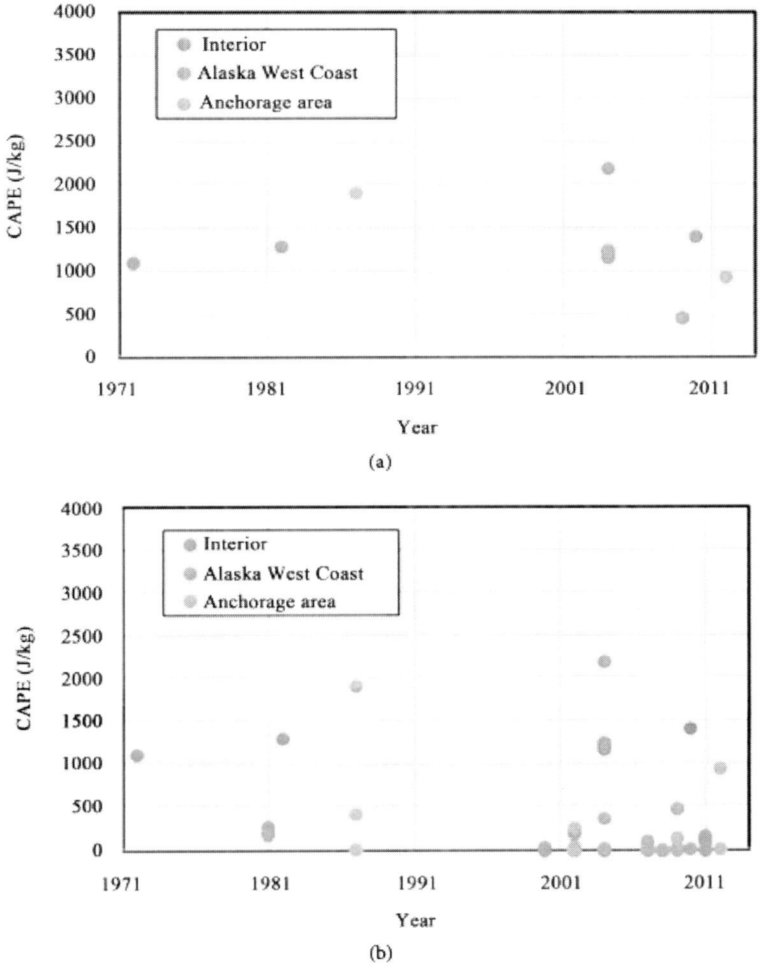

Figure 3. Frequency of observed funnel cloud events (circles) and their CAPE (a) within 2 hours of the radiosonde sounding or close to the site; (b) all events for which radiosonde data were available no matter of the temporal or spatial distance from the launch time and site. A circle corresponds to one event each. The gaps between 1989 and 1999 are due to erroneous dew-point temperature data and missing data.

the sites (Table 1). For atmospheric conditions to be considered as being like those during funnel cloud events occurring between the two sites, at each site, the profile had to fall within the limits identified for that respective site for "events between sites," and CAPE had to exceed 500 $J \cdot kg^{-1}$ at one site. For more details, see [1] .

3. FREQUENCY OF FUNNEL CLOUDS

Four funnel cloud events occurred between 1950s and 1960s, 12 in the 1980s and 1990s, and 26 between 2000 and 2014 [7] . All events that occurred close in space or time to the location of the radiosonde site or time of the sounding had CAPE of about or more than 500 $J \cdot kg^{-1}$ (Figure 3(a)). Most funnel cloud events occurred between May and September with a peak in July (Figure 3(b)), and between 11 am and 6 pm AKDT. Since funnel clouds occurred in various months and at different times in the diurnal cycle, the ambient conditions during these events varied over a broad range of air and dew-point temperatures (Table 1).

In the Interior, 13 funnel clouds were reported between 1955 and 2014. Near Fairbanks, three funnel events occurred in the 80s and two in the new Millennium. One event occurred in May, three in July, and one in August. Near McGrath, two funnel cloud sightings occurred in July. Six funnel cloud events, three in July, and August each occurred between Fairbanks and McGrath. Here, all events, but one were between 9 am and 6 pm AKDT.

In the Alaska West Coast area, five funnel cloud events occurred near Bethel in the new Millennium, in the afternoon to early evening, with one in June, three in August, and one in September. Two of the five sightings were on the same day late in August. Six events occurred between Bethel and Nome, of which all but one were in the new Millennium.

Eight funnel cloud events were reported between May and August near Anchorage. All events occurred between 7:30 am and 6 pm AKDT.

The similarity retrieval algorithm identified 2724, 442, and 744 profiles like those of funnel cloud events in the Interior, Alaska West Coast, and Anchorage region, respectively (Figure 4). In the Interior, 1303, 417, and 1004 potential funnel cloud events were found when searching the radiosonde data of Fairbanks, the combined data of Fairbanks and McGrath, and those of McGrath. In the Alaska West Coast region, 404 potential funnel cloud events were retrieved from the radiosonde data of Bethel, and 38 in the combined data of Bethel and Nome.

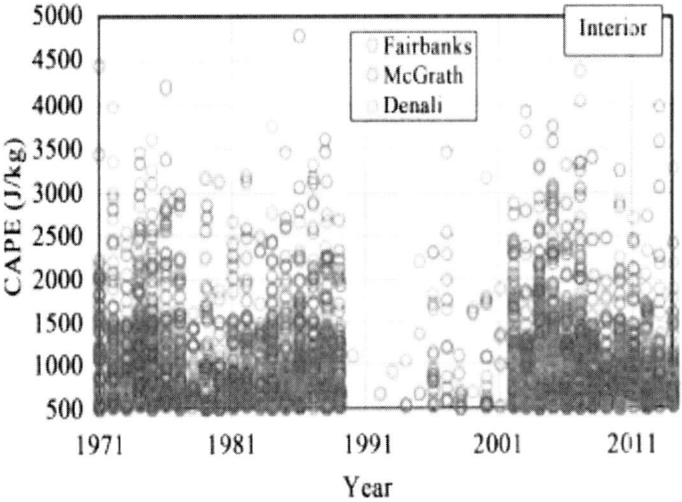
Interior
Fairbanks
McGrath
Denali
CAPE (J/kg)
5000
4500
4000
3500
3000
2500
2000
1500
1000
500
1971
1981
1991
2001
2011
Year

(a)

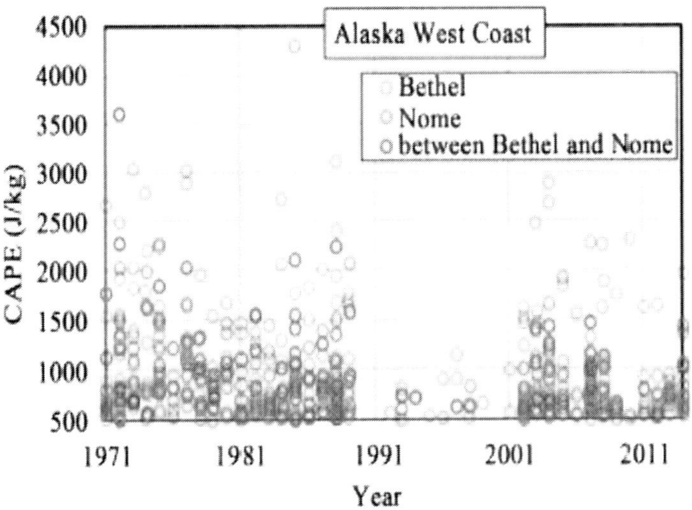
Alaska West Coast
Bethel
Nome
between Bethel and Nome
CAPE (J/kg)
4500
4000
3500
3000
2500
2000
1500
1000
500
1971
1981
1991
2001
2011
Year

(b)

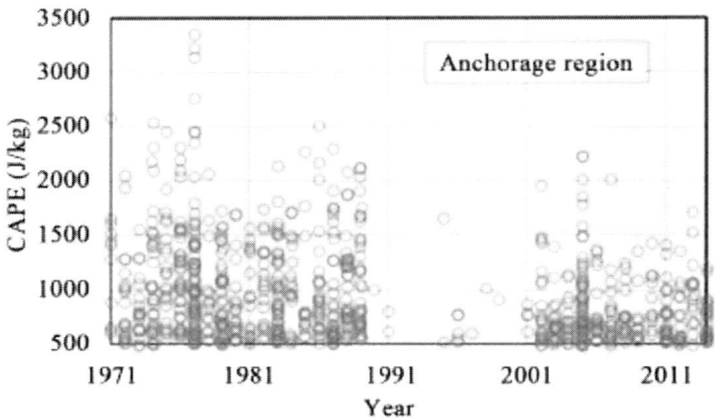

(c)

Figure 4. Frequency of retrieved funnel cloud events (circles) and their CAPE for the (a) Interior; (b) Alaska West Coast; and (c) Anchorage region between 1971 and 2014. A circle corresponds to one event each. The gaps between 1989 and 1999 are due to erroneous dew-point temperature data and missing data. Legends differ from Figure 3.

The retrieval algorithm identified more funnel cloud events prior to 1989 compared to after 2000. However, the number of observed funnel clouds has increased after 2000. If the retrieval algorithm works, these findings suggest that the NWS' funnel-cloud observation record is certainly biased by the increase in population and availability of devices for documentation.

No circulation pattern in the Doppler velocity could be found in the radar observations mainly due to the coarse temporal or spatial resolution of the radar and due to the blocking of the radar beam in the complex terrain. Nevertheless, the radar data indicated that no thunderstorms were present during the time and near the funnel cloud location.

4. ENVIRONMENTAL CONDITIONS DURING OBSERVED AND RETRIEVED FUNNEL CLOUD EVENTS

We analyzed the synoptic (mesoscale α) and mesoscale β/γ conditions during reported funnel cloud events and compared those to the synoptic and mesoscale β/γ conditions during retrieved funnel cloud events. The goal of this analysis was twofold: a) Evaluate the similarity retrieval algorithm and determine if the algorithm might be useful for identifying days of potential funnel cloud occurrence; and b) understand under which synoptic and mesoscale conditions funnel clouds might form, which could guide a more detailed analysis in the future.

4.1. Interior

During the observed funnel cloud events, the synoptic scale forcing was weak over the Interior throughout the troposphere. At 300 hPa, a common feature was a strong gradient in geopotential height south of the Alaska Range over the northern Gulf of Alaska and a weak gradient over the Interior (Figure 2(a)). The polar jet was located far south over the Gulf of Alaska with a west to east orientation parallel to the northern coast of the Gulf of Alaska. Per the reanalysis, at 500 hPa, the gradient winds came from the south, advecting air from the Gulf of Alaska across the Alaska Range towards the Tanana Valley, while near-surface winds blew west to east, up the Tanana Valley (Figure 5(a)). Conditions were zonal and quasi-stationary with respect to the temperature conditions above the ABL. Inversions were visible at different heights between 700 and 500 hPa in the Skew-T diagrams of all observed funnel cloud events (e.g. Figure 2(a)).

In all observed funnel cloud events, the lower troposphere was moist or saturated; while the upper troposphere was relatively dry (e.g. Figure 2(a), Table 1). The radiosonde profiles showed strong wind shear, but no common wind direction in the ABL and mid-troposphere. The latter is because the observed events occurred at different locations with different exposure and steepness of slopes.

The composite surface synoptic maps of retrieved and observed funnel cloud events were similar in a mesoscale sense (e.g. Figure 5(a) and Figure 5(b)). Individual mid- and upper air maps suggested two weak synoptic scale weather conditions were favorable for funnel cloud events in the Interior: a) A high- pressure system over the Beaufort Sea including the North Slope with low pres sure over the Gulf of Alaska or b) a low off the Pacific Northwest coast with marginal pressure differences north of the Alaska Range. Consequently, discrepancy between the composites from retrieved and observed events in the mid- and upper troposphere was expected (e.g. Figure 5(a) and Figure 5(b)).

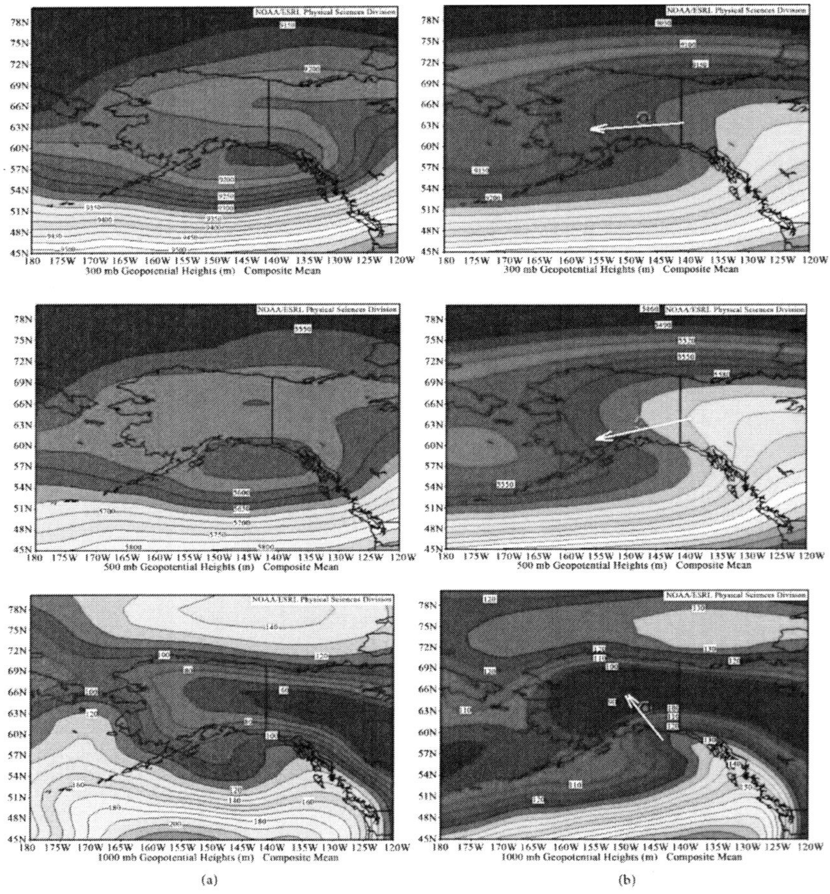

Figure 5. Composites of reanalysis data showing geopotential heights at 300 hPa (top panel), 500 hP (middle panel), and 1000 hPa (lower panel) on days of (a) observed; and (b) retrieved funnel cloud events in the Fairbanks region of the Interior. The synoptic situations for events around McGrath, and between McGrath and Fairbanks look similar in a mesoscale sense (therefore not shown).

4.2. Alaska West Coast

On days with funnel cloud events, all radiosonde temperature profiles, but one showed an inversion in the ABL (Figure 2(b)) suggesting the involvement of a sea-breeze system at least for events observed close to the coast (Figure 1). Unfortunately, the few surface meteorological sites in the

Alaska West Coast area were a) too far away from the coast to examine for landward penetration of a sea breeze, b) too far away from the funnel cloud event, or c) missing data during the event.

On all days with funnel cloud events, the synoptic scale forcing at 1000 and 500 hPa was weak (Figure 6(a)). The surface map showed high pressure over the Gulf of Alaska and low pressure over the Chukchi Sea. Depending on the position of these pressure systems, a northerly or southerly wind blew parallel to the Alaska West Coast. At 500 hPa, winds blew seawards, near-surface winds came from the South. Landward winds at 500 hPa coincided with near-surface winds from the North.

At all levels, individual and the composite synoptic maps of retrieved (e.g. Figure 6(b)) and observed funnel cloud events (e.g. Figure 6(a)) were similar in a mesoscale sense. This finding suggests that the retrieval algorithm can identify synoptic situations like those observed during observed funnel cloud events. This means the NWS could use NWP data and the retrieval algorithm to identify these synoptic situations.

4.3. Anchorage

The Skew-T diagrams for days with observed funnel cloud events suggested a 500 m to up to 2 km deep layer of close to saturation or saturated air and wind shear around this level in the ABL (Figure 2(c), Table 1). All reported funnel cloud events had synoptic scale forcing at 300 and 500 hPa over the region, the southern Yukon Territory and British Columbia (Figure 7(a)). The polar jet was located south of the Gulf of Alaska. At the surface, a low occurred in the Gulf of Alaska.

For the Anchorage region, for most retrieved profiles, the corresponding synoptic maps appeared like the composite synoptic maps of the observed funnel cloud events at all levels. This finding suggests that the algorithm captured situations of same synoptic influence. The composite surface synoptic maps of both retrieved (Figure 7(b)) and actual funnel cloud situations (Figure 7(a)) were similar in a mesoscale sense.

Only a few retrieved profiles showed a different synoptic situation at 300 and 500 hPa than the composites of observed funnel cloud events. However, their synoptic scale forcing was still weak like for the composites of observed funnel cloud events.

These few different synoptic situations may be false alarms. A comparison of the synoptic situation of each retrieved potential funnel cloud situation with the composite synoptic situation of observed funnel cloud events permits discarding them. Note that these few different synoptic situations cause some of the differ- ences between the composites of retrieved and observed synoptic conditions (Figure 7(a) and Figure 7(b)).

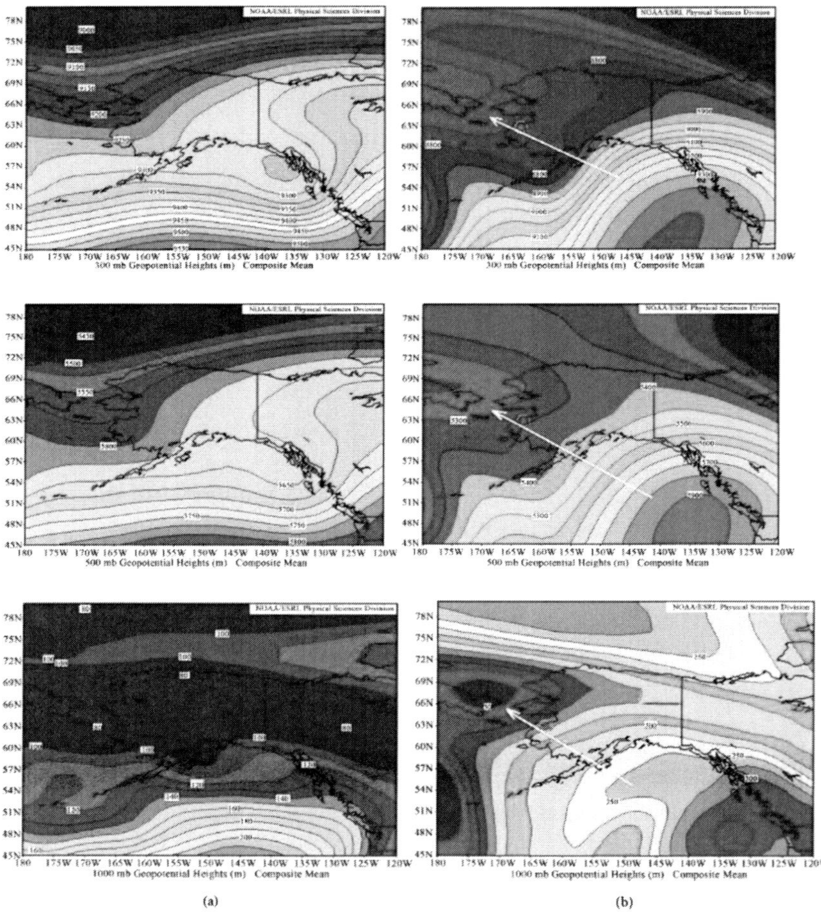

Figure 6. Composites of reanalysis data showing geopotential heights at 300 hPa (top panel), 500 hP (middle panel), and 1000 hPa (lower panel) on days of (a) observed; and (b) retrieved funnel cloud events around Bethel in the Alaska West Coast region. The synoptic situations for events between Bethel and Nome look similar in a mesoscale sense (therefore not shown).

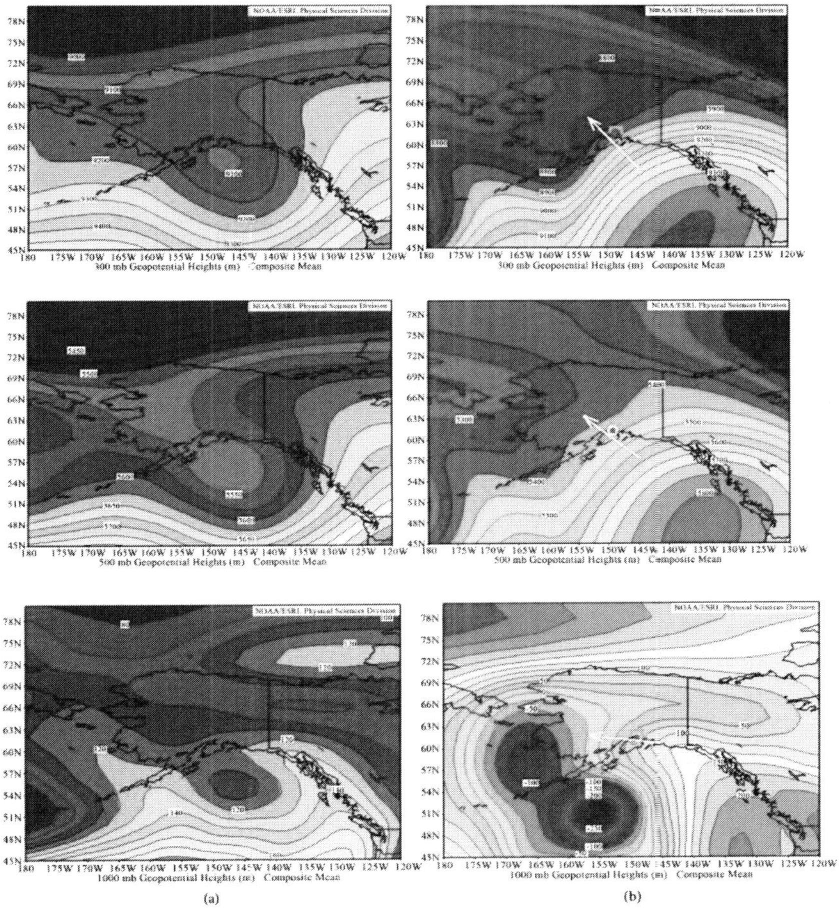

Figure 7. Composites of reanalysis data showing geopotential heights at 300 hPa (top panel), 500 hP (middle panel), and 1000 hPa (lower panel) on days of (a) observed; and (b) retrieved funnel cloud events for the Anchorage area.

5. DISCUSSION ON POSSIBLE FORMATION MECHANISMS

Since the observed funnel clouds were unlinked to severe thunderstorms, and occurred during weak synoptic conditions, we hypothesize that mesoscale dynamics might cause enough shear instability to generate vorticity for funnel cloud formation. Shear instabilities along the leading edge of density currents, such as frontal boundaries, sea-breeze fronts, thunderstorm outflow boundaries or gust fronts can produce small-scale vertical vorticity maxima and areas of enhanced updraft motion or "misocyclones" [14] - [24] . Studying 11 thunderstorm gust fronts with radar and rawinsonde data showed that the intensity of misocyclones was most closely related to the strength of horizontal wind shear across the gust front [22] . Here a series of small-scale (2 - 4 km) vertical vorticity maxima spaced at 3 - 7-km intervals occurred. Numerical studies on misocyclone development along thunderstorm outflow boundaries showed that misocyclones distort the horizontal wind field by producing local maxima of low-level convergence northwest and southeast of each circulation center [25] [26] . These low-level convergences can result in enhanced updrafts.

Alternatively to misocylones, as air descents the mountains into a valley with mountain-valley circulation, it generates horizontal vorticity rolls, which when tilted upward lead to vertical vorticity [5] . Such lifting may be due to an updraft, a barrier in the terrain like a hill, or the ascending branch of a sea breeze [24] .

5.1. Interior

Our analyses of available data suggest terrain-induced mesoscale β/γ flow interacting with large-scale synoptic flow under weak synoptic scale forcing (Figure 5(a)) as the vorticity producing mechanism for funnel clouds in the Interior.

All observed funnel cloud events showed weak to calm winds over the Interior up to 500 hPa in the reanalysis and radiosonde data (cf. Table 1, e.g. Figure 2(a)). Under such weak wind conditions, thermally-driven circulations such as up- and down-valley flow can develop [5] [13] [24] [27] . The landscape of the Interior permits differential surface heating due to local differences in vegetation, terrain elevation, slope orientation and surface properties [4] . Consequently, updrafts (small-scale), slope wind-systems (mesoscale γ), and mountain-val- ley wind systems (mesoscale β/γ) may form.

Figure 8 exemplarily shows the 10 m wind direction on 22 July 2010 at three sites (Fairbanks, Small Arms Range Alaska, Salcha; see Figure 1 for locations).

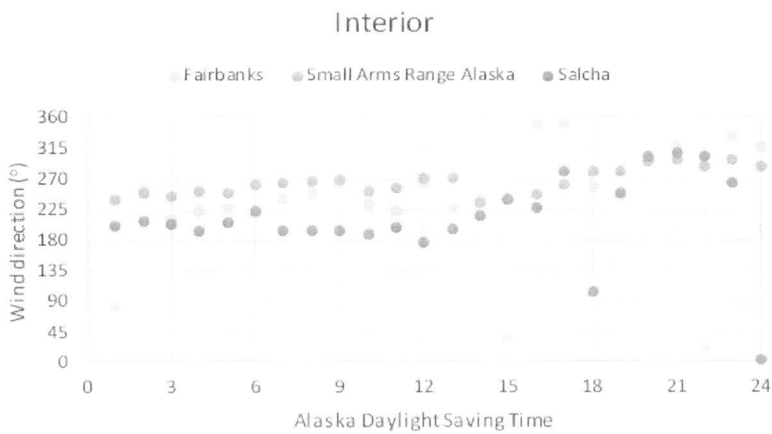

Figure 8. Diurnal course of wind direction (y-axis) in the Interior at three sites on July 22, 2010. The funnel cloud occurred at 9:47 am AKDT. See text for discussion.

The Salcha site is ~60 km up-valley from the Fairbanks site. The Small Arms Range Alaska site is ~15 km from Fairbanks. The valley has a West-East direction at Fairbanks and the range, and a West-Northwest to East-Southeast direction at Salcha. An up-valley flow developed after sunrise (4:17 am AKDT) blowing from westerly directions at Fairbanks and Small Arms Range (Figure 8). The funnel cloud occurred at 9:47 am AKDT over Fairbanks. The valley-mountain wind system broke down in the Fairbanks area around 3 pm. Up-valley flow was also found for other funnel cloud events in the Interior.

Given the weak to calm winds (cf. Table 1), any updraft would experience only slight offset from its source at the ground, which constrains the updraft. Chances are good for cloud development when saturation is reached [5] . The dew-point-temperature profiles of the observed funnel cloud events suggested high moisture in the ABL and cloud tops in the upper ABL to mid-troposphere (e.g. Figure 2(a)). The different direction of the wind systems was key in generating the vorticity for the funnel clouds (e.g. Figure 5(a), Figure 8), while the mixing of cool air with warm valley air became essential in cloud formation (Table 1).

The available data suggest the following mechanism for funnel cloud formation in the Interior (Figure 9). Moist air from the Gulf of Alaska is slowly lifted over the glacier-covered Alaska Range. It cools while flowing over the glaciers. Once this cold air descends to the valley floor, it warms, but remains cooler than the warm air that flows up the valley floor (Figure

5(a)). In addition, cold air from above the White or Kuskokwim Mountains (see Figure 1 for locations) may drain down the slopes into the valley (Figure 9). The valley near-surface

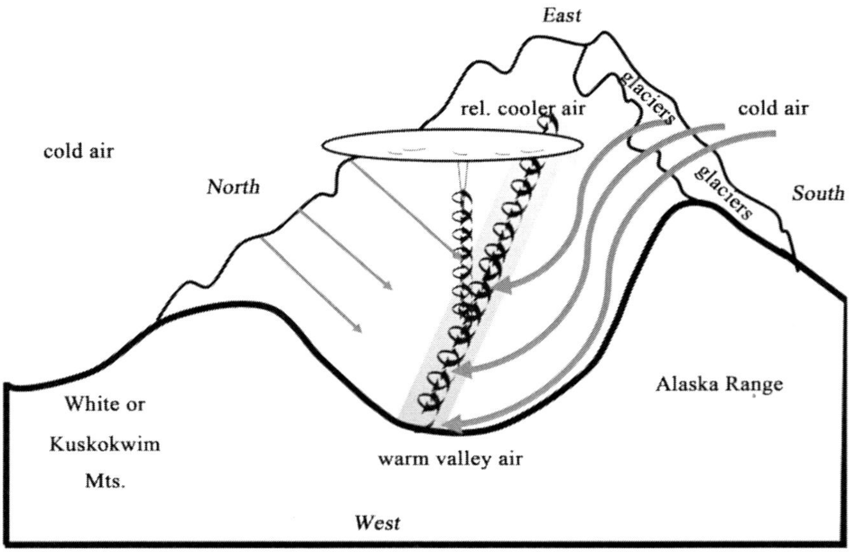

Figure 9. Potential mechanism for funnel cloud formation in the Interior. Gray letters indicate the near-surface conditions of air temperature. Black letters indicate the temperature conditions aloft. The red arrow indicates the up-valley flow with red to pink indicating relatively higher and lower temperatures along the valley. The thick blue arrows indicate the leading edge of the descending air. The thin blue arrows represent downslope winds. Black spirals indicate vorticity tubes.

winds and cool air moving downslope cause the wind shear. At the same time, air of different hydrothermal properties mixes, which modifies stability and may lead to cloud formation. The wind shear forces the air to rotate in horizontal tubes parallel to the direction of the valley (Figure 9). Buoyancy due to differential heating or slight terrain elevations lifts a rotating tube upward and a rotating updraft forms. Once the lifted air reaches saturation, the rotation becomes visible as funnels.

Alternatively, the cool descending air can act as a density current interacting with the ambient warm up-valley flow forming maxima in vertical vorticity or misocylones along the leading edge of the descending air (Figure 9).

5.2. Alaska West Coast

In the Alaska West Coast area, two different scenarios were found: a) An inland wind at the surface with northerly wind at 500 hPa, and b) a seaward near-sur- face wind with southerly wind at 500 hPa flowing parallel to the coast.

Temperature contrasts between land and water may induce a sea-breeze circulation under weak synoptic forcing during the day [5] [13] [24] [27] . Most radiosonde data suggested an inversion in the ABL (e.g. Figure 2(a)). Cold air drainage from mountains or hills can cause a frontal surface along their leading edge [5] [13] [24] .

The radiosonde profiles and reanalysis maps of funnel cloud events suggest two scenarios: Funnel clouds form when weak northerly flow exists at mid and upper tropospheric levels and near-surface winds blow from the sea. These conditions are favorable for sea-breeze development (Figure 10). The related wind shear creates the vertical vorticity along the leading edge of the sea-breeze front. During the long daylight hours (white nights), the circulation is barely disrupted.

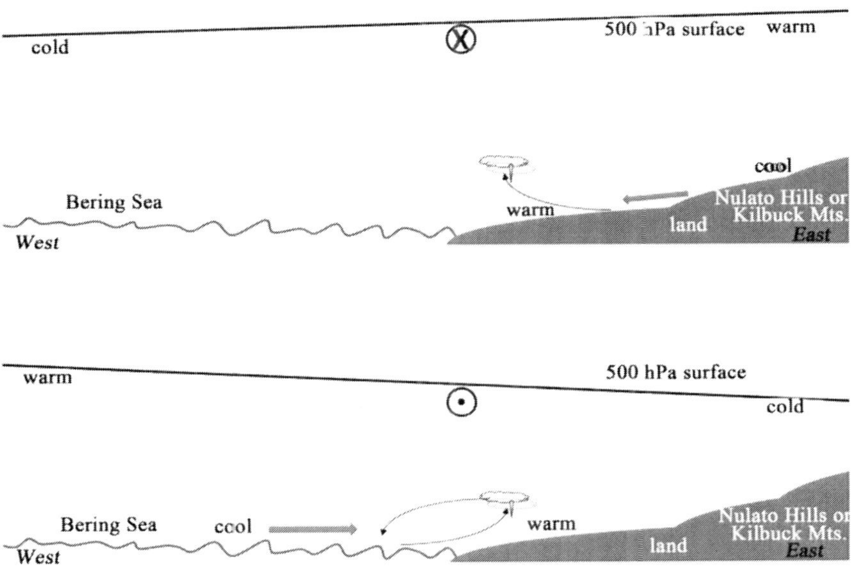

Figure 10. Potential mechanisms for funnel cloud formation in the Alaska West Coast area.

Funnel clouds also form under weak southerly flow in the mid and upper troposphere when cool air drains down the Nulato Hills or the Kilbuck Mountains and flows into the comparatively warm Yukon River Delta (see Figure 1 for locations). The density current interacts with the relatively lighter, warmer air in the Yukon River Delta forming wind-shear instability, which enhances vertical vorticity or misocylones along its leading edge. Once condensation occurs, funnel clouds can form in areas of enhanced vertical vorticity (Figure 10).

5.3. Anchorage Region

The Anchorage region has glacier-covered mountains to one side and ocean to the other. It combines the terrain features that produce horizontal and vertical vorticity in the Interior and Alaska West Coast area. As aforementioned, under weak synoptic scale forcing, mountainous terrain can support formation of slope-wind and mountain-valley systems including katabatic downslope winds [27] , while land-ocean contrasts can lead to sea-breeze systems [5] [13] [27] . These systems can interact [28] [29] .

Per the reanalysis data and their composites from the observed funnel cloud events (Figure 7(a)) weak synoptic scale forcing is an important pre-requisite for funnel cloud formation in the Anchorage region. Weak synoptic forcing permits development of local mesoscale wind systems that can create vorticity.

In the coastal regions, mountain valley circulations interact with the sea-breeze zone to generate horizontal vorticity. At the upward branch, the horizontal vorticity tube is tilted to form vertical vorticity. Farther inland, differential heating of the steep, often glacier-covered mountains causes slope winds and mountain-valley circulations. These wind systems create horizontal shear and vorticity. Cool descending air can act as a density current interacting with the ambient warm up-valley flow forming maxima in vertical vorticity or misocylones along the leading edge of the down-flowing air. Alternatively, terrain induced upward motions or buoyancy turn the horizontal vorticity tube vertical. During ascend cooling can lead to condensation, and rotation becomes visible as funnels.

Unfortunately, all reported funnel cloud events in the Anchorage area were far apart from each other in location (Figure 1) and far away from surface meteorological sites. Thus, examination of the diurnal behavior of wind direction was not possible.

6. CONCLUSIONS

Funnel cloud reports, surface meteorological, radiosonde, reanalysis and radar data were analyzed to determine the climatology, characteristics, and forming mechanism of funnel clouds observed in the Interior, Alaska West Coast and Anchorage area. Between 1955 and 2014, a total of 43 funnel clouds occurred. Most sightings occurred between 11 am and 6 pm AKDT between May and September with July having the highest number of events. All funnel cloud events close to the launch time and site had CAPE of ~500 $J \cdot kg^{-1}$ or more.

In all three regions, funnel clouds occurred under weak synoptic scale forcing throughout the troposphere over the respective region. All radiosonde Skew-T/ log-p diagrams and radar analyses indicated that Alaska funnel clouds are non-supercell thunderstorm clouds and owe their vorticity to wind shear in the ABL. Here, an approximately 500 - 2000 m thick layer of moist air was present while air was dry aloft.

Interaction between mesoscale dynamics of different scales created wind shear and the required vertical vorticity for funnel cloud formation in the ABL under specific weak synoptic scale (mesoscale α) forcing conditions. These synoptic situations depend on the region and can be identified from surface observations, mid- and upper-level synoptic maps, and radiosonde profiles.

Local scale dynamics creates horizontal vortex tubes either by a combination of a valley wind with slope winds (Interior) or a combination of slope or mountain-valley wind systems and sea-breeze circulation (Anchorage region). Differential heating due to the spatial patchiness of the land-cover, land-sea contrast, and different insolation of slopes creates buoyancy. Updrafts or forced lifting in uneven terrain or at surface discontinuities tilt the horizontal vortex tube into a vertical one. Adiabatic cooling during lifting and mixing of air cool, dry, descending air with moist, warm, ascending air lead to saturation and cloud formation at the top of the ABL, i.e. Alaska funnel clouds are boundary layer clouds.

In the Alaska West Coast area, two situations occur: a) The synoptic-scale wind field and land-sea contrast create the horizontal vortex tube that is tilted by the upward branch of the sea breeze. b) The synoptic-scale wind field and density flow from cold air drainage off the Nulato Hills or Kilbuck Mountains create the vertical vorticity.

A similarity retrieval algorithm was introduced by defining limits for temperature, dew-point temperature, and wind profiles and a minimum CAPE from soundings during funnel cloud events. Soundings that met the limits and requirement were considered retrieved funnel cloud events. In the Interior, 2% of the soundings between May to September 1971 to 2014 indicated potential funnel cloud events between Fairbanks and McGrath, a mostly unpopulated area. Potential for funnel cloud events was less in all other regions. Given the large fraction of unpopulated areas in Alaska, events may occur, but remain unobserved.

Alaska's population increased from 547,160 in 1989 to 736,732 in 2014 [30] . Based on the results of the similarity retrieval algorithm, the number of potential funnel cloud situations decreased after 2000 as compared to prior to 1989. This finding suggests that the increase in reported funnel cloud events after 2000 may be due to the increase in population and devices for documentation. Our results further suggest that the NWS record certainly under-represents past events.

The resolution of operational radars is too coarse to serve for funnel cloud warnings unless the event occurs close enough to the radar. However, the similarity of composites based on retrieved and observed funnel cloud events suggests that the retrieval algorithm can identify synoptic situations suitable for funnel cloud development. This finding is very important for development of funnel cloud forecasts and warnings. Forecasters could use radiosonde data together with synoptic analysis maps to identify days favorable for funnel cloud formation. In addition, the retrieval algorithm could be modified to run with NWP model forecasted vertical profiles to test whether the forecasted synoptic situation may be suitable for funnel cloud formation. However, as to what extent in lead-time such forecasts would be possible, would be subject to further research.

Future field studies should focus on vorticity development in the ABL to ascertain the exact forcing and winds necessary for funnel cloud development. Given the scarcity of reported and potential funnel cloud events, a targeted field campaign is rather difficult. Since such campaigns need at least a year or more in lead-time, we suggest establishing a monitoring network of at least nine wind profilers between Fairbanks and McGrath where most of the observed (Figure 1) and retrieved potential funnel cloud events occurred. For synoptic situations identified as supportive for funnel cloud formation in this area, the data could be screened for vorticity. Data from the cloud radar at Ft. Greely (Figure 1) that became available in 2015 could serve to examine cloud microphysical properties and cloud vertical extension where the beam is not blocked by mountains. Analyses of such data together with the radiosonde data at McGrath and Fairbanks could shorten the lead-time for the forecast of funnel clouds.

ACKNOWLEDGEMENTS

We thank G. Kramm, U.S. Bhatt, R.L. Collins and the anonymous reviewers for fruitful discussion and helpful comments, NOAA for access to the radiosonde and radar data, the National Center for Atmospheric Research for access to the reanalysis data, the National Science Foundation (NSF), the SOARS program, the Gwichyaa Zhee Gwich' in Tribal Government, and SLOAN for financial support.

CITE THIS PAPER

Edwin, S.G., Mölders, N., Friedrich, K., Schmidt, S. and Thoman, R. (2017) Conditions Supporting Funnel Cloud Development in Alaska. Atmospheric and Climate Sciences, 7, 223- 245.

REFERENCES

1. Edwin, S.G. (2016) Climatology and Forcing Mechanism of Funnel Clouds in Alaska. MS Thesis, Department of Atmospheric Sciences, University of Alaska Fairbanks, Fairbanks.
2. von Storch, H. and Zwiers, F.W. (1999) Statistical Analysis in Climate Research. Cambridge University Press, Cambridge, 484 p.
3. Shulski, M. and Wendler, G. (2007) The Climate of Alaska. University of Alaska Press, Fairbanks, 216 p.
4. Mölders, N. and Kramm, G. (2007) Influence of Wildfire Induced Land-Cover Changes on Clouds and Precipitation in Interior Alaska—A Case Study. Atmospheric Research, 84, 142-168.
5. Lin, Y.-L. (2007) Mesoscale Dynamics. Cambridge University Press, Cambridge.
6. Houze, R.A. (2014) Cloud Dynamics. Vol. 104, Academic Press, Cambridge.
7. Department of Commerce (2014) Stormevents. http://www.ncdc.noaa.gov/stormevents/listevents.jsp?eventType=%28C%29+Funnel+Cloud&beginDate_mm=05&beginDate_dd=01&beginDate_yyyy=1971&endDate_mm=08&endDate_dd=30&endDate_yyyy=2014&county=ALL&hailfilter=0.00&tornfilter=0&windfilter=000&sort=DT&submitbutton=Search&statefips=2%2CALASKA
8. Bieniek, P.A., Walsh, J.E., Thoman, R.L. and Bhatt, U.S. (2014) Using Climate Divisions to Analyze Variations and Trends in Alaska Temperature and Precipitation. Journal of Climate, 27, 2800-2818.
9. Kalnay, E., Kanamitsu, M., Kistler, R., Collins, W., Deaven, D., Gandin, L., et al. (1996) The NCEP/NCAR 40-Year Reanalysis Project. Bulletin of the American Meteorological Society, 77, 437-471.
10. Kistler, R., Collins, W., Saha, S., White, G., Woollen, J., Kalnay, E., et al. (2001) The NCEP-NCAR 50-Year Reanalysis: Monthly Means CD-Rom and Documentation. Bulletin of the American Meteorological Society, 82, 247-267.
11. Office of the Federal Coordinator for Meteorological Services and Supporting Research (OFCM) (1997) Federal Meteorological Handbook No. 3: Rawinsonde and Pibal Observations. FCM-H3-1997, Washington DC, 191 p.

12. Pielke, R.A. (2002) Mesoscale Meteorological Modeling. Academic Press, New York, 676 p.
13. Mölders, N. (2011) Land-Use and Land-Cover Changes: Impact on Climate and Air Quality. Vol. 44, Springer Science & Business Media, Berlin, 193 p.
14. Fujita, T.T. (1981) Tornadoes and Downbursts in the Context of Generalized Planetary Scales. Journal of Atmospheric Sciences, 38, 1511-1534.
15. Carbone, R.E. (1982) A Severe Frontal Rainband. Part I: Stormwide Hydrodynamic Structure. Journal of Atmospheric Science, 39, 258-279.
16. Mueller, C.K. and Carbone, R.E. (1987) Dynamics of a Thunderstorm Outflow. Journal of Atmospheric Sciences, 44, 1879-1898.
17. Wakimoto, R.M. and Wilson, J.W. (1989) Non-Supercell Tornadoes. Monthly Weather Review, 117, 1113-1140.
18. Weckwerth, T.M. and Wakimoto, R.M. (1992) The Initiation and Organization of Convective Cells Atop a Cold-Air Outflow Boundary. Monthly Weather Review, 120, 2169-2187.
19. Wilson, J.W., Foote, G.B., Crook, N.A., Frankhauser, J.C., Wade, C.G., Tuttle, J.D., et al. (1992) The Role of the Boundary-Layer Convergence Zones and Horizontal Rolls in the Initiation of Thunderstorms: A Case Study. Monthly Weather Review, 120, 1785-1815.
20. Kingsmill, D.E. (1995) Convection Initiation Associated with a Sea-Breeze Front, a Gust Front, and Their Collision. Monthly Weather Review, 123, 2913-2933.
21. Arnott, N.R., Richardson, Y.P., Wurman, J.M. and Rasmussen, E.M. (2006) Relationship between a Weakening Cold Front, Misocyclones, and Cloud Development on 10 June 2002 during IHOP. Monthly Weather Review, 134, 311-335.
22. Friedrich, K., Kingsmill, D.E. and Young, C.R. (2005) Misocyclone Characteristics Along Florida Gust Fronts during CAPE. Monthly Weather Review, 133, 3345-3367.
23. Murphey, H.V., Wakimoto, R.M., Flamant, C. and Kingsmill, D.E. (2006) Dryline on 19 June 2002 during IHOP. Part I: Airborne Doppler and LEANDREII Analyses of the Thin Line Structure and Convection Initiation. Monthly Weather Review, 134, 406-430.
24. Mölders, N. and Kramm, G. (2014) Lectures in Meteorology. Springer, Heidelberg, 591 p.
25. Lee, B.D. and Wilhelmson, R.B. (1997) The Numerical Simulation of Non-Supercell Tornadogenesis. Part I: Initiation and Evolution of Pretornadic Misocyclone Circulation along a Dry Outflow Boundary. Journal of Atmospheric Science, 54, 32-60.
26. Lee, B.D. and Wilhelmson, R.B. (1997) The Numerical Simulation of Non-Supercell Tornadogenesis. Part II: Evolution of a Family of Tornadoes along a Weak Outflow Boundary. Journal of Atmospheric Science, 54, 2387-2415.
27. Bluestein, H.B. (1985) An Observational Study of a Mesoscale Area of Convection under Weak Synoptic-Scale Forcing. Monthly Weather Review, 113, 520-539.
28. Kondo, H. (1990) A Numerical Experiment on the Interaction between Sea Breeze and Valley Wind to Generate the So-Called "Extended Sea Breeze". Journal of the Meteorological Society of Japan, 68, 435-446. https://www.jstage.jst.go.jp/article/jmsj1965/68/4/68_4_435/_pdf
29. Kurita, H., Ueda, H. and Mitsumoto, S. (1990) Combination of Local Wind Systems under Light Gradient Wind Conditions and Its Contribution to Long-Range Transport of Air Pollutants. Journal of Applied Meteorology, 29, 331-348.
30. US Census (2017). https://www.census.gov/

CHAPTER 6

A Review of Some Recent Studies on Buoyancy Driven Flows in an Urban Environment

Bodhisatta Hajra

Department of Mechanical Engineering, The University of Hong Kong, Pokfulam Road, Hong Kong

ABSTRACT

This paper reviews some recent studies (after 2000) pertaining to buoyancy driven flows in nature and thier use in reducing air pollution levels in a city (city ventilation). Natural convection flows occur due to the heating and cooling of various urban surfaces (e.g., mountain slopes), leading to upslope and downslope flows. Such flows can have a significant effect on city ventilation which has been the subject of study in the recent times due to increased pollution levels in a city. A major portion of the research reviewed here consists of natural convection flows occurring along mountain slopes, with a few studies devoted to flows along building walls. The studies discussed here primarily include field measurements and computational fluid dynamics (CFD) models. This review shows that for densely populated cities with high pollution levels, natural convection flows (mountain slope or building walls) can significantly aid the dispersion of pollutants. Additional studies in this area using CFD and water channel measurements can explain the physical processes involved in such flows and help improve CFD modelling. Future research should focus on a complete understanding of the mechanisms of buoyancy flows in nature and developing design guidelines for better planning of cities.

1. INTRODUCTION

Global air pollution has been a subject of study for several decades. For instance, global warming has caused the melting of ice in the west Antarctic region, thereby increasing the sea-levels by several centimetres [1]. However, pollution levels within a city may increase only due to the surrounding topography and local environmental conditions. For example, the London smog in 1952 consisted of large amounts of dusty fumes from nearby factories causing poor visibility and resulting in the deaths of several people during the time [2]. Lack of available land and increased population have led to the construction of high density and high rise buildings in many cities like New York and Hong Kong, thereby increasing city pollution. This situation has been aptly described by Fernando [3] as "The future of humankind is sure to evolve in large urban areas that proffer the highest quality of life. Accelerated growth, however, has overstressed some urban environments, thus calling for sound planning tools for sustainable growth. In this context, of cardinal importance is the prediction of the urban atmosphere (i.e., key meteorological variables and turbulence) and air-quality indicators (e.g., pollutant levels, comfort index)." One of the key issues with urban megacities like Hong Kong is their geographic location and complex topography. These cities are surrounded by mountains and skyscrapers, where wind penetration is greatly reduced causing accumulation of pollutants [4]. Furthermore, during low wind speeds the buoyancy driven flows caused by heating and cooling of surfaces (mountain slopes and building walls) can greatly help disperse these pollutants [5, 6]. Another problem associated with city level pollution is the urban heat island (UHI) effect, which causes an increase in urban air temperatures as opposed to the surrounding rural region [7–9]. Recent studies indicate that buoyancy driven flows (along mountains and building walls) can help reduce urban air pollution levels and improve the thermal environment in a city. Based on detailed water tank measurements, Reuten et al. [10] stated "Upslope flows are a crucial mechanism in the transport of air pollutants in complex terrain, both as separate flow systems and as part of other thermally driven flows."

Urban ventilation studies have been carried out through field measurements, CFD, and analytical methods. A well-known field study on the dispersion of tracers on nocturnal drainage flows was carried out under the Atmospheric Studies in Complex Terrain (ASCOT) program (see [11] for details) in the Brush Creek valley in Western Colorado, USA. According to

[12], the upslope flows generated immediately after sunrise aided in the dispersion of the tracers in the ASCOT study. Similarly, a CFD study on buoyancy driven flows along building walls of a street canyon was carried out by [13], and they found that the ratio of building height to street width played a major role in changing the air and pollutant flow characteristics in the street canyon. Buoyancy driven flows along building walls and mountain slopes were quantified by using the Air Changes per Hour (ACH) concept by [14]. Their study which was primarily based on air temperature measurements using a thermal camera showed that the contribution of building wall flows was twice as high as mountain slope flows due to greater surface area of buildings than in mountains. More recently, CFD simulations to assess city ventilation using the age of air and ACH concept were carried out. It was found that for buildings taller than 100 m, buoyancy driven flow along building walls is sufficiently high and even comparable to that produced by mountain slope flows [15].

This review focuses on some important recent studies (after 2000) on the application of mountain slope and building wall flows on urban ventilation. A great majority of the studies reviewed in this paper are about mountain slope flows with only a small portion of literature on building wall flows, since most of the studies in the past have considered air pollution in a city primarily due to buoyancy flows along a mountain as opposed to building walls. Only more recent studies (e.g., [16]) have shown that considering flows along building walls is equally important since building surfaces are a major contributor to buoyancy flows, particularly in cities like Hong Kong. Therefore, the present review presents some of the major urban buoyancy flow studies in the recent times and stresses the need to carry out additional buoyancy flow studies on building walls.

Initially, a brief discussion on the mechanism of buoyancy driven flow along vertical plates and mountain slopes is presented (Section 2), followed by the description of field measurements (Section 3) and analytical models (Section 4). Thereafter, CFD simulations (Section 5) and discussions on some key topics (Section 6) are presented. Future developments that are likely to occur (Section 7), followed by conclusions from this review (Section 8), form the concluding parts of this paper.

2. BUOYANCY DRIVEN FLOWS

According to Jaluria [17], some of the examples of natural convection flows include, "buoyant flow arising from heat or material rejection to the atmosphere, heating and cooling of rooms and buildings, recirculating flow driven by temperature and salinity differences in oceans." Unlike forced convection flows, natural convection flows are generally associated with a difference in density, which is caused by differences in temperature. In fact, plenty of types of literature exist on natural convection flows over a vertical plate [18]. For instance, Eckert and Jackson derived a formula for heat transfer coefficient for Grashof numbers in the range of 10^{10} to 10^{12} for a vertical flat plate, for aerospace engineering applications [19]. However, the focus of this paper is on natural convection flows occurring in nature (along building walls and mountain slopes) and how they aid the dispersion of pollutants within a city consisting of closely-spaced, high rise buildings. There is more literature on mountain slope flows and less on building wall flows because the latter has been investigated more recently for UHI issues.

2.1. Buoyancy Driven Flows along Mountains

Plenty of types of literature exist on the buoyancy driven flows along mountain slopes [20], and therefore, only a brief description of the physics of these flows is mentioned here. Figure 1 shows the nature of the winds that develop along mountain slopes [21]. The wind velocity profiles generally tend to be zero closer to the surface owing to friction and gradually increase to about 1–4 m/s, at a height of about 15 m during the night (Figure 1(a)). However, the direction is reversed during the day time (Figure 1(b)) and the magnitude of the maximum wind velocity is about 1–5 m/s, occurring at a height of about 10–50 m. In fact, the depth of the wind profile extends up to 100 m during the night, as opposed to 200 m during the day.

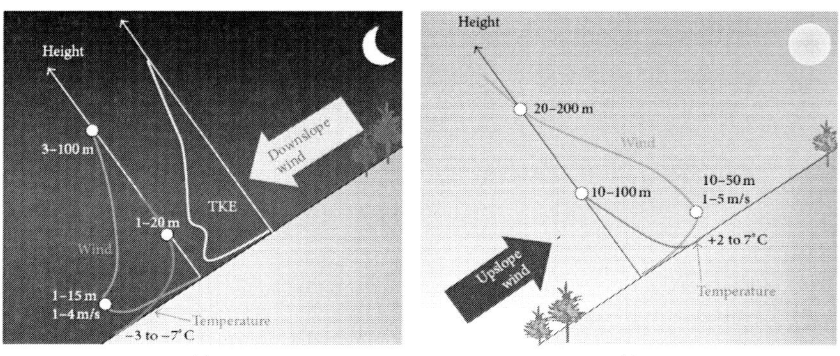

Figure 1: Mountain wind system: (a) downslope winds; (b) upslope winds (from [21]).

Diurnal mountain wind systems have been studied by many researchers in the past [22, 23]. The momentum equation, energy equation, and continuity equations, by incorporating the Boussinesq's assumption (neglecting the density in the continuity equation) are solved numerically. Zardi and Whiteman [21] state that "A fundamental obstacle to rapid progress in mountain meteorology is that there are almost infinitely many possible terrain configurations. So, any field measurement or numerical experiment that is valid for a specific situation does not automatically have greater significance beyond that case."

This statement aptly describes the problems faced by most CFD modellers and city planners (described later), involved in improving thermal and wind conditions in cities located amidst mountains.

2.2. Buoyancy Driven Flow along a Vertical Plate

Analogous to mountain slope flows is the buoyancy driven flow along a vertical plate (Figure 2). Natural convection flows over a flat vertical surface develop due to the differences in density, induced by temperature changes. For instance, when the wall temperature is higher than the ambient air temperature , the warmer air close to the wall rises and the air surrounding it gets entrained (Figure 2(a)), resulting in a net upward flow, whilst the opposite occurs when . At a given height the velocity of flow tends to increase up to a certain point, before reducing to zero, whilst the temperature reduces from higher values to zero gradually away from the wall (Figure 2(b)). These flows are described by several nondimensional numbers, such as Grashof number (ratio of buoyancy to viscous forces), Rayleigh number (product of Grashof and Prandtl numbers), and Nusselt number (ratio of convective to conductive heat transfer coefficients). There are many approximate methods available to estimate the velocity and temperature distributions [24], although the application of these techniques for geophysical flows is not fully validated.

Detailed reviews on natural convection flows for vertical and inclined surfaces can also be found from [25, 26]. Buoyancy flows along a building wall can be considered equivalent to flows along a vertical plate with high Rayleigh number. Although there are studies on measurement of building wall surface temperatures (e.g., [6]), detailed studies on the thermal boundary layer along a building wall must be carried out in future. In fact, an excellent review of natural convective heat transfer coefficients on horizontal and vertical surfaces and their application to buildings is provided in Khalifa [27].

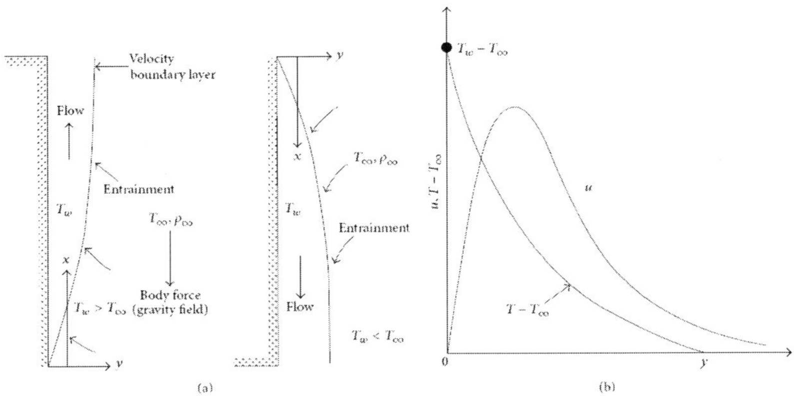

Figure 2: Natural convection over a flat vertical surface: (a) velocity boundary layer; (b) velocity and temperature profiles (from [17]).

3. FIELD STUDIES ON BUOYANCY DRIVEN FLOWS IN THE URBAN ENVIRONMENT

There are quite a few studies on buoyancy flow measurements in nature, with most of them focused on mountain slope winds and very few on building walls. Only a few are described here, with a more detailed list available in Table 1.

3.1. Atmospheric Studies in Complex Terrain (ASCOT), Colorado, USA

One of the earliest studies on the application of mountain slope flows on the dispersion of pollutants in a valley was carried out in western Colorado, USA, under the ASCOT program (see [12] for details). Perfluorocarbon tracers were released in the valley and its dispersion under nocturnal conditions was studied (see Figure 3). It may be mentioned that although this review primarily covers the progress following 2000, this particular study carried out in late eighties is of great significance to mountain slope flow, and, hence, described here. The experiments consisted of releasing the tracers from midnight to early morning, and later these tracers were sampled using about 90 surface samplers. Additionally, some air crafts flying at a height of 1980 m were also utilised to collect the samples [28].

From the results the authors concluded that the ventilation of the tracers out of the valley started early in the morning after sunrise, when upslope flows gradually start to dominate.

Table 1: Various physical modelling studies of buoyancy driven urban flows (2000 and later).

Study type and name	Year* and Reference	Characteristic features
Field study: Vertical Transport and Mixing (VTMX), Salt Lake, USA	2000; [29]	Experiments to assess vertical transport and mixing of various parameters for mountain slope flows (e.g. heat, air pollution, etc.)
Field study: URBAN 2000	2000; [74]	(i) Conducted simultaneously with the VTMX study in Salt,Lake, USA. (ii) Assess tracer dispersion in street canyons due to synoptic and thermally driven winds (part of the data from URBAN 2000 was also used for the VTMX study).
Field study in Lower Fraser valley, British Columbia, Canada,	2001; [30]	(i) Water channel measurements were also performed for some cases. (ii) Return flows that opposed upslope flows were observed.
Field study at PICO-NARE observatory on Pico mountains, Azores island, North Atlantic Ocean	2004; [75]	(i) Study of buoyancy driven flows and their influence on mountain top observatories during low synoptic winds. (ii) Isoprene concentrations (generated from vegetation at the bottom) on the mountain top reduced due to upslope flows.
Field study at Moravian-Silesian Beskydy Mountains, Czech Republic	2004; [76]	(i) Night time slope flows over a mountain with a dense forest at the foot of the hill were studied. (ii) Katabatic flows measured using sonic anemometers, with flow reversal (similar to [28]), were observed.
ADVEX field study, Bolzano, Italy	2005; [77]	(i) Study was carried out in a mountainous terrain surrounded by forests at the bottom; measurements were carried out using sonic anemometer (wind) and infrared gas analyzer (CO_2 and H_2O). (ii) Synoptic winds above the canopy affected the classical slope wind phenomenon (upslope in daytime and downslope at night) significantly.
Field study: Meteor Crater Experiment (METCRAX), Winslow, Northern Arizona, USA	2006; [78]	(i) SODAR, Radar wind profiler used for the measurements of downslope flows. (ii) The flows were not influenced by soil moisture and surface roughness.
Field study: Transition flow experiment (TRANSFLEX), Phoenix, USA	2006; [34]	(i) Study of transition mechanism in mountain slope flows (flow reversal from upslope to downslope). (ii) Main focus was to assess pollutant concentrations in two of the sites in Phoenix, USA.
Field study: Stable Atmospheric Boundary Layer Experiment in Spain (SABLES), Valladolid, Spain	2006; [36]	(i) Study of katabatic winds due to shallow slopes and their effect in reducing pollution levels. (ii) Use of wind vanes, tethered balloons, and microbarometers to measure meteorological variables, stability, and turbulent parameters.
Water channel measurements of upslope flows over an inclined surface, USA	2007; [43]	(i) Water channel measurements of upslope flows on an inclined surface were conducted. (ii) Relationship between Prandtl number and slope flows were developed, besides a critical slope angle beyond which the flow sustains.
Field study of buoyancy flows in a street canyon at COSMO, Japan	2008; [16] 2010; [6]	(i) Studies on buoyancy driven flows along building walls of a street canyon were carried out. (ii) Height and spacing between buildings were found to be major factors affecting the flow.
Field studies at The University of Hong Kong, SAR China	2008; [14]	(i) Use of infrared camera to assess surface temperatures. (ii) Study of both slope and wall flows and their influence on city ventilation. (iii) Use of air changes per hour concept to quantify city ventilation.
Wind tunnel study of buoyancy driven flows in a street canyon, ETH, Zurich, Switzerland	2012; [79]	Heating of building walls and ground surface induces two vortices within the street canyon and their positions change for different windward wall temperatures.

* Refers to the year in which the experiments were performed.

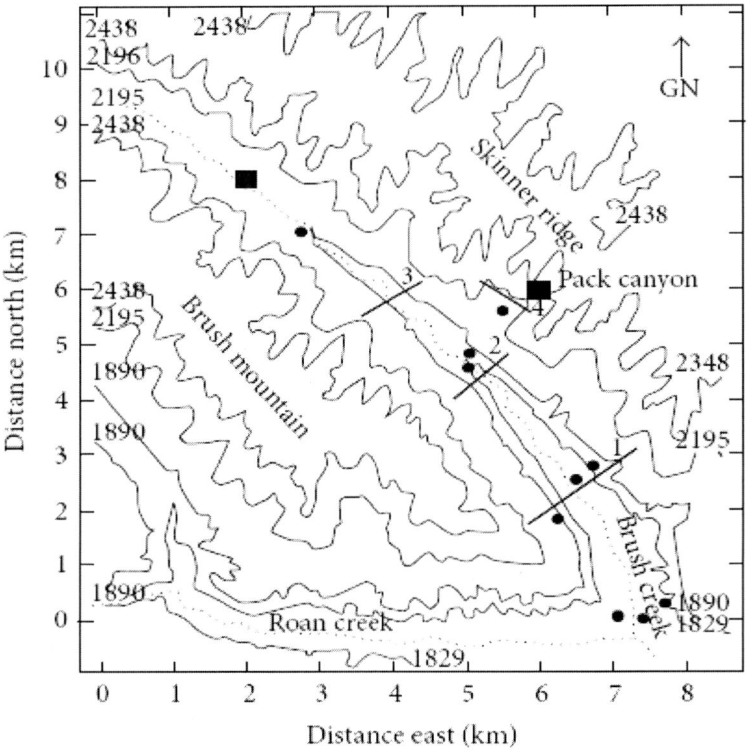

Figure 3: Map of the Brush Creek Valley showing the tracer release sites (squares) and the locations of the vertical profile samplers (circles). The thick straight lines denote the positions of the surface sampling arcs (from [12]).

Additionally, this study showed that the solar intensity, location of the tracer, and ambient meteorology played a major role in the ventilation process. This is one of the most important field studies on estimating the effectiveness of slope winds in ventilating a valley.

It is worth noting that although this study was carried out in the eighties, it has a lot of significance today, since global air pollution has gone up and there is plenty of human settlement along valleys/areas located at the foot of the mountains. This makes it even more important to utilise buoyancy driven winds to reduce pollution levels, especially in the absence of synoptic winds.

3.2. Vertical Transport and Mixing (VTMX), Salt Lake Valley, USA

A major field campaign titled VTMX, sponsored by the US Department of Energy through their Environmental Meteorology program took place in October 2000 in Salt lake valley, USA [29]. The key goals of this field study were to assess the vertical mixing and transport of various parameters, such as heat, air pollution, and momentum in the Salt Lake Valley, and formation of cold pools. Furthermore, studying the effect of nocturnal stability on air pollution concentrations in the valley and how this could affect an urban area located at the foot of the mountain was also a part of the investigation. A number of instruments including SODARS and LIDARS (wind velocity and direction), Radio Acoustic Sounding System (temperature), and tethered balloons (wind velocity, temperature, and humidity), were deployed. The authors reported "a complicated pattern of local flows influenced by lake effects, heating and cooling of the valley and the nearby mountains, and interactions with synoptic systems." The idea behind this field study was to incorporate the results of this study in numerical models to better simulate the physical processes involved in mountain flows. Most previous studies including the ASCOT study were mainly focused on understanding the physics behind slope flows. However, the VTMX study also utilised their data for numerical simulations and theoretical model development.

3.3. Upslope Flows in Lower Fraser Valley, British Columbia, Canada

Reuten and his associates [30] carried out field measurements along the slopes of the Lower Fraser Valley in British Columbia, Canada (Figure 4). According to Reuten et al. [30], "The Lower Fraser Valley is nearly flat, mostly below 100 m above sea level, and has an approximately triangular shape narrowing from about 100 km width at the Strait of Georgia in the west to about 2 km approximately 90 km inland to the east. It is bounded by the Coast Mountain Ranges to the north and the Cascade Ranges to the south-east, which have heights of about 2000 m and 1000 m above mean sea level (m.s.l.), respectively." The instruments used for measurement included a Doppler SODAR for measuring wind speeds on the slopes; besides a scanning LIDAR system and a tethersonde were used to measure specific humidity, wind speed and direction, temperature and particulate matter. Their field measurements focused on upslope (anabatic) flows, as opposed to most previous studies that were mostly focused on ventilating a city due to katabatic (downslope) winds. Reuten and his associates showed, from their study, that upslope flows helped reduce pollutant concentrations by dispersing them into the atmosphere, besides the existence of a strong "return flow" due to synoptic winds opposing the upslope flows.

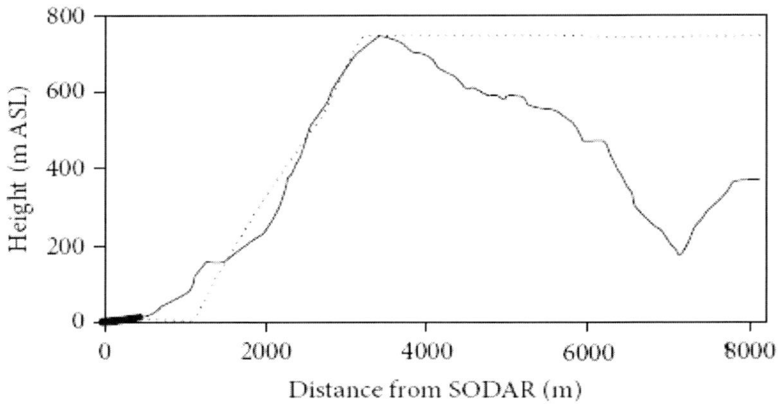

Figure 4: Vertical cross-section of the slope; the dashed line is a simplified topography that includes a 19° angle reaching up to a plateau at 780 m above the SODAR site, which was 6 m above sea level (from [28]).

This was an interesting observation, which was not made in any of the earlier field measurements on mountain slope flows. The field measurements were later simulated using water channel measurements by Reuten et al. [10] who reported that "Non-dimensional boundary-layer depths near the base of slope in atmosphere and water tank agree within the measurement uncertainties of the field observations (20%)."

3.4. Measurements of Urban Thermal Environment in Tokyo, Japan

Buoyancy driven flows are related to urban thermal environment and due to high levels of pollution in Tokyo, a field measurement was conducted by Ooka and his associates at the University of Tokyo [31] to asses wind velocity, temperature (surface and air), and humidity. This data was used to develop a model that can take account of urban convection flows to assess the thermal environment in a densely packed city environment. Additionally, the model could be incorporated within any existing air pollution dispersion model in order to predict precise urban air quality information. In fact, Ooka et al. [32] also used the results from this study to develop a multiobjective genetic algorithm to assess the outdoor thermal environment, besides researching on the effect of tree plantation on mitigating UHI effects. Some of the key aspects of the studies conducted in Tokyo are the assessment of humidity in addition to air velocity and surface temperature measurements. Additionally, the impact on human comfort (pedestrian comfort) due to UHI effects was also part of these studies, while most other studies on urban buoyancy flows were mostly focused on the physical processes involved in these flows. Wind and thermal environments at the pedestrian level are closely related to buoyancy driven flows through urban surfaces. More recently, CFD simulations of these flows using previous field measurements are gradually being used to develop computational modelling as discussed in detail by Mochida and Lun [33].

3.5. Transition Flow Experiment (TRANSFLEX), Phoenix, USA

The TRANSFLEX experiments were conducted between January 7 and 17, 2006, in Phoenix, Arizona, USA, to understand the transition mechanism in the evening [34]. The experiments were supported by the Arizona Department of Environmental Quality due to concerns on poor air quality in the region; very often the PM_{10}limit of 50 μg/m^3 (yearly average) was exceeded. For this purpose, two sites, namely, Mountain View High School and another site in Southern Phoenix, were chosen since these areas reported high PM_{10}concentrations. The two main issues of interest in this study were to(a)study flow reversal due to change in buoyancy;(b)assess the formation of a transition front.

For this purpose, concentration sensors and tethered balloons were deployed. A general observation during this study was the arrival of the transition period which varied from one site to another, but generally after sunset, a flow reversal (upslope to downslope) was observed. An MM5 urbanised model [35] was used to obtain a general qualitative agreement with measured data. From this study, the authors concluded that a local Richardson number may be suitable for quantifying such events. This study showed that stable stratification may prevent the dispersion of the particulate matter, and generally a transition front could allow entrainment of dust particles in some regions to increase particle concentrations. Unlike most previous studies that were focused on any one particular phenomenon (upslope/downslope flows), this field study is one of the most important studies carried out in recent times, with a particular emphasis on "transitional flows" along mountain slopes and its effect on urban ventilation.

3.6. Stable Atmospheric Boundary Layer Experiment in Spain (SABLES), 2006

The SABLES field study [36] was conducted from June 19 to July 5, 2006, at Valladolid, Spain. Although this location consists of a relatively flat terrain, it is surrounded by two small slopes of 1 : 6000 and 1 : 1660, respectively [37]. The instruments employed for this study were similar to some of the previous field measurements, namely, sonic anemometers and wind vanes (wind velocity) and microbarometers to measure the absolute pressure at various heights and tethered balloons for night time wind conditions. Although some of the main objectives of this field study were to assess mean micrometeorological variables, stability, and turbulent parameters, the effects of katabatic flows were mainly observed from some of their results. For instance, Figure 5 shows the inversion strength between 2 m and 35 m from June 21 to June 30, 2006. Results show that the temperature changes on certain days (e.g., June 28 and June 29) are markedly different from other days (e.g., June 21 and June 22). According to [36], "Katabatic winds (radiatively cooled air descending along local slopes) are the most probable cause for this behaviour." In fact, shallow slopes are capable of producing drainage flows of about 1-2 m/s, as reported in Wangara, Australia, and the Great Plains, USA [38].

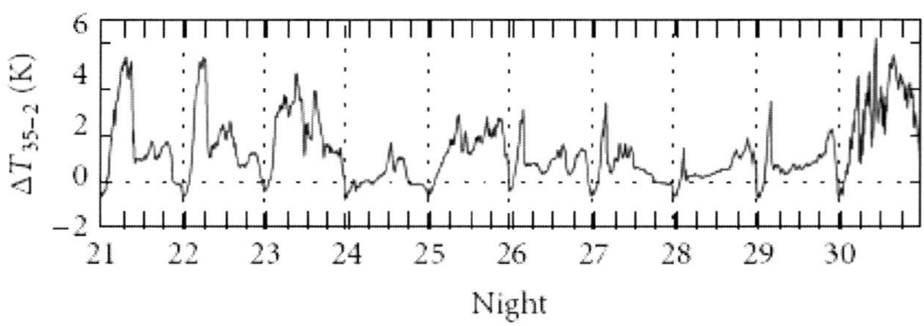

Figure 5: Inversion strength between 35 m and 2 m (from [36]).

The significance of buoyancy driven flows, especially nocturnal drainage (katabatic) flows, in diluting pollutant concentrations and cooling regions located at the foot of the mountains can be observed from this study. One of the important features that distinguishes this study from the rest is that the SABLES study consisted of mountains with shallow slopes (earlier studies had steeper slopes) which produced significant slope winds to ventilate a city. However, most previous studies tried to quantify the ventilation rate in terms of wind speeds along slopes and observed reduction in concentration of pollutants (e.g., during katabatic winds), although the field study at the University of Hong Kong (described next) is uniquely different.

3.7. Field Studies at The University of Hong Kong, SAR China

Compared to most other field studies that employed tethered balloons and anemometers to assess mountain slope winds, the field study at The University of Hong Kong (HKU) utilised a thermal infrared camera [14]. Yang and Li carried out field measurements using an infrared camera to assess surface temperatures on building surfaces. They used the concept of air changes per hour (ACH) to quantify urban ventilation rates due to buildings and mountain slopes in Hong Kong and also considered the effects of Sky View Factor, which was not a part of previous experimental studies [14]. According to their study (Figure 6), the heating of the building wall surfaces and mountain slopes during the day produced upward buoyancy driven flow, whilst the air from the Victoria harbour replaced this upward air.

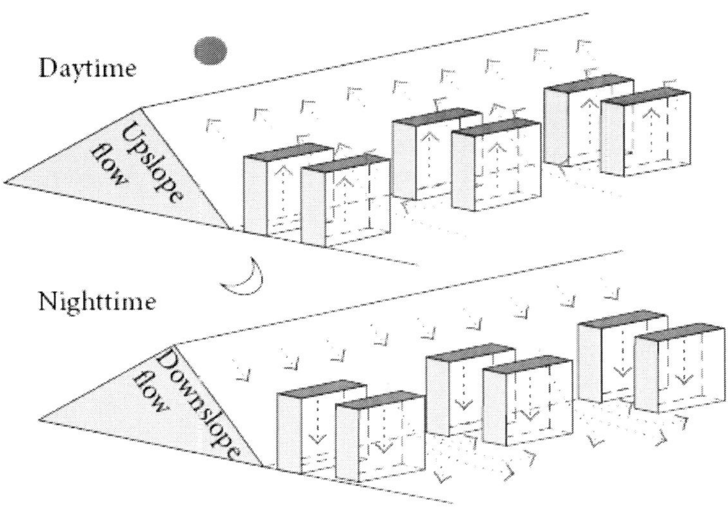

Figure 6: Schematic 3D illustration of the slope wind and wall flow system (from [14]).

At night the opposite happens (Figure 6) and there is a downward flow of air due to the cooling of the building and mountain surfaces. Yang and Li. [14] state that "The slope flows and wall flows dominate the city ventilation when the large scale flows (or mesoscale sea-land breezes) are weak, that is, no-wind conditions." In fact, this study also showed that the ACH values of wall flows were significantly higher than slope flows. Most previous studies focused only on katabatic/anabatic mountain flows, without considering the contribution of the building surfaces (vertical walls), and, therefore, the findings of this study are very important in the context of city ventilation.

Figure 7: COSMO facility, Japan (from [16]).

3.8. Field Measurements of Buoyancy Flow on a Building Wall in Japan

A comprehensive outdoor scale model (COSMO) facility to study the natural convection flows along building walls in a street canyon was carried out at the Nippon Institute of Technology, Japan, in December 2008 (Figure 7), by Onomura et al. [16]. The windward and leeward wall sections were heated alternately and particle image velocimetry (PIV) was used to visualise the flow patterns at night. According to Onomura et al. [16], "Shear stress is strong at the top of the street canyon in the heating cases. In the windward wall heating case, a heated wall enhances turbulence within the canyon. Reynolds stress is distinguished at the top of the canyon in the leeward wall heating case."

An extension of this study was carried out by Nottrott et al. [6] (see Figure 8) using sonic anemometers and thermocouples.

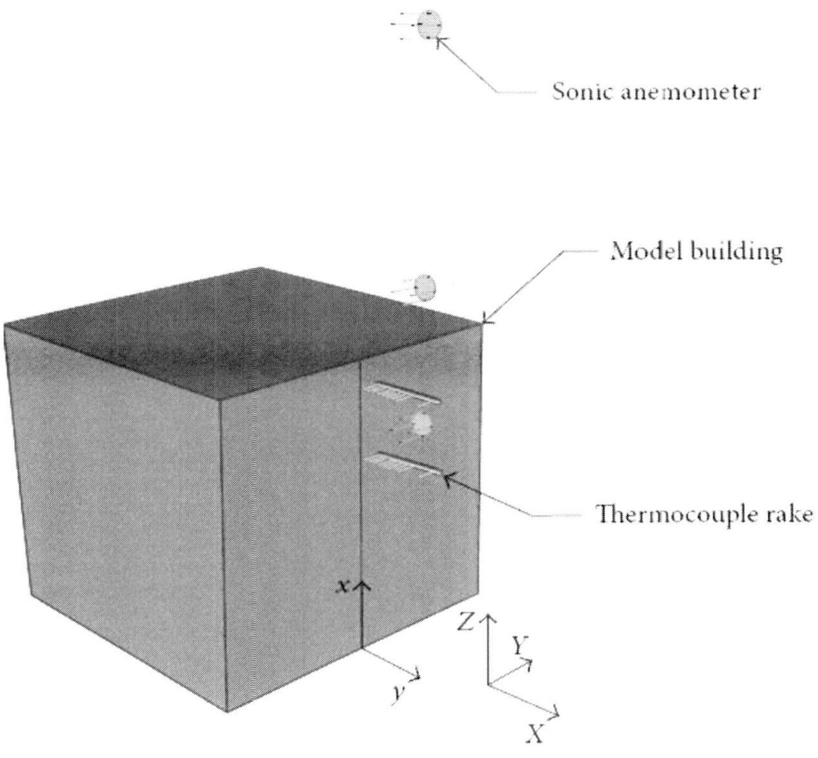

Figure 8: A schematic drawing of the experimental setup at COSMO (from [6]).

This study was conducted in July 2010 at the COSMO test facility to assess thermally induced buoyancy flows within a street canyon. Many studies in the past have investigated the natural convection flows along vertical plates [39], including water channel measurements by Tsuji and Nagano [40]. However, no full scale measurements on buildings were conducted, although CFD simulations in street canyons have been carried out by Xie et al. [13] as described later. The sonic anemometers and thermocouples were used to obtain 30-minute continuous datasets for 3 days. They found that turbulent natural convection occurred even in low values of Grashof and Reynolds numbers. Based on different values of building height (H) and distance between buildings (S), Onomura et al. [16] concluded that "decreasing the ratio H/S in cities located in warm climates has the potential to increase convection on leeward building walls, thus reducing overall building cooling load." Table 1 summarises the various physical modelling studies covered in this section, besides additional studies. It is important to realise that in an urban setting surrounded by mountains, buoyancy flows occur due to both buildings and mountains. Therefore, it is necessary for any theory/model to integrate both aspects, which has generally not been the case as discussed further.

4. THEORETICAL MODELS DEVELOPED/SUGGESTED BASED ON FIELD/LABORATORY STUDIES

Based on field measurements, many theoretical models were developed. Some of the models were used to assess the flow characteristics of buoyancy flows, whilst some of them tried to establish relationships between various parameters. For instance, Monti et al. [41] utilised the data from the VTMX (described earlier in Section 3.2) and found that mean gradient Richardson number and mixing efficiency (flux Richardson number) were closely related. Similarly, Haiden and Whiteman [42] utilised the data from the VTMX experiments and found that the acceleration of katabatic flows at two different sites was different from those predicted from observed buoyancy. This departure in results was attributed to the unevenness of the slopes, which otherwise appeared smooth to any observer. Based on these observations, Haiden and Whiteman suggested that the momentum and heat budget equations must be revisited.

Upslope buoyancy driven flows were studied by Princevac and Fernando [43] using water channel measurements on a heated inclined slope (Figure 9).

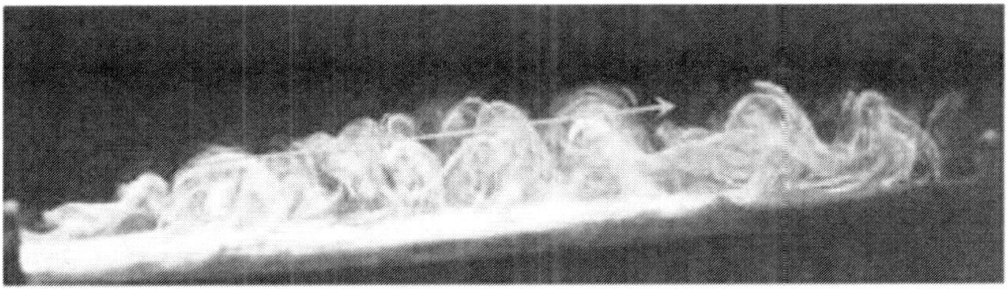

Figure 9: Experimental visualization of convection above the sloped heated plate (from [43]).

They found that the heated thermals had a tendency to travel upwards along an inclined slope at a particular Prandtl number (Pr) and slope angle and established a relationship for upslope flow to sustain at (Pr). Based on the laboratory measurements, Princevac and Fernando concluded that for small slope angles (about 0.1°), the upslope flows were possible and this was found to be consistent with findings from the field study at Phoenix, USA (see Section 3.5), and SABLES study (Section 3.6), where even slopes of about 0.18° also produced anabatic flows.

Similarly, Princevac and his coresearchers utilised the results from VTMX study (Section 3.2) and developed a relationship for the downslope wind velocity, slope length and angle, and buoyancy jump between current and background atmosphere, for different stability conditions [44]. They showed that the katabatic flows exhibited pulsations with a time period proportional to the buoyancy frequency of the background atmosphere. Their theory which was supported by the VTMX study suggested that the internal wave oscillations were partly responsible for weak sustained turbulence, often observed in nocturnal flows.

There are other studies on analytical models that have tried to model mountain slope flows [45]. For instance, Zammett and Fowler [46] modified the classical Prandtl model to simulate katabatic flows on mountain slopes with glaciers and ice sheets. Their study showed that the improved model compared well with observations and other available regional models. A model to simulate nocturnal low level jets over slopes, such as the Great plains in USA, was developed by Shapiro and Fedorovich in 2009 [47], "to include up- and downslope motion in the boundary layer within a stably stratified environment."

5. CFD SIMULATIONS OF BUOYANCY DRIVEN FLOWS IN NATURE

Most CFD simulations have utilised k-ε based model. For instance, Xie et al. [48] used the renormalisation group (RNG) k-ε model to assess the effect of solar radiation on building walls and the resulting buoyancy flows in a street canyon. Their simulations showed that the air exchange rate increased with heating intensity, but decreased with increasing Reynolds number. Furthermore, they also developed a regression based model relating the air exchange rate and heating intensity.

The authors extended their study on street canyons (Figure 10) by varying the street canyon aspect ratio (building height H to street width W) using the RNG k-ε model as described in [13]. They found that "In contrast to single surface heating, the multi-surface-heating configuration has greater influence on the flow field and pollutant transport in the street canyons."

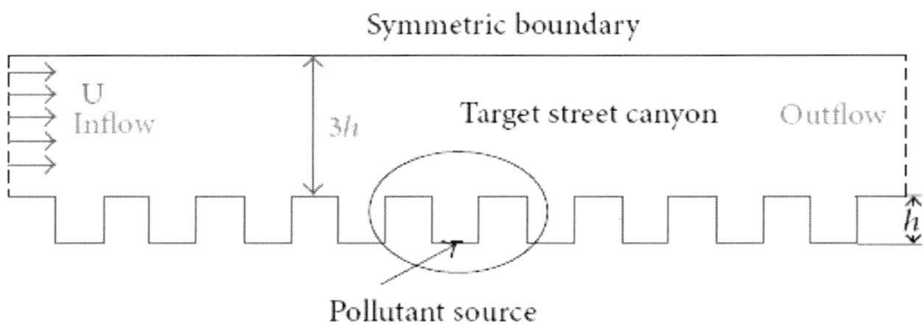

Figure 10: Schematic diagram of the computational domain and boundary conditions (from [13]).

For instance, Figure 11 shows normalised pollutant concentration profiles for two different cases.

Additionally, they also found that the street canyon aspect ratio was a dominant factor in altering the flow characteristics. A similar CFD study on the flow and dispersion of pollutants within a street canyon was also carried out by Wang et al. [49] and they found that "When the windward wall is warmer than the air, an upward buoyancy flux opposes the downward advection flux along the wall, and divides the flow structure into two counter-rotating vortices indicating a clockwise top vortex and a reverse lower vortex within the canyon." Due to advancements in computer speeds, Large Eddy Simulations (LES) of mountain slope flows have also been carried out recently. For instance, Skyllingstad [50] carried out LES simulations of a steep (20°) and a gentle slope (1°) and stated that

"Comparison of the gentle slope case with a flat terrain simulation indicates that drainage winds can effectively prevent the formation of very stable boundary layers, at least near the top of sloping terrain." Other examples include Axelsen and Van Dop [51, 52] who also performed LES simulations of katabatic winds and found reasonable agreement with some previous simulations from Schumann [53].

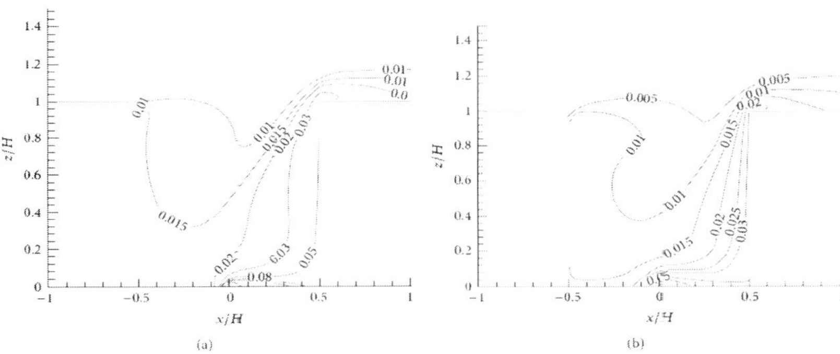

Figure 11: Spatial contours of normalised pollutant concentration in street canyon with different surfaces heated in aspect ratio 1: (a) ground heated; (b) leeward façade and ground heated (from [13]).

A more recent CFD study using the k-ε model that includes the buoyancy effects of buildings and mountain slopes on city ventilation was carried out by Luo and Li [15] (Figure 12). The ventilation of the city of Hong Kong was evaluated in terms of air changes per hour and age of air concepts. From their studies they stated that "When the building is lower than 60 m, the slope wind dominates. When the building is as high as 100 m, the contribution from the urban wall flow cannot be ignored."

Luo and Li [15] found from their simulations that slope flows (particularly katabatic flows) are highly beneficial in ventilating a city. However, due to increased urbanisation in Hong Kong (high rise buildings), building walls significantly increased the buoyancy driven flows as opposed to mountains. Compared to most previous studies which mainly focused on any one aspect (mountain slope or building wall in street canyon), this study dealt with the effects of both building walls and mountain slopes which is more realistic.

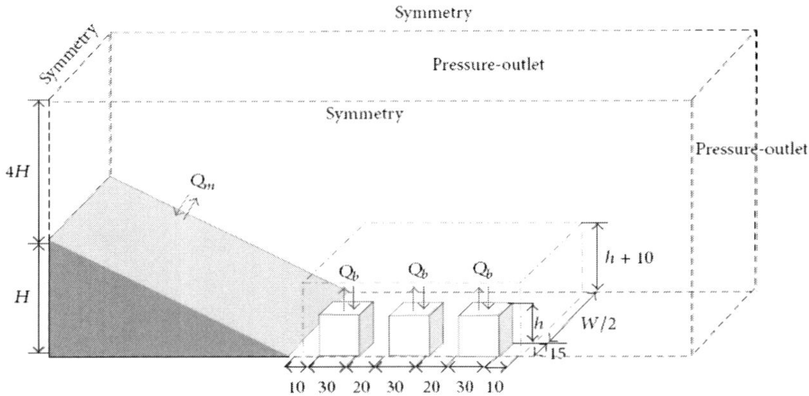

Figure 12: Computational model of mountain slope and building blocks (from [15]).

6. DISCUSSION

This review mainly covered some recent studies (after 2000) on buoyancy driven flows in nature (mountain slopes and buildings) and its influence on reducing pollution levels in a city. Some issues that are worthy of discussion from this review are as follows.

6.1. Instrumentation and Quantification of Buoyancy Flows

Most studies discussed in this paper (e.g., VTMX and TRANSFLEX studies) have mainly utilised tethered balloons, LIDARS, SODARS, and sonic anemometers to assess mountain slope flows, whilst the field study at the University of Hong Kong (Section 3.7) utilised an infrared camera. This is a significant improvement in technology, because infrared cameras were originally designed for military use [54], but these have been used only recently for estimating surface temperatures (e.g., [14]). This makes measurements much more convenient since these thermal cameras can be installed almost anywhere to measure the temperature of a particular surface (ground, building, or mountain slope), besides being cost effective (e.g., tethered balloons are quite expensive). Furthermore, using a tethered balloon has certain disadvantages for meteorological measurements and as aptly described by Parker et al. [55] "the familiar pressurized balloons made of latex are easily ruptured by sharp objects in canopies, or because the sizes of the canopy spaces that accommodate balloons might limit the payload." Additionally, strong winds may affect the measurements made by tethered balloons. In recent times, unmanned aerial vehicles [56] have been used for meteorological measurements over volcanic mountains (e.g., [57]) and may be an alternative for tethered balloons in future. Obviously, these problems are overcome by using an infrared camera. More recently, Onomura et al. [16] have even used an infrared camera to measure 2D wind velocity distribution on a building surface exposed to solar radiation. They used the thermal image velocimetry (TIV) concept of tracking brightness temperature images to measure the wind velocities and reported excellent comparisons with measurements from sonic anemometers. Indeed, gradually new technology to facilitate the measurement of buoyancy driven flows is being developed and is very important in the context of city ventilation.

Some studies on UHI have also estimated the air temperatures in rural and urban areas of Hong Kong using satellite images (e.g., [58]), although the data from those studies have actually never been used to study buoyancy driven flows (slope and wall flows and improving wind conditions in the city), but one should expect satellite imaging to be a promising tool of the future for city ventilation studies.

Furthermore, most studies have tried to quantify the effect of buoyancy induced flows in nature using wind speeds (along slopes) or tracer concentrations in valleys (e.g., VTMX and ASCOT studies). However, more recent studies by [14] and [15] have utilised the air changes per hour concept that was originally used in building ventilation [59] and extended it to urban ventilation flows. Of particular importance is the successful application of empirical relations (e.g., the velocity and thickness of the thermal boundary layer on a wall by [19] utilised by Yang and Li [14] for urban flows), which is a significant improvement with regards to the quantification of urban ventilation.

In general, the field measurements of buoyancy driven flows should adopt a systematic approach depending on the parameter being measured and the location of measurement.(a)Locations that are difficult to access: while measuring surface temperatures on a mountain slope or high rise building, it may be difficult to fix thermocouples on these surfaces. Therefore, in such cases, an infrared camera could be most useful since this can provide greater ease of measurement with high accuracy [14]. In order to measure buoyancy driven wind velocities, sonic anemometers as well as infrared images to assess 2D wind speeds may be utilised.(b)Larger areas of measurement: while covering larger areas (such as New York), satellite imaging technology to gather urban surface temperature data (e.g., [58]) can be useful, since infrared cameras cannot cover such larger areas of interest. On certain locations on the urban surface, thermocouples can be used in order to compare with satellite data.(c)Measuring parameter: if the parameter being measured is wind velocity, sonic anemometers are extremely accurate and convenient due to their light weight (compared to the conventional weather stations) making it easier to install them at any location. A weather station should be used only if additional parameters (e.g., humidity, air temperature etc.) are being measured, although in such cases only limited measurements at certain critical locations may be carried out.

6.2. Flow Processes in the Urban Environment

It is understandable that the buoyancy driven flows in the urban environment are too intricate and are not fully understood yet. One of the key issues is modelling the flow appropriately. For a populated city surrounded by skyscrapers and mountains (e.g., Hong Kong), it is not realistic to simulate any one process in particular. For instance, most previous studies (e.g., VTMX and ASCOT field measurements) have mostly focused on mountain slope flows, whilst some of them have tried to simulate buoyancy flows only due to building walls in a street canyon (e.g., [48]). But in reality, an urban area consists of many sources of buoyancy driven flows, which include buildings, mountains and other structures (e.g., transmission line towers, cooling towers and industrial chimneys, and trees that are sufficiently tall) and humans. Most studies particularly, numerical simulations, generally consider only one aspect (only mountain slopes or only building walls), the only exception being the study by Luo and Li [15] which considered both aspects. For instance, Gao and Niu found that the Rayleigh number (product of Grashof and Prandtl numbers) at the head level of a human being is approximately , indicating a turbulent nature of flow [60]. In a highly populated city like Hong Kong with a population of above 7 million, each person living in the city can greatly contribute to the buoyancy driven flow. Furthermore, additional structures (e.g., industrial stacks and tall trees) could also contribute to natural convection flows, and therefore future studies must take account of this for realistic cases.

It is worth noting that most studies have tried to estimate the dispersion of tracers (e.g., in TRANSFLEX study) due to katabatic/downslope winds, with very few studies on anabatic/upslope winds (e.g., Section 3.3). Buoyancy driven flows become particularly important for city ventilation when the synoptic winds are weak and interestingly most studies (particularly, field measurements) have not highlighted the difficulty encountered in measuring city ventilation rates for each case (presence of synoptic winds and absence of synoptic winds). In fact, the study by Reuten et al. [30] is the only study that has highlighted this problem. For instance, in one of their measurements in the convective boundary layer (CBL), they found that during the day the upslope flow along a mountain had a tendency to develop a "clockwise rotation" after a certain distance up the slope. Reuten et al. [30] stated that "We cannot give a conclusive explanation for the observed wind directions but suspect a superposition of the synoptic wind with an upslope flow in the bottom part of the CBL and with a return flow in the upper part of the CBL." However, this difficulty can be countered using CFD simulations, since each effect (buoyancy and synoptic winds; buoyancy without synoptic winds) can be studied separately or as a combination.

6.3. Computational and Analytical Models

Since, field measurements may not always be possible due to financial and time constraints, computer based simulations of buoyancy driven flows will dominate the future and the results obtained from previous field measurements will be useful for validation. The main problem with any model is that none of them can actually represent the physics of the entire process accurately. For instance, conventional mesoscale models like MM5 [35] use the "roughness approach" to replicate urban elements (buildings) and although MM5 was found to be useful in the TRANSFLEX study of mountain slope flows (Section 3.5) its application for an urban site consisting of building and mountains may be less accurate. Similarly, other empirical/analytical models (e.g., [61]) that were developed from laboratory measurements may not be applicable for modelling urban flows.

It may be necessary to analyse the problem in a slightly different manner. For instance, the concept of ACH which was initially used for building ventilation was later used for city ventilation studies [14]. It is worth noting that a study carried out by Xi et al. [62] has reported many "plume-vortex" and "plume-plume" interactions, based on Rayleigh-Benard turbulent convection experiments carried out in a convection cell, with the bottom and top plates being heated and cooled, respectively. This experimental set up is very much representative of a large city which is heated at the bottom (ground surface) and cooled at the top (top of the urban canopy layer). Although their study was for Rayleigh numbers of about 10^8, whilst the natural convection flows in nature are of much higher Rayleigh numbers, the results from [62] may be applied to the study of city ventilation, although this is just a speculation and the effectiveness of the application of this concept can only be verified by future researchers.

From the various studies reviewed, it is clear that even today most people use the k-ε based models to simulate buoyancy flows [15]. However, Direct Numerical simulation and LES may be better alternatives given the recent improvements in high speed computers, and, in fact, some LES studies on mountain slope flows have yielded excellent results [50]. It is understandable that CFD using LES may be a widely used tool in future, but experiments and numerical modelling must be done in parallel to improve the latter.

6.4. Designing a City for Better Wind Conditions and Reduced Pollution Levels

For a developed city like Hong Kong consisting of a large built up area (which reduces wind penetration) and numerous structures (mountains and buildings contributing to the buoyancy flows), besides high population and acute shortage of land, very few options to improve the city ventilation conditions exist. One option may be to use very large floating structures (VLFS), as described in [63]. These structures have been employed (Figure 13) in the past, particularly in coastal regions where there is shortage of land.

Figure 13: Floating restaurant in Yokohama, Japan (from [63]).

This can avoid the further construction of skyscrapers which have otherwise lead to reduced wind speeds in cities like Hong Kong [64, 65].

Another widely used method discussed by many researchers is the planting of trees that can reduce UHI effects, improve day-lighting and wind conditions in a city [66]. Similar concepts were earlier used to develop "green roofs" [67], and now the same is being extended to the city scale. Another example is the numerical simulations carried out by Hong et al. [68] for a district in Beijing to assess thermal and wind conditions due to trees for various "parameters" (tree locations, distance between trees and buildings, type of tree, etc.).

7. FUTURE DEVELOPMENTS

As human civilization expands, particularly in cities located amidst mountains (e.g., Hong Kong and Phoenix), air pollution (particulate matter from industries and vehicles; anthropogenic heat, etc.) will also rise. Obviously, the developments that arise in future will stem from the deficiencies encountered currently. Some of the key issues faced by researchers today are as follows.

(a) Understanding of the problem: Fernando [3] has rightly stated that "Fluid mechanics is inextricably related to fundamental developments in urban meteorology," and therefore it is important to realise that any model (theoretical/empirical) must capture the complete mechanism in the urban scenario. For instance, previous studies exclusively focusing on any one aspect (wall or slope flow) are not fully accurate.

(b) Modelling issues and coordination between experimentalists and physicists: once the problem is fully understood, modelling it using CFD, particularly using LES, is preferred. Although k-ε models do not require high speed computers, as opposed to LES, the latter is much more accurate for modelling these flows. More importantly, CFD results must be validated either from previous field measurements or from laboratory simulations and this will require collaborations between physical modellers and computational scientists. This collaboration is essential, since computational modelling is expected to make an impact in future, considering the time and financial factors involved in field studies, although it is expected that laboratory simulations (e.g., water channel measurements) will certainly continue in future and will possibly be a major source of experimental data for CFD scientists. However, accurate modelling of these flows in water channels is essential. For instance, Reuten et al. [10] considered three different "Pi" groups for simulation of slope flows in a water tank and found that despite equality of the "Pi" groups, "similarity of maximum upslope flow velocity in atmosphere and water tank was violated by a factor of more than two." Reuten et al. [10] further stated that "We speculate that differences in Reynolds number and roughness height between tank and atmosphere may be the cause for the velocity similarity violation." Therefore, it is important to have an accurate simulation of urban buoyancy flows in laboratory measurements and this may require revisiting of some of the existing similarity criteria being currently followed.

(c) Better instrumentation for field measurements: it has often been observed that most studies have tried to estimate the wind velocities using various measurement techniques along slopes with very little effort to estimate whether these "velocities" were caused by a combination of buoyancy flow and synoptic winds or due to any one individual "component." Therefore, future instrumentation should be capable of distinguishing each "component" of flow. With so much wireless technology rising, it may be possible to have this technology applied to the procurement of data, especially in remote locations which may be difficult to access [69]. Similarly, unmanned vehicles [70] with global positioning system may be used for meteorological observations.

(d) Designing a city for improved wind and thermal conditions: it is important for designers to come up with design codes/guidelines that can help upcoming cities that are located amidst mountains where buoyancy driven flows could play a major role in city ventilation. Some of these criteria should include building dimensions (max. allowable height/width to enable wind penetration), width of streets, allowable emissions (from vehicles, industries, etc.), and location of these industries within the city. Other options such as cool roofs and tree plantation, besides using specific building construction materials (e.g., [71, 72]) to provide a cooler environment, should also be investigated in detail.

8. CONCLUSIONS

The following may be concluded based on the review of recent work (after 2000) on buoyancy driven flows in nature.(a)Most field studies were focused on mountain slope flows, with limited research on buoyancy driven flows along a building wall. Additional "urban buoyancy flow" studies focusing on the combined effect of "building wall" and "mountain slope" are necessary.(b)Field measurement techniques to assess buoyancy driven flows in the urban environment have gradually improved from tethered balloons to the use of infrared camera for estimating surface temperatures [14] and wind velocities [73].(c)CFD simulations have mostly focused on k-ε based models of wall flows [13], with limited studies applying LES modelling of slope flows [50], and few studies focused on the combined effect of building wall and slope flows [15].(d)Analytical and empirical relations based on laboratory measurements have not been successfully applied for geophysical scales, and, hence, a proper understanding of the physical processes associated with buoyancy driven flows in nature is necessary.(e)In future, collaboration between physical modellers and computational scientists to improve CFD modelling is necessary. Furthermore, design guidelines to generate improved wind conditions in a city, taking account of buoyancy flows to mitigate urban air pollution, must be developed.

CONFLICT OF INTERESTS

The author declares that there is no conflict of interests regarding the publication of this paper.

REFERENCES

1. M. Oppenheimer, "Global warming and the stability of the West Antarctic ice sheet," Nature, vol. 393, no. 6683, pp. 325–332, 1998.
2. D. L. Davis, M. L. Bell, and T. Fletcher, "A look back at the London smog of 1952 and the half century since," Environmental Health Perspectives vol. 110, no. 12, pp. A734–A735, 2002.
3. H. J. S. Fernando, "Fluid dynamics of urban atmospheres in complex terrain," Annual Review of Fluid Mechanics, vol. 42, pp. 365–389, 2010.
4. J. Hang, M. Sandberg, and Y. Li, "Effect of urban morphology on wind condition in idealized city models," Atmospheric Environment, vol. 43, no. 4, pp. 869–878, 2009.
5. J. C. R. Hunt, H. J. S. Fernando, and M. Princevac, "Unsteady thermally driven flows on gentle slopes," Journal of the Atmospheric Sciences, vol. 60, pp. 2169–2182, 2003.
6. A. Nottrott, S. Onomura, A. Inagaki, M. Kanda, and J. Kleissl, "Convective heat transfer on leeward building walls in an urban environment: measurements in an outdoor scale model," International Journal of Heat and Mass Transfer, vol. 54, no. 15-16, pp. 3128–3138, 2011.
7. R. Giridharan, S. Ganesan, and S. S. Y. Lau, "Daytime urban heat island effect in high-rise and high-density residential developments in Hong Kong," Energy and Buildings, vol. 36, no. 6, pp. 525–534, 2004.
8. S. Di Sabatino, B. C. Hedquist, W. Carter, L. S. Leo, and H. J. S. Fernando, "The phoenix, urban heat island experiment: effects of built elements," in American Meteorological Society, 89th Annual Meeting, pp. 10–16, Phoenix, Ariz, USA, January 2009.
9. H. J. S. Fernando, D. Zajic, S. Di Sabatino, R. Dimitrova, B. Hedquist, and A. Dallman, "Flow, turbulence, and pollutant dispersion in urban atmospheres," Physics of Fluids, vol. 22, no. 5, pp. 1–20, 2010.
10. C. Reuten, D. G. Steyn, and S. E. Allen, "Upslope flows in atmosphere and water tank, part I: scaling," The Open Atmospheric Science Journal, vol. 4, pp. 178–187, 2010.
11. W. E. Clements, J. A. Archuleta, and P. H. Gudiksen, "Experimental design of the 1984 ASCOT field study," Journal of Applied Meteorology, vol. 28, no. 6, pp. 405–413, 1989.
12. P. H. Gudiksen and D. L. Shearer, "The dispersion of atmospheric tracers in nocturnal drainage flows," Journal of Applied Meteorology, vol. 28, no. 7, pp. 602–608, 1989.
13. X. Xie, C.-H. Liu, and D. Y. C. Leung, "Impact of building facades and ground heating on wind flow and pollutant transport in street canyons," Atmospheric Environment, vol. 41, no. 39, pp. 9030–9049, 2007.
14. L. Yang and Y. Li, "City ventilation of Hong Kong at no-wind conditions," Atmospheric Environment, vol. 43, no. 19, pp. 3111–3121, 2009.
15. Z. Luo and Y. Li, "Passive urban ventilation by combined buoyancy-driven slope flow and wall flow: parametric CFD studies on idealized city models," Atmospheric Environment, vol. 45, no. 32, pp. 5946–5956, 2011.
16. S. Onomura, S. Takimoto, and M. Kanda, "Influence of a heated wall on urban canopy flow using PIV measurements," in Proceedings of the 7th International Conference on Urban Climate, Yokohama, Japan, June–July 2009.
17. Y. Jaluria, Natural Convection Heat and Mass Transfer, Pergamon Press, Oxford, UK, 1980.

18. P. Chandran, N. C. Sacheti, and A. K. Singh, "Natural convection near a vertical plate with ramped wall temperature," Heat and Mass Transfer, vol. 41, no. 5, pp. 459–464, 2005.
19. E. R. G. Eckert and T. W. Jackson, "Analysis of Turbulent Free-Convection Boundary Layer on Flat Plate," NACA Report 1015, 1951.
20. C. D. Whiteman, Mountain Meteorology: Fundamentals and Applications, Oxford University Press, New York, NY, USA, 2000.
21. D. Zardi and C. D. Whiteman, "Diurnal mountain wind systems," in Mountain Weather Research and Forecasting: Recent Progress and Current Challenges, pp. 35–119, Springer, New York, NY, USA, 2012.
22. F. Defant, "Local winds," in Compendium of Meteorology, pp. 655–672, American Meteorological Society, Boston, Mass, USA, 1951.
23. C. D. Whiteman, "Observations of thermally developed wind systems in mountainous terrain," AMS Meteorological Monographs, vol. 45, pp. 5–42, 1990.
24. S. Ostrach, "An analysis of laminar free convection flow and heat transfer about a flat plate parallel to the direction of the generating body force," NACA Technical Report 1111, 1953.
25. J. S. Turner, Buoyancy Effects in Fluids, Cambridge University Press, Cambridge, UK, 1973.
26. B. Gebhart, Y. Jaluria, R. L. Mahajan, and B. Sammakia, Buoyancy Induced Flows and Transport, Hemisphere Publishing, New York, NY, USA, 1988.
27. A.-J. N. Khalifa, "Natural convective heat transfer coefficient—a review I. Isolated vertical and horizontal surfaces," Energy Conversion and Management, vol. 42, no. 4, pp. 491–504, 2001.
28. M. M. Orgill, "Early morning ventilation of a gaseous tracer from a mountain valley," Journal of Applied Meteorology, vol. 28, no. 7, pp. 636–651, 1989.
29. J. C. Doran, J. D. Fast, and J. Horel, "The VTMX 2000 campaign," Bulletin of the American Meteorological Society, vol. 83, no. 4, pp. 537–551, 2002.
30. C. Reuten, D. G. Steyn, K. B. Strawbridge, and P. Bovis, "Observations of the relation between upslope flows and the convective boundary layer in steep terrain," Boundary-Layer Meteorology, vol. 116, no. 1, pp. 37–61, 2005.
31. H. Huang, R. Ooka, and S. Kato, "Urban thermal environment measurements and numerical simulation for an actual complex urban area covering a large district heating and cooling system in summer," Atmospheric Environment, vol. 39, no. 34, pp. 6362–6375, 2005.
32. R. Ooka, H. Chen, and S. Kato, "Study on optimum arrangement of trees for design of pleasant outdoor environment using multi-objective genetic algorithm and coupled simulation of convection, radiation and conduction," Journal of Wind Engineering & Industrial Aerodynamics, vol. 96, no. 10-11, pp. 1733–1748, 2008.
33. A. Mochida and I. Y. F. Lun, "Prediction of wind environment and thermal comfort at pedestrian level in urban area," Journal of Wind Engineering and Industrial Aerodynamics, vol. 96, no. 10-11, pp. 1498–1527, 2008.
34. H. J. S. Fernando, B. Verhoef, S. di Sabatino, L. S. Leo, and S. Park, "The phoenix evening transition flow experiment (TRANSFLEX)," Boundary-Layer Meteorology, vol. 147, no. 3, pp. 443–468, 2013.
35. S. Dupont, T. L. Otte, and J. K. S. Ching, "Simulation of meteorological fields within and above urban and rural canopies with a mesoscale model," Boundary-Layer Meteorology, vol. 113, no. 1, pp. 111–158, 2004.
36. C. Yague, S. Viana, G. Maqueda, M. F. Lazcano, G. Morales, and J. M. Rees, "A study on the nocturnal atmospheric boundary layer: SABLES2006," Física de la Tierra, vol. 19, pp. 37–53, 2007.
37. J. Cuxart, C. Yague, G. Morales et al., "Stable atmospheric boundary-layer experiment in Spain (sables 98): a report," Boundary-Layer Meteorology, vol. 96, no. 3, pp. 337–370, 2000.
38. R. B. Stull, An Introduction to Boundary Layer Meteorology, Kluwer Academic Publishers, Dordrecht, The Netherlands, 1988.

39. R. Cheesewright, "Turbulent natural convection from a vertical plane surface," Journal of Heat Transfer, vol. 90, pp. 1–8, 1968.
40. T. Tsuji and Y. Nagano, "Characteristics of a turbulent natural convection boundary layer along a vertical flat plate," International Journal of Heat and Mass Transfer, vol. 31, no. 8, pp. 1723–1734, 1988.
41. P. Monti, H. J. S. Fernando, M. Princevac, W. C. Chan, T. A. Kowalewski, and E. R. Pardyjak, "Observations of flow and turbulence in the nocturnal boundary layer over a slope," Journal of the Atmospheric Sciences, vol. 59, no. 17, pp. 2513–2534, 2002.
42. T. Haiden and C. D. Whiteman, "Katabatic flow mechanisms on a low-angle slope," Journal of Applied Meteorology, vol. 44, no. 1, pp. 113–126, 2005.
43. M. Princevac and H. J. S. Fernando, "A criterion for the generation of turbulent anabatic flows," Physics of Fluids, vol. 19, no. 10, Article ID 105102, 2007.
44. M. Princevac, J. C. R. Hunt, and H. J. S. Fernando, "Quasi-steady katabatic winds on slopes in wide valleys: hydraulic theory and observations," Journal of the Atmospheric Sciences, vol. 65, no. 2, pp. 627–643, 2008.
45. R. G. Barry, Mountain Weather and Climate, Cambridge University Press, Cambridge, UK, 3rd edition, 2008.
46. R. J. Zammett and A. C. Fowler, "Katabatic winds on ice sheets: a refinement of the Prandtl model," Journal of the Atmospheric Sciences, vol. 64, no. 7, pp. 2707–2716, 2007.
47. A. Shapiro and E. Fedorovich, "Nocturnal low-level jet over a shallow slope," Acta Geophysica, vol. 57, no. 4, pp. 950–980, 2009.
48. X. Xie, C. H. Liu, D. Y. C. Leung, and M. K. H. Leung, "Characteristics of air exchange in a street canyon with ground heating," Atmospheric Environment, vol. 40, no. 33, pp. 6396–6409, 2006. View at Publisher •
49. P. Wang, D. Zhao, W. Wang, H. Mu, G. Cai, and C. Liao, "Thermal effect on pollutant dispersion in an urban street Canyon," International Journal of Environmental Research, vol. 5, no. 3, pp. 813–820, 2011.
50. E. D. Skyllingstad, "Large-eddy simulation of katabatic flows," Boundary-Layer Meteorology, vol. 106, no. 2, pp. 217–243, 2003.
51. S. L. Axelsen and H. Van Dop, "Large-eddy simulation and observations of slope flow," Acta Geophysica, vol. 57, pp. 803–836, 2008.
52. S. L. Axelsen and H. van Dop, "Large-eddy simulation of katabatic flow over an infinite slope," Acta Geophysica, vol. 57, no. 4, pp. 837–856, 2009.
53. U. Schumann, "Large-eddy simulation of the up-slope boundary layer," Quarterly Journal of the Royal Meteorological Society, vol. 116, no. 493, pp. 637–670, 1990. View at Publisher • •View at Scopus
54. N. C. Currie, F. J. Demma, D. D. Ferris Jr., B. R. Kwasowsky, R. W. McMillan, and M. C. Wicks, "Infrared and millimeter-wave sensors for military special operations and law enforcement applications," International Journal of Infrared and Millimeter Waves, vol. 17, no. 7, pp. 1117–1138, 1996.
55. G. G. Parker, P. J. Stone, and D. Bowers, "A balloon for microclimate observations within the forest canopy," Journal of Applied Ecology, vol. 33, no. 1, pp. 173–177, 1996.
56. K. Nonami, F. Kendoul, S. Suzuki, W. Wang, and D. Nakazawa, Autonomous Flying Robots: Unmanned Aerial Vehicles and Micro Aerial Vehicles, Springer, 1st edition, 2010.
57. D. Pieri, J. A. Diaz, G. Bland et al., "In situ observations and sampling of volcanic emissions with NASA and UCR unmanned aircraft, including a case study at turrialba volcano, Costa rica," Geological Society Special Publication, vol. 380, no. 1, pp. 321–352, 2013.
58. J. E. Nichol, W. Y. Fung, K.-S. Lam, and M. S. Wong, "Urban heat island diagnosis using ASTER satellite images and "in situ" air temperature," Atmospheric Research, vol. 94, no. 2, pp. 276–284, 2009.

59. D. Etheridge and M. Sandberg, Building Ventilation-Theory and Measurement, John Wiley & Sons, Chichester, UK, 1996.
60. N. Gao and J. Niu, "Modeling the performance of personalized ventilation under different conditions of room air and personalized air," HVAC & R Research, vol. 11, no. 4, pp. 587–602, 2005. View at Publisher •
61. A. J. Wells and M. G. Worster, "A geophysical-scale model of vertical natural convection boundary layers," Journal of Fluid Mechanics, vol. 609, pp. 111–137, 2008. View at Publisher • • View at Zentralblatt MATH
62. H. D. Xi, S. Lam, and K. Q. Xia, "From laminar plumes to organized flows: the onset of large-scale circulation in turbulent thermal convection," Journal of Fluid Mechanics, no. 503, pp. 47–56, 2004.
63. E. Watanabe, C. M. Wang, T. Utsunomiya, and T. Moan, "Very large floating structures: applications, analysis and design," Tech. Rep. 2004-02, Centre for Offshore Research & Engineering (CORE), National University of Singapore, 2004.
64. W.-L. Ginn, C. Lee, and K.-Y. Chan, "Past and future changes in the climate of Hong Kong," Acta Meteorologica Sinica, vol. 24, no. 2, pp. 163–175, 2010.
65. E. Ng, Designing High-Density Cities for Social and Environmental Sustainability, Earthscan, Sterling, Va, USA, 2009.
66. W. Hongbing, Q. Jun, H. Yonghong, and D. Li, "Optimal tree design for daylighting in residential buildings," Building and Environment, vol. 45, no. 12, pp. 2594–2606, 2010.
67. R. Kumar and S. C. Kaushik, "Performance evaluation of green roof and shading for thermal protection of buildings," Building and Environment, vol. 40, no. 11, pp. 1505–1511, 2005.
68. B. Hong, B. Lin, L. Hu, and S. Li, "Optimal tree design for sunshine and ventilation in residential district using geometrical models and numerical simulation," Building Simulation, vol. 4, no. 4, pp. 351–363, 2011.
69. M. Zennaro, A. Bagula, D. Gascon, and A. Bielsa, "Planning and deploying long distance wireless sensor networks: the integration of simulation and experimentation," in Proceedings of the 9th International Conference (ADHOC-NOW '10), pp. 191–204, Edmonton, Canada, August 2010.
70. D. E. Cook, P. A. Strong, S. A. Garrett, and R. E. Marshall, "A small unmanned aerial system (UAS) for coastal atmospheric research: preliminary results from New Zealand," Journal of the Royal Society of New Zealand, vol. 43, no. 2, pp. 108–115, 2013.
71. P. G. Tuohy, L. B. McElroy, and C. Johnstone, "Thermal mass, insulation and ventilation in sustainable housing—an investigation across climate and occupancy," in Proceedings of the 9th International IBPSA Conference on Building Simulation (BS '05), pp. 1253–1260, Montreal, Canada, August 2005. View at Scopus
72. F. Aldawi, F. Alam, I. Khan, and M. Alghamdi, "Effect of climates and building materials on house wall thermal performance," Procedia Engineering, vol. 56, pp. 661–666, 2013. View at Publisher •
73. A. Inagaki, M. Kanda, S. Onomura, and H. Kumemura, "Thermal image velocimetry," Boundary-Layer Meteorology, vol. 149, no. 1, pp. 1–18, 2013.
74. K. J. Allwine, J. H. Shinn, G. E. Streit, K. L. Clawson, and M. Brown, "Overview of Urban 2000: a multiscale field study of dispersion through an urban environment," Bulletin of the American Meteorological Society, vol. 83, no. 4, pp. 521–536, 2002.
75. J. Kleissl, R. E. Honrath, M. P. Dziobak et al., "Occurrence of upslope flows at the Pico mountaintop observatory: a case study of orographic flows on a small, volcanic island," Journal of Geophysical Research D: Atmospheres, vol. 112, no. 10, Article ID D10S35, 2007. View at Publisher • •View at Scopus
76. P. Sedlák, M. Aubinet, B. Heinesch et al., "Night-time airflow in a forest canopy near a mountain crest," Agricultural and Forest Meteorology, vol. 150, no. 5, pp. 736–744, 2010.

77. C. Feigenwinter, L. Montagnani, and M. Aubinet, "Plot-scale vertical and horizontal transport of CO2modified by a persistent slope wind system in and above an alpine forest," Agricultural and Forest Meteorology, vol. 150, no. 5, pp. 665–673, 2010.
78. L. S. Savage, S. Zhong, W. Q. Yao, T. Horst, and W. O. Brown, "A regional-scale downslope wind observed during the METCRAX Experiment in Northern Arizona," in Proceedings of the 18th Symposium on Boundary Layer and Turbulence, Stockholm, Sweden, June 2008.
79. J. Allegrini, V. Dorer, and J. Carmeliet, "Wind tunnel measurements of buoyant flows in street canyons," Building and Environment, vol. 59, pp. 315–326, 2013.

CHAPTER 7

Emissions and Possible Environmental Implication of Engineered Nanomaterials (ENMs) in the Atmosphere

Astrid C. John [1,*], Miriam Küpper [1], Astrid M.M. Manders-Groot [2], Bruno Debray [3], Jean-Marc Lacome [3] and Thomas A.J. Kuhlbusch [4,5]

[1]*Institute for Energy and Environmental Technology (IUTA), Bliersheimer Str. 58-60, 47229 Duisburg, Germany*

[2]*Netherlands Organisation for Applied Scientific Research TNO, PO Box 80015, 3508 TA Utrecht, The Netherlands*

[3]*Institut National de l'Environnement Industriel et des Risques (INERIS), Parc ALATA, 60550 Verneuil-en-Halatte, France*

[4]*Federal Institute for Occupational Safety and Health (BAuA), Friedrich-Henkel-Weg 1-25, 44149 Dortmund, Germany*

[5]*CENIDE, University Duisburg-Essen, Carl-Benz-Str. 199, 47057 Duisburg, Germany*

ABSTRACT

In spite of the still increasing number of engineered nanomaterial (ENM) applications, large knowledge gaps exist with respect to their environmental fate, especially after release into air. This review aims to summarize the current knowledge of emissions and behavior of airborne engineered nanomaterials. The whole ENM lifecycle is considered from the perspective of possible releases into the atmosphere. Although in general, emissions during use phase and end-of-life seem to play a minor role compared to entry into soil and water, accidental and continuous emissions into air can occur especially during production and some use cases such as spray application. Implications of ENMs on the atmosphere as e.g., photo-catalytic properties or the production of reactive oxygen species are reviewed as well as the influence of physical processes and chemical reactions on the ENMs. Experimental studies and different modeling approaches regarding atmospheric transformation and removal are summarized. Some information exists especially for ENMs, but many issues can only be addressed by using data from ultrafine particles as a substitute and research on the specific implications of ENMs in the atmosphere is still needed.

Keywords: engineered nanomaterials; ENMs; release; transformation processes; aerosols; ultrafine particles; atmospheric transport

1. INTRODUCTION

Engineered nanomaterials (ENMs) are one of the subgroups of "nanomaterials" comprising also natural and unintentionally produced nanomaterials [1]. They are commonly defined as materials containing particles or having structures with at least one dimension within 1–100 nm, or powders with 50% or more of the particles by number having one or more external dimensions in the size range 1–100 nm [1]. Other definitions, e.g., based on a threshold in volume-specific surface area [2,3] or concentrating on properties differing from those of bulk samples rather than on a generalized size range [4], have also been proposed. These unique properties are inter alia caused by the large percentage of atoms at the surface and an increasing number of crystalline defects with decreasing particle size [4]. This may lead to increased reactivity, high conductivity and exceptional optical properties [5,6,7] making them valuable for numerous applications.

Developments of new materials and technologies, here ENMs, also require the investigation of possible implications for humans and the environment [8,9,10]. These investigations become more important with the increased use of ENMs [11,12,13]. Conservative global market estimates for e.g., metal oxide nanomaterials added up to 270,000 tons in 2012, rising to 1,663,000 tons by 2020 [14]. As a consequence, an exposure risk to ENMs may become a concern for the population [15,16].

Health effects of anthropogenic and naturally emitted UFP (ultrafine particles <100 nm) as well as ENMs are still under investigation, with regulatory concerns in air quality control [11]. However, since physicochemical characteristics of intentionally manufactured ENMs might differ from those of unintentionally produced UFP in numerous ways, such as aspect ratio, reactivity, chemical composition and toxicity [17,18], further progress has to be made assessing alternative metrics, such as airborne particle number or particle surface area concentration, and their link to health and environmental effects of the multiplicity of ENMs [19,20]. Furthermore, it is necessary not just to target the pristine ENMs, as done in the REACH guidance ("Regulation concerning the Registration, Evaluation, Authorisation and Restriction of Chemicals" [21], but the specific formulation of the ENM comprising e.g., coatings or adherent parts of the matrix they have been embedded in [22].

In addition to health impacts, ENMs may—analogously to UFP—also have implications on global climate if used in huge amounts. Especially, secondary UFP formed by oxidation of SO_2, NO_x or VOCs to $H_2SO_4/(NH_4)_2SO_4$, HNO_3/NH_4NO_3 and secondary organic aerosols (SOA) followed by nucleation or condensation processes (e.g., [23,24]) are important for climate due to their impact on atmospheric chemistry, cloud formation and the radiation budget of the earth (e.g., [25]).

The majority of the literature dealing with fate and transformation processes of ENM in the environment focuses on aquatic systems and soils (e.g., [26,27]), since the major part of ENMs is known to end up in these two compartments, whereas only 0.1–1.5% of produced ENMs are estimated to be released into the atmosphere during their life cycle [28]. Calculations conducted in studies by Gottschalk et al. [29] resulted in concentrations of several pg of ENMs/m^3 in the Danish atmosphere, whereas Müller and Nowack [30] modelled quantities of 1.7 ng/m^3 nano-Ag, 1.5 ng/m^3 nano-TiO_2 and 1.5 ng/m^3 carbon nanotubes (CNT) for the atmospheric compartment above Switzerland (realistic release scenario based on substance flow analysis), respectively. Though, no concentrations are published which are measured around production facilities where higher concentrations can be expected. In any case, monitoring concentrations, transformation processes and fate of airborne ENMs is relevant as atmospheric transformations can have a crucial impact on ENM's role in atmospheric chemistry and further fate in the environment [5]. In addition, most importantly, altered ENM characteristics caused by these processes may influence the toxic potential [31,32,33,34].

We currently lack sufficient knowledge of the types, rates, and extent of transformations expected for ENMs in environmental and biological systems [35,36]. Especially cross-media effects (i.e., transformations in the atmosphere that may affect fate in the aquatic or terrestrial environments or vice versa) have been largely overlooked in discussions of the environmental impacts of ENMs [37], but making predictions about ENM's fate, transport, toxicity and implications on atmospheric chemistry requires significant, multidisciplinary advances [38,39].

Part of the ENM transformation occurs shortly after release and depends on the ambient aerosol. Therefore it is highly important to be precise on the state in which ENMs are released, and in practical cases emission preprocessors may be necessary to account for such transformation processes close to the source before emissions can be used in models.

This review summarizes the current state of knowledge with regard to the emission, transformation, transportation and possible implications of ENMs in ambient air, to give an overview of which information is already present and which aspects might need further investigation.

2. HOW CAN ENMS BE RELEASED AND HOW MUCH IS EMITTED INTO AMBIENT AIR DURING THE WHOLE LIFE CYCLE?

So far most published measurement campaigns concentrated on assessing workers' exposure by measuring or calculating ENM concentrations in air at the workplace [40,41,42] whilst there has been little study of the release of engineered nanomaterials to the environmental air compartment during the product life cycle. Indeed, field measurements of ENMs in the environment are scarce and often measurements of releases of ENMs to air from industrial plants are done by the industry under strict confidential conditions. The following section aims at detailing the current knowledge of the various release scenarios that can lead to emission of ENMs to ambient air.

2.1. ENM Life-Cycle

Each step of the ENM life cycle from production, processing and use to the end-of-life (Figure 1) may lead to emissions into ambient air [43]. The emissions can be continuous, e.g., resulting from the evacuation of polluted air flows from production, transformation or waste treatment plants, the emission from the secondary material being processed (e.g., release from various matrixes following mechanical, thermal, chemical or combined stresses) either during the production of ENMs or during their use. ENMs can also be released accidentally, resulting from a sudden uncontrolled loss of containment, a fire or an explosion [44,45].

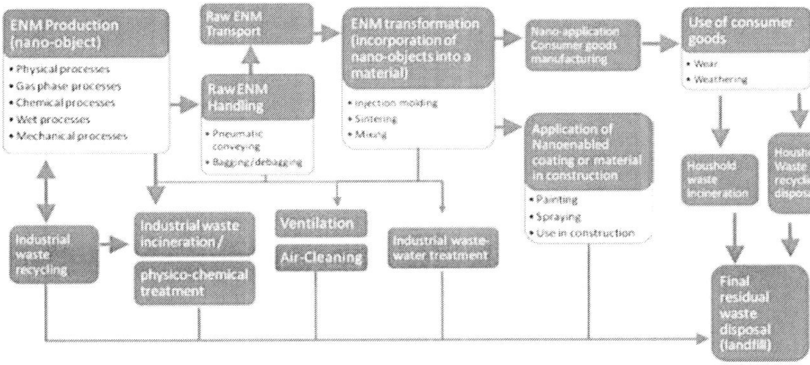

Figure 1. ENM life cycle.

An overall estimate of ENM release to different compartments by Keller et al. [28] considers that currently, around 63–91% of the ten most produced ENMs are disposed in landfills, whereas releases into water are around 0.4–7% of total global production. Air emissions represent the smallest flow to the environment for the studied ENMs, with an estimate of less than 1.5% of production when considering the highest release estimates. Yet, considering the potential impact of these emissions at the local scale and the major concerns about the potential health effects of ENMs when inhaled, the relative significance of the emissions to air should not be underestimated. At the local scale, emissions to air may become the major concern both with respect to environmental impacts and human health.

2.2. Production of ENM: A Qualitative Overview

Gottschalk and Nowack [46] estimated in 2011 as an initial approach from 0 to 2% release to the environment during manufacturing. More recently a study produced as support for 3rd regulatory review on nanomaterials for European Commission DG Environment has proposed updated emission factors during manufacturing [47] yet still mostly based on expert judgement ranging from 0.7% to 1.6% of which 6% is estimated to be emitted through air.

There exists a variety of ENM manufacturing processes with very variable emission potential. In many cases, the major emissions are not stemming from the process step generating the nanomaterial, but at a subsequent phase such as recovering particles from the reactor, milling, drying or further handling. Indeed, several production processes are closed as they are based on the chemistry of gas mixtures sometimes under low or high pressure. Other processes such as spray drying or milling are more prone to leaks as they are often performed in air at atmospheric pressure conditions.

The emission potential is also dependent on the form of the ENMs, i.e., whether it is produced as free powder, suspension in a liquid or already as solid material (e.g., master batch). Powders have a higher emission potential than suspensions or undivided solid forms.

As an illustration, Table 1 lists a series of ENM production processes [48] and indicates whether the process is potentially highly emissive [49]. The classification is qualitative and mainly based on expert judgement considering the intrinsic degree of tightness of the reactor in which the main chemical synthesis reaction occurs. It is considered here that synthesis reaction that require vacuum or controlled atmosphere are intrinsically less emissive during the synthesis phase, to the work environment and ambient air, than processes performed under atmospheric conditions. Yet it is important to underline the fact that tightness can always be reinforced to limit exposure and that opening a closed reactor to recover the produced particles or clean the reactor can be a source of emission. Emission control systems can also be added to reduce releases at each step of the process [50,51,52,53]. More precise figures on measured emissions from manufacturing processes in work environment can be found in [54] supporting the high variability of emissivity and the dependence on activities and processes involved as well as engineering controls in place.

Table 1. ENM production processes and their potential emissivity in the absence of controls: H = potentially high emissivity during the production phase, L = Low emissivity in normal conditions during the production phase.

Process Type	Process	Production Conditions	Emissivity
Physical processes	Electrospinning	Closed	H
	Gas phase processes based on Heating and fast cooling	Controlled atmosphere	L
	Laser pyrolysis	Controlled atmosphere	L
	Spray drying	Atmospheric conditions	H
Chemical processes	Liquid phase processes: precipitation, sol gel, exfoliation, sonochemistry	ENM produced as liquid suspension; in case of drying, it can be emissive	H
	Chemical vapor deposition	Controlled atmosphere	L
	Flame synthesis	Atmospheric conditions	H
	Hot plasma synthesis	Controlled atmosphere	L
Mechanical processes	High energy milling	Atmospheric conditions	H
	Severe plastic deformation	Atmospheric conditions	H
	Shock wave processes	Atmospheric conditions	H

2.3. Internal Handling and Transport of ENMs

After being produced, the ENM will be handled. Internal transport and handling within the plant, including packaging operations, can be very important sources of accidental or continuous ENM emissions. For instance, pneumatic transport involves the movement of particles in an air flow from which they need to be separated by filters. Depending on the efficiency of this separation process, emissions can occur to the environment. Additionally, accidental loss of containment due to a breach or the piercing of a filter can lead to a spillage and air transport of particles. Falling of powder into a hopper or filling of a big bag are also potentially emissive processes [42].

However, certain measures have been developed to reduce ENM emissions in the workplace, such as containment or ventilation and air treatment [40,51].

2.4. Emission from Stacks

The main point sources at production plants are emissions from stacks which either stem from process flows from reactors and process units, or ventilation flows from working areas or specific working steps. Usually, process flows carry much more particles than workshop ventilation flows. The rate of emission is highly dependent on the rate of emission from processes, either directly to process flow or indirectly to the ambient air in the premises, and the efficiency of filtering devices used to prevent emissions to the environment in collecting nanoparticles. EN 1822-1 Standard [55] specifies the filtration efficiency for a variety of particle filters used in ventilation and climatization systems ranging from 85% for EPA E10 filters up to 99.999995% for ULPA U17 filters. The filtration efficiency is measured at the "Most Penetrating Particle Size", which is determined specifically for each filter, ranging usually between 0.1 μm and 0.5 μm [56,57]. Collection efficiencies varying from 90% to 99.999% were reported for fabric filters used in municipal solid waste incineration plants with a minimum efficiency around 0.4 μm [58].

2.5. Diffuse Emission from Building

If the ventilation and air cleaning of buildings is not efficient enough, diffuse emissions can also occur from the buildings through openings such as doors or windows. Such fugitive emissions were observed and documented for microscale (PM_{10}) particles in conventional industrial activities such as steel production [59] and ceramic industry [60].

Monfort et al. [60] determined for ceramic industry that if the appropriate corrective measures are applied, high efficiencies can be achieved in reducing PM_{10} diffuse emissions from the storage, handling, and transport of raw materials, particularly in semiclosed and closed facilities, in which the overall efficiency obtained was 75–80% and >95%, respectively. In such closed facilities emission factors between 7 to 11 g of PM_{10} per ton of product were estimated. So far, such a study still needs to be undertaken for the production and transformation of ENMs.

Material deposited on the ground by accident (leak, accidental spillage during handling) can remain in place for some time and be resuspended to form a new continuous emission.

2.6. ENM Emissions during Manufacturing and Processing of Composite Materials

There is a wide variety of ENMs. Their most frequent application involves the incorporation of nanoobjects into matrix material (polymer, ceramic or metal) to form a composite. The processes and the associated handling operations are potentially emissive, especially if the nanoobjects are incorporated as powder. The most common process for their incorporation in polymers is injection molding, but other processes such as spray-drying can also be used. The impact is usually limited to the workplace, although, due to improper filtration of local ventilation, particles might also be conveyed to the surrounding environment. Nowack et al. [61] described nine release scenarios along the lifecycle of a CNT bearing nanocomposite. The first one concerns the manufacturing of the material by injection molding. The second is relative to the processing of the composite material (cutting, sawing, drilling and sanding of raw nanocomposites). Machining operations like sanding, cutting and drilling are based on high energy input and may lead in each case to a considerable generation of ENMs in workplaces [61]. Unfortunately, these scenarios are only described in a qualitative way and quantitative estimates are lacking.

As for the initial manufacturing, engineering controls can lower the emission rates from plants in which these operations take place.

2.7. Continuous Emission Rates from Industrial Plants

Emission rates from industrial plants are rather difficult to apprehend due to the variety of industrial processes and industrial plant configurations. They are all the more difficult for ENMs as there is no specific dedicated regulation that imposes the measurement and declaration of ENM releases to the environment.

The IED directive is the European framework for the control of industrial emission. As far as particle emissions into air are concerned, the IED directive only focuses on PM_{10} and $PM_{2.5}$ [62]. The IED directive requires plants to use best available technologies (BAT) in their processes to minimize the impact on the environment. Separately from the IED, the UNECE Convention on Access to Information, Public Participation in Decision-making and Access to Justice in Environmental Matters (better known as Aarhus Convention) grants the public rights to access environmental information. Under this Convention, the European Pollutant Release and Transfer Register (E-PRTR [63],) was established through Regulation (EC) No 166/2006 where pollutant emitters above a certain threshold have to report annually their pollutant release to air and to water. For particulates, only PM_{10} is required for reporting in E-PRTR, and ENMs are not considered at all. Although included in PM_{10}, their emission rate cannot be deduced from PM_{10} data.

France was the first country to impose a declaration of the amounts of produced or imported ENMs [64]. Since the beginning of this legal requirement three annual reports were published with a progressive increase of the declaration rate. The total amount of ENMs declared as produced in France in 2014 [65] was 300,822 t. The total amount of ENMs declared as imported was 114,951 t. Table 2 presents the main categories of ENMs produced or imported in France in 2014. Due to confidentiality rules applicable to the use of declarations, only the broad tonnage interval is displayed. It is important to note that some ENMs of concern such as carbon nanotubes, although present in the declaration data, do not appear in this table because it presents only the ENM produced or imported in amounts above 100 t. In total, more than 300 different chemical species were identified among the declarations.

Although these data improve greatly the vision of the ENM market, they do not directly inform on release rates. The published data do not allow to differentiate the forms in which the ENM is produced: powder, suspension in a liquid or in a solid, nor do they mention the production and transformation process. Yet they can be used in a first approach to identify the ENMs present on the market in large quantities and have been used in [47] as a basis to estimate global releases by nanomaterials family.

Table 2. Ten categories of ENM produced or imported in France in amounts higher than 1000 t [65].

Generic Name	Tonnage Band
Carbon black	>100,000 t
Silicon dioxide	>100,000 t
Calcium carbonate	10,000–100,000 t
Titanium dioxide	10,000–100,000 t
Boehmite (Al(OH)O)	1000–10,000 t
Vinylidene chloride copolymer	1000–10,000 t
Silicic acid, magnesium salt	1000–10,000 t
Aluminium oxide	1000–10,000 t
Vinyl polychloride	1000–10,000 t
Mixture of cerium dioxide and zirconium dioxide	1000 10,000 t

2.8. Emission During Use

During the ENM use phase, particles embedded in a matrix material can be emitted due to abrasion induced by the user or material degradation under various types of ambient solicitations (rain, wind, sun). Such phenomena have been studied for a variety of materials, such as construction materials [66], composites [67,68] and surface coating [69,70]. Also, some reviews on release mechanisms [71,72] and on protocols simulating aging are available [73]. The emission rate is highly variable and depends on the ENM and on the conditions of use. Values ranging from 0 to 90% release of the ENM to the environment have been estimated by Keller et al. [28], with only low fractions (0–5%) of these values released into air. For most ENMs the use phase is the most important source of ENM release to the environment, however release to air is only a small fraction of this (e.g., [29]). Therefore, far, although several authors have developed conceptual and computational material flow analysis models, reliable data are missing to estimate the release rate from objects containing ENMs during their use phase [43].

2.9. Transportation of ENM

During transportation, if proper packaging is used, the risk of emission is mostly accidental. Yet, so far there is no specific requirement regarding the transportation of ENMs. Decisions are based on requirements applicable to any powder material and depend mainly on proven hazardous properties determined through regulatory testing. Thus, it cannot be ruled out that transportation of ENMs is done in a way that allows for particles to be emitted.

At least one road accident involving spilling of ENM (TiO_2) powder was reported [45]. This accident occurred when big bags fell on the road due to improper tightening of the truck load. However, it is possible that several others occurred and were not identified as such due to the absence of specific labeling of nanomaterials.

The emission that results from a transportation accident is difficult to assess. It is partly composed of a first emission due to the shock and initial loss of containment, but it is then followed by a continuous lifting by air of particles from the spill until the spill is completely cleaned-off.

2.10. Emission into Air during Waste Treatment and Disposal

Various waste treatment technologies can be envisaged for ENMs such as recycling, landfill disposal, and incineration. The emission of ENMs from incineration has been studied in various projects [74,75]. Ounoughene et al. [76] showed that the fate of ENMs is highly dependent on their composition. Carbonaceous and organic ENMs have a tendency to burn and therefore to be completely eliminated. Other ENMs such as nanoclays were shown to remain in the combustion residues [75]. When burning silicone polymers with a SiO_2 load, nanoparticles were found in the fly ashes, but it was concluded that a large part of these particles were in fact the result of the silicon combustion and not only the release of the incorporated ENMs [76]. Most of these particles can be collected by usual smoke treatment processes in state of the art incinerators.

In a quantification approach Mueller and Nowack [30] estimated rates of release from waste incineration plants of 0.05–1% into air, 1–50% into slag, and 50–98% being captured by filters. For comparison, waste water treatment plants are estimated to remove between 75% and 97% of ENMs from the inflow waters [77,78] and transfer them to biosolids which are then disposed off through incineration, application to agricultural land, or landfills.

2.11. Emissions during Abnormal Conditions along the Life Cycle: Accidental Release

Beside continuous emissions, incidents or accidents can also lead to significant emissions. The main scenario is an accidental or at least unwanted loss of containment which results in a flow of particles in the air, which will directly be transported to the surrounding environment or will form a depot on the ground that will be resuspended and transported until the spill is removed. Fire and explosions can also be the cause of emission of ENMs to air. In terms of quantity, the amount of ENMs released to the environment during one single accidental event could, in some cases, exceed the continuous emissions from plants, especially when very stringent measures are applied to limit continuous emissions.

3. HOW CAN ENMS BE TRANSFORMED DURING THE AIRBORNE PHASE?

While a lot of research has been done on transport and alterations of ENMs in aquatic media during the last years, information on ENM transformations within the atmosphere is still quite scarce. However, information and studies about the emission and behavior of unintentionally produced UFP are readily available, which can be used as starting point for the ENM assessment. While engineered nanomaterials consist of primary particles, UFP can be released as primary particles (e.g., soot) or can be formed in the atmosphere as secondary particles from gaseous precursors (e.g., formation of $(NH_4)_2SO_4$). Modest differences between the characteristics and behavior of atmospheric ENMs and UFP have been observed, originating from the higher ENM uniformity in shape, size and functionality [79]. However, projecting knowledge of UFP behavior on ENMs seems to be justified in most cases. In some cases, though, there can be significant discrepancies when e.g., ENM cannot coagulate due to specific surface functionalities like coatings [80].

We can differentiate between the physical transformation processes of hetero- and homo-coagulation and condensation, physico-chemical dissolution as well as chemical transformation processes such as oxidation or chemisorption during cloud processing and photo-induced reactions that might result in removal of coatings and change from hydrophobic to hydrophilic behavior. Usually, the fate of ENMs is affected by a combination of several of these processes.

3.1. Physical Transformations

3.1.1. Coagulation

Generally, particles in the nanometer size range tend to coagulate due to their large diffusion coefficients. However, since the coagulation rate for simple monodisperse coagulation (homo-coagulation) is proportional to the square of the particle number concentration [81], this process is negligible at low concentrations of ENMs. For example, Maynard and Kuempel [82] calculated that nanometer-diameter particle concentrations below 10^{12} particles/m^3 can be assumed to be relatively stable in an occupational environment. Due to the diminutive atmospheric concentrations of ENMs (e.g., [30]), homo-coagulation seems to be negligible in the atmosphere. This process could be important at higher concentrations found in accidental industrial releases or potentially occurring near the emission source, e.g., close to production lines. However, homo-agglomerates might be formed in local environments with higher concentrations and persist at a background scale, as predicted by Quik et al. [83] for aqueous media.

Polydisperse (hetero-)coagulation is enhanced when small particles with their high diffusion coefficients diffuse to a large particle with its large surface area [84]. Hetero-coagulation is therefore most effective for particles of different sizes. Hence, hetero-coagulation is assumed to dominate over homo-coagulation between ENMs in the atmosphere, as shown for example by Seipenbusch et al. [85], who conducted experiments in an aerosol chamber to investigate the agglomeration behavior of metallic nanoparticles (Pt). They simulated background aerosol by introducing appropriate concentrations (typically $\sim 10^3$ to 10^4 cm^{-3}) of oil droplets and monodisperse silica particles (particle sizes in the range of 0.1–1 μm) and found almost the entire primary ENM peak to vanish by attachment onto the larger background particles.

It is important to keep in mind that although the particle number concentration is diminished by coagulation, the ENM itself is not removed from the atmosphere. However, the change in size of the particle does have an impact on the lifetime of the ENM as discussed in Section 3.1.4.

In aquatic media polymeric-like molecules such as humic substances may be capable of imparting strong electrostatic charges or steric stability to the surface of ENMs preventing coagulation upon sorption [86,87]. Although there has been some work on humic-like substances [88], there have been no studies to assess if the presence of large organic molecules in the atmosphere can play a similar role.

3.1.2. Condensation

Airborne particles down to ca. 40 nm may act as nuclei for the condensation of water and semi-volatile compounds—most frequently sulfate, nitrate, ammonium and organic compounds. For UFP typical particle growth rates in the range 1–20 nm h^{-1} have been identified by Kulmala et al. [24] in mid-latitudes depending on the temperature and the availability of condensable vapors.

Once released into the atmosphere, ENMs might also serve as condensation nuclei. Their condensational growth is generally assumed to occur within a similar time frame. However, this property depends not only on particle size, but also on surface characteristics meaning that these processes may be hampered by particular surface coatings.

In a study performed by Lee and Donahue [89] the authors showed that polyacrylic acid coatings on TiO_2 and CeO_2nanoparticles act as retardant for secondary organic aerosols, generated from α-pinene ozonolysis under atmospheric conditions. This assumption is confirmed by studies relating measurement efficiencies of Water-Based Condensation Particle Counters (WCPC) to those performed by butanol-based Condensation Particle Counters (CPC) which revealed lower counting efficiencies for some hydrophobic types of particles, such as candle-generated aerosol (paraffin) and emery oil [90], as well as for Di(2-ethylhexyl) sebacate (DEHS) particles [91]. Therefore it can be assumed that such kinds of ENMs or ENMs coated with hydrophobic substances might exhibit lower atmospheric growth rates.

In addition to particle enlargement, condensation processes can also cause ENMs to get more spherical, e.g., due to compaction of the agglomerates, as shown for soot particles by Slowik et al. [92]. This is especially relevant for CNT, since their aspect ratios might be significantly reduced during condensation, resulting in modified behavior in the environment.

Some case-studies highlighting exceptional coagulation-properties of particular ENM have been published so far. Tiwari and Marr [37] compiled such literature focusing on atmospheric transformations of C_{60} fullerenes and CNT. They suspect fractal and non-spherical ENMs, such as aggregated fullerenes and tangles of CNT, to convert to more spherical structures during atmospheric processing, due to analogous findings in laboratory experiments for fractal agglomerates of carbon particles exhibiting branched chains (e.g., [93]).

3.1.3. Dissolution

Solubility of ENMs depends not only on solvent characteristics such as ionic strength, saturation state, pH, etc. [94], but also on several parameters of the particles themselves, such as size, surface, shape and aggregation. Atmospheric particles are known to traverse multiple condensation cycles [95] influencing their dissolution behavior: Laboratory experiments conducted by Desboeufs et al. [96] and Spokes et al. [97], in which evaporation chambers have been applied to simulate cloud processing, revealed elements dissolved during each condensation/evaporation cycle (e.g., aluminum, iron, and manganese), to deposit back on the surface possessing an altered structure increasing their dissolution within the next cycle. Desboeufs et al. [96] state this mechanism to depend strongly on pH of the aqueous phase, which can in turn be influenced by the particle itself. Consequently, dissolution of particles (including ENMs) within the atmosphere is not straightforward.

Nevertheless, some progress in tackling this problem has already been made. For instance, Kaptay et al. [98] concentrated on the impacts of size and shape of ENMs on their solubility, and by Mihranyana and Strømme [99] investigating dissolution behavior of fractal ENMs. Furthermore, Quadros and Marr [100] referring to a study by McMahon et al. [101] assume Ag-nanoparticles to dissolve into cloud or fog droplets under slightly acidic conditions, releasing silver ions. Apart from silver, dissolution of iron and zinc oxide ENMs was observed within aqueous media and might also occur at high humidities within air, releasing Fe^{2+}, Fe^{3+} [102], and Zn^{2+} [103] whereas the frequently applied TiO_2-, SiO_2-ENMs as well as unfunctionalized fullerenes and CNT exhibit very low solubilities [28,104,105].

3.1.4. Removal from Atmosphere

ENMs will ultimately deposit on soil or surface water [106] by dry or wet deposition. This deposition may either happen directly or follow growth of ENMs via coagulation and/or condensation.

According to Seinfeld and Pandis [107] the vertical dry deposition flux is dependent on local particle concentration in a specific height over the ground, since it does not act on the full column (except for gravitational settling of large particles) but is related to processes close to the earth's surface where it includes turbulent diffusion and collection of particles by surface and vegetation. The efficiency of dry deposition is relatively high for small particles and for high friction velocities [81]. Removal reaches a minimum for a particle size range of 100 to 1000 nm resulting in extended atmospheric lifetimes for ENMs that have grown into this size range or are attached to larger co-emitted particles [37].

Within regions of precipitation, wet deposition (scavenging by precipitation elements) can be the dominant removal process. Wet deposition depends on the amount, type and intensity of the precipitation and on the properties of the scavenged material. For aerosols, the scavenging coefficient is dependent on the particle size and other particle properties (for example, whether the aerosol is hydrophilic or hydrophobic).

For all these deposition processes, theoretical frameworks exist but uncertainties are large. Measurements on dry and wet deposition and deposition velocities only cover a limited set of atmospheric conditions and both measurements and the theoretical framework show uncertainties of a factor 10 to 100. Deposition of UFP has been investigated in some case studies, e.g., [108,109], who conducted experiments within test chambers to determine deposition rates of NaCl particles and gasoline vehicle exhaust, respectively.

3.2. Chemical Transformations

Literature on chemical transformation of ENMs within the atmosphere is scarce and within the existing studies most authors concentrated on a single type of nanomaterial [9,110,111]. Generally, chemical transformation affects the ENM's surface first and might even be restricted to the outer layer depending on chemistry of the particles themselves and of the coating if present. These implicated alterations of surface chemistry are most important, since they can influence the fate of the particle in the environment. Most prominently, oxidation processes or photo-induced chemical reactions may result in active sites on particle surfaces to be passivated or coated with time [18,82]. For instance, gas-phase oxidation of carbonaceous particles is known to cause an increase in hygroscopicity [112], and other atmospheric processes, such as condensation of organic molecules or the incorporation of carbonaceous particles into internally mixed aerosols, might have similar impacts [37] potentially influencing ENM's fate and transport within the atmosphere.

3.2.1. Oxidation in Ambient Air

Upon release into the atmosphere metallic ENMs, such as nano-iron or nano-silver, are transformed via chemical reactions with atmospheric oxidants [113,114], whilst common nano-oxides such as TiO_2, SiO_2 and CeO_2 will not undergo further oxidation after being released into the atmosphere.

Atmospheric organic aerosols are well known to be oxidized by reactions with hydroxyl radicals, ozone, and other oxidants, leading to the formation of oxide, carbonyl, hydroxyl, carboxylic acid, and other functionalities that make the particles more hydrophilic [115]. First studies have proven that the same seems to apply for carbonaceous ENM [37].

For instance, Tiwari et al. [116] investigated reactions between aerosolized fullerenes (C_{60}) and O_3 at atmospherically relevant O_3 mixing ratios with reaction rates ranging from 9×10^{-6} to 2×10^{-5} s^{-1}. Results indicate that C_{60} is oxidized by O_3 and forms a variety of oxygen-containing functional groups on the aerosol surface, which are expected to reduce hydrophobicity of the material, since this effect was observed for studies of the same reaction within aqueous media [117]. Additionally, the authors found the extent of this reaction to be influenced by relative humidity (surface oxygen content rises with relative humidity), whilst kinetics may be influenced by photolysis (see section "Photo-induced reactions").

So far, ambivalent results have been published concerning the reactivity of CNT with O_2. Collins et al. [118] found CNT exposure to air or oxygen to dramatically influence their electrical resistance, thermoelectric power, and local density of states, as determined by transport measurements and scanning tunneling spectroscopy. Giannozzi et al. [119] report triplet oxygen to retain its spin-polarized state when interacting with a nanotube leading to the formation of a weak physisorptive bond with essentially no charge transfer between the molecule and the CNT.

A possible explanation for these ambivalent results is provided by Goldoni et al. [120], who performed investigations of single-wall CNT processed in buckypaper using Electron Energy Loss-Spectroscopy (EELS). They revealed the presence of residual contaminants in purified SWCNT to be possibly responsible for their claimed sensitivity to oxygen. However, the authors cannot rule out for sure that their treatment of SWCNT at high temperatures under ultrahigh-vacuum conducted for purification purposes might have induced the insensitivity of the CNT towards O_2.

3.2.2. Reactions with Other Substances Apart from Oxygen

To identify and quantify possible transformation processes affecting nano-silver, McMahon et al. [101] measured surface plasmon-polariton spectra upon exposure to laboratory air. They observed sulfur to be chemisorbed at the surface of the particles and determined a growth rate of this tarnishing product (Ag_2S) of 3 nm per day, being 7.5 times higher than that of bulk Ag under the same conditions. Consequently, atmospheric nano-silver might be assumed to transform even more quickly, since larger parts of the ENM's surface area are expected to be exposed to air compared to the experimental setup by McMahon et al. [101], where particles were placed onto a substrate.

In a study performed by Goldoni et al. [120] that has already been described briefly in the previous section, chemical alterations of SWCNT due to reactions with NH_3, NO_2 and SO_2 were confirmed to take place, whereas no changes in electronic spectra of "clean" SWCNT were observed after exposure to O_2, CO, H_2O and N_2, indicating that these gases interact with clean SWCNT via weak London dispersion forces only.

3.2.3. Photo-Induced Transformation Processes

Apart from chemical reactions such as oxidation, certain ENMs can be photoactivated upon exposure to light of specific wavelengths [82]. This photoactivation can result in photolysis, photooxidation or photoreduction, which can act to change the structure and properties of ENMs [18].

Probably the most prominent example for photo-induced reactions are TiO_2 particles: Depending on the type of TiO_2applied (photoreactive or photostable), photochemical transformation on the surface of TiO_2 nanoparticles takes place due to its known photoreactivity (e.g., [22,121]) which is discussed in detail in Section 4.1.

For the aqueous phase photolysis of C_{60} clusters has been observed [122]. Since this process has been shown to depend on aggregation state, this transformation may be marginal in natural waters, but might play a major role in the atmosphere, where aggregation is less likely under typical conditions [37]. However, atmospheric processing and thus functionalization of C_{60} may influence fullerenes' potential for undergoing photolysis in the natural environment [37]. As mentioned by Tiwari et al. [116], photolysis may also influence the C_{60}–O_3 reaction in terms of rate, distribution of products, and/or reactive oxygen species (ROS) generation potential, suggesting that atmospheric transformation of C_{60} fullerenes would alter them substantially and may result in the introduction of highly oxygenated and cross-linked fullerenes into the environment. However, the extent to which photolysis of carbonaceous ENMs in the atmosphere may occur is unknown [37] since the study of photochemistry at interfaces, specifically the air—particle one, is still in its infancy [123].

Sunlight exposure might also represent one major pathway causing the loss of coatings, as sunlight-catalyzed redox reactions are suspected to be involved in the degradation of some polymeric coatings, such as gum Arabic, present on ENMs [124]. The loss of engineered polymeric coatings may inter alia induce instability and facilitate (hetero-

)aggregation [18] and can thus alter the ENM's fate in the environment crucially.

4. WHAT ARE THE POSSIBLE ATMOSPHERIC IMPLICATIONS OF ENMS?

Besides the effects the atmosphere has on ENM transformation, ENMs also might affect atmospheric chemistry and physics. Wiesner et al. [125] figured out that ENMs also show high absorption coefficients, and many of them can act as catalysts [80]. However, no information is currently available on the interactions between ENMs and the chemicals they absorb, and how this interaction might influence atmospheric chemistry.

4.1. Photo-Catalytic Properties of TiO_2

As mentioned before, TiO_2 owns special properties: When excited with light of wavelengths equal to or greater than the band gap energy (3.0 and 3.2 eV equal to lights wavelength of 290 and 390 nm for rutile and anatase, respectively), electron hole pairs are generated in the conduction and valence bands, respectively (e.g., [126]). These electron-hole pairs can then follow several pathways: (i) recombination of electrons and holes on the surface or in the bulk of the semiconductor; (ii) reduction of electron acceptors by photogenerated electrons; and (iii) oxidation of electron donors by photo-generated holes [127,128]. As a consequence, TiO_2 surfaces facilitate interfacial photochemical reactions and these reactions have the potential to change the chemical balance of the atmosphere [129,130].

The formation of reactive oxygen species (ROS, e.g., $\cdot OH$, $HO_2\cdot$, $\cdot O^{2-}$) is known to be catalyzed by TiO_2 [131] and ROS again might also initiate additional photo-catalytic reactions, which is described as remote photo-process (e.g., [132]), or can get involved in redox-reactions with other atmospheric constituents, such as VOC (e.g., [133]). Accordingly, adsorption of VOC on TiO_2 surfaces and subsequent degradation into CO_2, H_2O, and other low-molecular weight compounds acts as a potential removal pathway.

CO_2 and SO_2 are known to adsorb on TiO_2 surfaces resulting in formation of adsorbed HCO^- [134] and adsorbed sulfite or sulfate (e.g., [135]), respectively.

TiO_2 has been reported as an effective NO_x removal material by reducing NO_2 into gaseous HONO and minor amounts of N_2O as well as into surface adsorbed NO^-, $N_2O_2^{2-}$, and under atmospheric conditions most frequently nitrate [130,136]. However, TiO_2 has also been observed to produce NO_x by photo-oxidation of Ammonia under atmospheric conditions [137]. Additionally, Monge et al. [138] revealed the production of O_3 to be induced by TiO_2 upon exposure to NO_x under illumination, and the deactivation of TiO_2 surfaces by nitrate coating formation was observed ([130] and references therein) resulting in a more complex role of TiO_2 in atmospheric chemistry than previously assumed.

Photo-catalytic decomposition of O_3 on TiO_2 under atmospheric conditions has been observed by various authors (e.g., [139]) and might represent an important sink of ozone in the atmosphere. By combining experimental observations with modeling studies, it has furthermore been suggested that heterogeneous removal of H_2O_2 by mineral dust aerosol may have a significant effect on tropospheric HO_x chemistry, OH budgets, and, consequently, ozone levels [130].

These processes are influenced by different parameters, such as relative humidity, concentration of reactants present, the relative binding of water versus reactants to the surface, and the mechanism of the photocatalytic reactions. Especially water plays a major role since apart from being involved in generating OH radicals by photo-chemical reactions on TiO_2surfaces, H_2O can also act as inhibitor/promotor by blocking/establishing surface reactive sites for adsorption or reaction of reactant molecules on TiO_2 particles [131]. For instance, Bedjanian and El Zein [136] investigated the interaction of $NO_2$with solid TiO_2 films in a flow reactor via mass spectrometric and ion chromatographic detection and measured the following yield of the products of NO_2 interaction with pure TiO_2 under real atmospheric conditions: 0.65 ± 0.10, 0.05 ± 0.05, and 0.15 ± 0.05 for HONO, NO, and N_2O, respectively. The distribution of the products between NO and HONO was found to be independent of temperature in the range T = 280–320 K and was governed by relative humidity (RH) with an increase in RH leading to lower NO and higher HONO yield at a maximum of nearly 65% reached at 5% RH [136].

As implied by experiments conducted by Egerton and Mattinson [126], silica seems to block the photo-activity of TiO_2, since they observed the photo-oxidation rate of propan-2-ol to be reduced when TiO_2 particles were coated with silica compared to uncoated particles. Hence, Majestic et al. [113] conclude that a quick deactivation of TiO_2 might take place in arid areas with high concentrations of wind-blown dust. However, this hypothesis is not confirmed by any study.

Although nanoscale TiO_2 particles are representing only a small portion of the mass of PM, ENMs could contribute significantly to TiO_2 in the atmosphere [140,141] due to their widespread use (see Table 2). Despite lacking literature, TiO_2ENM particles are expected to display high photocatalytic activity due to their large surface area to volume ratios, potentially playing an important role in atmospheric chemistry. For instance, Auffan et al. [4] state that the anatase phase which is greater for very small TiO_2 particles, is more effective in the production of hydroxyl radicals and the subsequent decomposition of organic compounds than the rutile phase. This size-dependent photo-catalytic activity was not observed to increase monotonically with decreasing size but rather pass through a maximum significantly below 100 nm ([4] and references therein).

4.2. Atmospheric Relevant ENM Other Than TiO_2

Although TiO_2 is seen to be the most prominent example for ENMs influencing atmospheric chemistry, other ENMs might have impacts on the atmosphere as well. As the concentration of the different ENMs in the atmosphere is key, measurements are needed.

Inorganic ENMs other than TiO_2 (e.g., ZnO) and photosensitive fullerenes have also been associated with oxidative stress since light stimulation was proven to catalyze ROS production at the ENM's surfaces [141,142,143,144]. In addition to the ENMs that have been mentioned in the literature as affecting atmospheric chemistry, all substances being capable of producing ROS in the atmosphere or in water are assumed to be relevant. For instance, the ROS generation of nano-Au suspended in water has been shown upon exposure to X-ray and UV irradiation [145]. Further exemplary data on that topic can be found in Wohlleben et al. [73]. However, ENMs might also have the contrary effect regarding ROS: For instance, ceria ENMs have been shown to scavenge superoxide [146].

In addition, ENM can be involved in chemical atmospheric reactions not involving ROS. For example, Patakfalvi et al. [147] have shown SO_2 to adsorb onto the surface of silver ENMs as well as the Ag catalyzed reduction of NO in aquatic media, which might also be relevant processes upon release into the atmosphere at high RH.

4.3. Other Effects of ENMs on the Atmosphere

UFP are known to be important for climate studies as they can act as cloud condensation nuclei (CCN) and therefore impact cloud formation and in consequence the radiation budget of the earth (e.g., [148]). This indirect radiation effect is also possible for ENMs. However, due to low ambient concentrations, the impact of ENMs is expected to be much lower compared to conventional particles.

In polluted air, light scattering by particles is relevant especially for those containing sulphate, organic carbon and nitrate species, whilst black carbon particles (e.g., soot from Diesel engines and coal combustion) are relevant for light absorption. Although particles <100 nm are too small to be visible and thus do not directly interact with radiation, ENMs might have an impact for this direct radiation effect when being present as aggregates/agglomerates >300 nm.

Visibility impairment is generally caused by a build–up of suspended particles in the atmosphere. It increases with relative humidity and atmospheric pressure and decreases with temperature and wind speed [149]. At high relative humidity, the light scattering cross–sectional areas of particles enlarge by uptake of water. For an ammonium sulphate particle, the increase may be by a factor or five or more above that of the dry particle [150]. Light scattering may cause 60% to 95% and light absorption 5% to 40% of overall visibility reduction [151,152].

At this point in time, no further relevant implications that ENMs could exert on the atmosphere have been identified.

5. HOW LONG WILL ENM STAY IN THE ATMOSPHERE, HOW FAR ARE THEY TRANSPORTED AND WHERE DEPOSITED?

Transport and deposition of species in the atmosphere have been studied intensively over the past 50 years. Drivers of these studies were concerns about air quality and nuclear safety. Most of the research was devoted to describing mass concentrations and deposition, and state-of-the-art chemistry-transport models describe ambient concentrations fairly well. Over the past ten years ultrafine particles have gained attention in view of health and climate effects. For these particles, nucleation, agglomeration and condensation have to be taken into account. Although this area of research can build on the expertise for conventional particles, several aspects for UFP are currently still described less accurately. This holds even more strongly for ENMs. On the other hand, modelling of atmospheric fate of ENM is less uncertain than environmental fate in water and soil systems.

As pointed out in Section 3 on ENM transformation, not much is known about the deposition behaviour of ENM and analogy is sought with ultrafine particles (UFP) from other origins. A difference between UFP and ENMs which may affect their removal rates is the fact that fresh

ENMs may have more pronounced shapes (e.g., nanotubes, fibers). On the other hand, ENMs are quite often not released at nanoscale size but as larger aggregates so that they can be described as conventional particles with respect to wet and dry deposition. Ding et al. [40] have compiled an overview of ENM release at different workplaces, which can be used as a first estimate of ENM sizes. For most processes (cleaning, packing, sonication, mixing and spraying) ENMs are released as aggregates at sizes larger than 100 nm and mostly in the order of 1 μm. Only for machining and to a lesser extent production, ENMs were found to be released at particle sizes between 10 and 100 nm. Another peculiarity for assessing the fate of particles is the accidental release of ENMs. Accidental release of nano-powders to air can happen by explosion, a fire or a major loss of containment, and is associated with high energy and high concentrations as compared to regular release during production, manufacturing and in the use and disposal phase described in chapter 2. Such scenario already occurred in France with carbon black being released from a pneumatic transport pipe after its rupture, leading to a massive jet release of nanomaterial [153]. This is the most critical scenario as the quantity of released ENMs is maximized. So far, literature studies concerning the modeling of massive release of powders are rather scarce and were devoted to the study of micropowders [154,155]. In order to fully understand the behavior of nanoparticles when they are released with high concentrations in a jet, observations and CFD-modelling including particle dynamics should be combined to identify contours of high concentrations and determine transformations in the state of the material.

For more general environmental fate assessment (production and use phase of material) lifetime and transport distance are dominated by deposition and transformations. Specific models are not yet available to simulate the spatial and temporal ENM concentration field in the air at a local to middle scale [76]. However, models predicting UFP concentrations on a regional scale (e.g., LOTOS-EUROS) are available. A problem is that the material is poorly classified at emission. Most important information needed for accurate fate description are size, attachment to other material, coating, reactivity, volumes, continuous or intermittent emission and the emission height.

Another difficulty to determine lifetime and transport distances of ENMs is caused by their wide variety of particles and release forms (release as raw material, tailpipe emission of motor oil additives together with UFP and condensables, or abrasion of ENM-containing material like tire and break wear), that undergo transformations in the atmosphere as described in Section 3. Properties also depend on concentration at release and presence and concentrations of co-emitted and ambient particles, and subsequent dilution. This means that it is necessary to explicitly investigate a specific material and release form in order to come to a quantitative expression on transport distances and life time, and these values cannot be simply extrapolated to other ENMs. Spatially resolved emission inventories with good characterization for such particles are still in their earliest phase (see Section 2) and are a prerequisite for spatially resolved concentration and deposition.

Also the release height and vertical mixing play a role. In this paper, we only consider release and processes relevant for the atmospheric boundary layer, which will be the most relevant case for ENMs (surface emissions from traffic, abrasion of materials and low stack emissions from plants).

5.1. Study Results on Atmospheric Processes Influencing Fate of UFP

5.1.1. Near-Source Processes: Coagulation Versus Dry Deposition

As already mentioned in Section 3.1, both, coagulation (especially hetero-coagulation) and dry deposition are explicitly relevant and competing processes regarding the fate of ENM. Close to the source, relatively high concentrations of particles may lead to rapid particle growth due to coagulation. On the other hand, deposition velocities for the smallest particles are relatively high. The balance between the two is important to determine the lifetime and transport distances. Since both processes lead to a decrease in number of particles they are difficult to separate in field observations. Here we present results from field studies of UFP as an indication for ENM behavior.

In particular traffic exhaust emissions have been the subject of several recent studies that are relevant here, for their general conclusions regarding behavior of UFP right after release, but also with respect to emissions of e.g., motor oil additives that may be co-emitted. Ketzel and Berkowitz [156] conclude for exhaust emissions that deposition is faster than coagulation for particle sizes >50 nm while coagulation is faster for smaller particle sizes in their modelling study. Karl et al. [157] performed a modelling study with aerosol dynamics (MAFOR model) for traffic emissions and compared results to observations in various cities. Dry deposition and coagulation of particles were identified to be the most important aerosol dynamic processes that control the evolution and removal of particles. However, CFD calculations on a street canyon for different wind directions by Nikolova et al. [158] revealed that the overall effect of deposition and coagulation on the UFP is negligible for the cases considered. They conclude that overall, deposition is more efficient and faster than coagulation in removing particles, and especially those in the nucleation mode. Choi and Paulson [159] performed a field study on freshly emitted vehicle exhaust particles in an early-morning experiment, at a time at which dilution has a smaller (but still dominant; ~70–80%) impact on particle concentrations. They observed that concentrations in the smaller size bins (5.6–23.7 nm) peaked further downwind than the reference particles (42.1–562 nm), and decayed significantly faster than larger particles particularly in the area 100–400 m downwind. Comparisons of the cumulative contributions of van der Waals enhanced coagulation, dry deposition, and dilution and the observed decay curves, imply that for up to the first 50–100 m there is nucleation and/or growth of particles smaller than 5.6 nm. In contrast, in the ~100–400 m region, some of the smaller particles evaporate, which would not apply for non-volatile ENMs but would be relevant for the interaction of ambient particles with ENMs. In the further downwind areas (>400 m) the particles all appear to decay at rates consistent with the sum of dilution, coagulation, and deposition. Keuken et al. [160,161] performed measurements around Schiphol Airport (The Netherlands). Airplanes are strong emitters of UFP, compared to vehicles, producing more and smaller particles. Elevated UFP (10–40 nm) concentrations were detected at stations around Schiphol Airport, even up to 40 km distance, but with a shift in size from 10–20 nm close to Schiphol to 20–40 nm at 40 km distance. It should be noticed that field studies for UFP are particularly difficult to interpret since it is difficult to separate effects of change in size due to coagulation and removal of particles due to deposition, since both result in loss of total particle number. This implies that models are needed to interpret the results of field studies, whereas field studies are needed to constrain models.

5.1.2. Dry Deposition

Parameterizations for dry deposition of particles over a surface are often based on field experiments. Since these parameterizations work best for the conditions and species and circumstances for which they were derived they may not be accurate when extrapolating to other species and environments. Petroff and Zhang [162] show the deposition velocities from a compilation of observations for different surface types, with large scatter and hardly any observations for particles smaller than 100 nm for most surface types, so that it is difficult to validate their new parameterization. Their new parameterization has lower deposition velocities for the particles <1 μm than in the parameterization by Zhang et al. [163] that was and is still used in many chemistry-transport models, implying longer transport distances/lifetimes or higher relevance of wet deposition as removal mechanism. When such parameterizations are used in chemistry-transport models, additional uncertainties arise from the quantification of surface properties that are always a generalization in which regional and seasonal variations due to e.g., growing season, crop characteristics can easily induce variations in vegetation height of a factor 5, which has an impact on the deposition velocity. New parameterizations are still being developed, addressing specific aspects of the process (e.g., for deposition on rough surface of simple shape by Zhang and Shao, [164]). Table 3 shows typical values calculated by and used in the LOTOS-EUROS model (using deposition scheme by [163]) for three representative land use types. Values exhibit a strong daily cycle and day-to-day variability, depending on weather conditions, and the values differ by an order of magnitude over the range of particle sizes.

Table 3. Typical deposition velocities in m/s for the LOTOS-EUROS model [165].

Particle Size	Desert	Grass	Forest
330 nm	10^{-4}–1.5×10^{-3}	10^{-4}–2.5×10^{-3}	10^{-4}–5×10^{-3}
700 nm	10^{-4}–10^{-3}	10^{-4}–1.5×10^{-3}	10^{-4}–3×10^{-3}
3 μm	3×10^{-4}–8×10^{-4}	3×10^{-4}–1×10^{-3}	3×10^{-4}–3×10^{-3}
5μm	10^{-3}–2×10^{-3}	10^{-3}–1.5×10^{-3}	10^{-3}–0.013
8 μm	2×10^{-3}–7.5×10^{-3}	2×10^{-3}–4×10^{-3}	2×10^{-3}–0.065

5.1.3. Wet Deposition

Wet deposition can be separated into in-cloud processes and below-cloud processes, but often these processes are taken together as a bulk process for simplicity and the availability of observations (typically concentrations in rainwater). Andronache et al. [166] describe the role of several physical processes for scavenging of UFP in the boundary layer and combine the parameterization of several processes to come to an effective scavenging coefficient. Both below-cloud scavenging and in-cloud processing of UFP are important. Scavenging ratios are dependent on the particle size and composition on the one hand and cloud droplet/rain droplet size on the other. Below-cloud scavenging, entrainment in the cloud, in-cloud nucleation scavenging, in-cloud collection and coagulation scavenging are included in this effective scavenging coefficient. Observations above a boreal forest [167] show the largest scavenging coefficients for the smallest particles (10 nm, order 10^{-4} s^{-1}) with a minimum around 100 nm (10^{-5} s^{-1}), median values between 7×10^{-6} and 4×10^{-4} s^{-1}. Andronache et al. [166] find values in the same range, based on theoretical considerations, but with the minimum scavenging coefficient around 50 nm. The theoretical results indicate the largest sensitivity (differences > factor 2) to the in-cloud collection efficiency for particles >50 nm, which is dependent on the solubility of the aerosol. Andronache et al. [166] state that for typical mixing during stratiform rain events, in-cloud processes dominate the UFP scavenging. Little is known about scavenging due to snowfall and the use of detailed microphysical models is required to study the interaction between particles and hydrometeors. Also diffusophoresis, thermophoresis and electric charging may play a role. Wet deposition is estimated to be of small importance for very small particles. For example, Andronache et al. [166] reported that particles need to grow to a diameter of ~100 nm to become activated as cloud droplets. This means that ultrafine particles need to grow to this diameter. For nanoparticles in the nucleation mode size in range (~10 nm) this would take a few days for typical growth rate of the particles of about 5 nm h^{-1} [24].

Wang et al. [168] have reviewed several theoretical and empirical size-resolved parameterizations of the wet scavenging coefficient, and compared them with available field and laboratory measurements. They investigated the uncertainties in several parameters and their impact (raindrop particle collection efficiency: 2 orders of magnitude in 0.01–3 μm range, formulations of raindrop number size distribution: factor of 3 to 5 for all particle sizes, droplet terminal velocity: factor 2). Still, these uncertainties were not enough to overcome differences with observed values, and point at the importance of processes that were not included in these parameterizations (i.e., turbulent transport and mixing, cloud and aerosol microphysics) that influence field data but that are not considered in current theoretical parameterizations. For models that do not include explicit in-cloud processing, Wang et al. [168] recommend to use scavenging coefficients in the upper range of the values, but to be more conservative when in-cloud processing is included. This is in line with the findings of Andronache [166] who point at the importance of entrainment in clouds and cloud processing. According to Wang et al. [168], the net effect of the uncertainties is a factor 2–10 for the size-resolved particle concentrations and a factor 2 for the bulk mass and number concentrations. For snowfall additional research would be needed.

5.2. Evaluation of Numerical Modelling in Assessing the Fate of UFP and ENMs

For the determination of residence time and deposition of ENMs, it is difficult to rely on rules of thumb since large variations in time and space occur due to the intermittent character of clouds and rain and large differences in dry deposition velocities due to different surface roughness and atmospheric stability.

Still, to get an overall estimate of lifetime and concentrations in compartments (air, soil, and water), box or flow models can be used in which average conditions are assumed. Such a model was developed [169] and applied by e.g., Gottschalk et al. [29] using a probabilistic material flow. A different approach is to start from a materials flow model as used for the assessments of chemical substances for REACH as done by Meesters et al. [170]. In these models, all compartments are taken into account. For a broad overview of case studies we refer to Gottschalk et al. [171]. At the other side of the range, local models have been used. Computational Fluid Dynamics has been used to investigate behavior and concentrations of ultrafine particles close to the source [158]. Also Gaussian plume models, treating ultrafine particles as inert tracers [160,161] or including particle dynamics [172] have been used to make estimates close to the sources.

To arrive at spatial detailing for specific regions, chemistry-transport models can be used which explicitly resolve the timing of emissions, interaction with ambient particle concentrations, transport, cloud processes and deposition. Several models are available, all with different model formulations and parameterizations of the processes. Global models have been used with aerosol dynamics modules, mainly from the perspective of climate impact of aerosol. They resolve particle formation, emissions of natural (sea spray, desert dust) and anthropogenic aerosol (combustion, biomass burning) and quantify the impact of (ultrafine) particles on cloud formation and scattering and absorption of radiation. However, they cannot resolve high concentrations in source areas. Several parameterizations for aerosol dynamics were developed for these global models, both in sectional approaches and modal approaches. Both approaches have their advantages and disadvantages in terms of level of detail and computational effort. Examples are MAFOR [172], UHMA (Korhonen et al. [173]), M7 (Vignati et al. [174]), SALSA [175].

Such parameterizations have not only been applied in global, but in regional modelling as well. For regional-scale modelling over Europe, a special version of the model CAMx with a particle dynamics module (PMCAMX-UF) has been applied [176] with a new emission inventory for UFP. Also LOTOS-EUROS with the M7 [174] aerosol dynamics module has been applied [177] with an update of the emission inventory used by Fountoukis et al. [176]. Large-scale background concentrations were combined with Gaussian plume models to describe ultrafine particle concentrations originating from local traffic at city-level [177]. It should be noted that both, Fountoukis et al. [176] and Kukkonen et al. [177], found that although total particle number concentrations were represented reasonably well (within factor two) as compared with measurements, their size distribution showed too many small particles and too little large particles. This may be due to the use of the emission data, which may be representative for tailpipe conditions but not for conditions at short distance to the tailpipe. This cannot be resolved with a regional-scale model, although Fountoukis et al. [176] used a preprocessor for the emission to compensate for this. Other reasons for the observed discrepancy might be a lack of interaction with semi-volatile organic particles [178,179], interaction with other secondary aerosols (e.g., ammonium nitrate), and large uncertainties in the parameterization of nucleation that may be region-specific but is applied over larger areas. Whilst the uncertainties in nucleation are assumed to be less relevant for the ENM fate modelling, taking account of condensables and near-source processes is important for regional-scale models. The existing models should be extended with new tracers to take ENMs into account. Special attention should be paid to surface properties of ENM, that may influence their behavior regarding condensation of species, water uptake, coagulation efficiencies, and behavior in clouds with electric charges.

Despite all uncertainties, a simple estimate of the lifetime and transport distances of particles can be carried out, based on explicit calculations with the LOTOS-EUROS model. This model has been used to trace emissions of a passive tracer with assumed particle diameter of 700 nm [180] (model characteristics identical to elemental carbon), released at constant emission rate of 10 kg/h at a point source in the Netherlands. Concentrations and depositions were calculated for the year 2012, and annual average concentration and annual total deposition are displayed in Figure 2.

Elevated concentrations were found up to distances of the order of 50 km in the annual average, but absolute transport distances are larger. Deposition maps show that nearly the whole model domain receives some of the tracer, albeit in small quantities. In particular, remote mountain areas receive significant amounts of the tracer due to larger amounts of rain related to orography. This example should be considered as a very rough indication of the order of magnitude study, since aerosol dynamics were not taken into account, the diameter that was taken is certainly not representative for all types of ENMs, and uncertainties in wet and dry deposition velocity are easily a factor 2. However, these results indicate that for modelling the fate of ENMs, a reasonable modelling domain should cover at least a circle of 100 km around the source to describe concentrations in air. When looking at the overall fate of ENM, including deposition to soil and water, the scale of a large river catchment (e.g., the Rhine) would be an appropriate scale. It should be noted that for the atmosphere, removal is relatively efficient and a simulation of one year would be fairly representative. For soil and water, processes are slower so that the spatial scale would match but longer periods must be modelled to come to representative concentrations in soil and water.

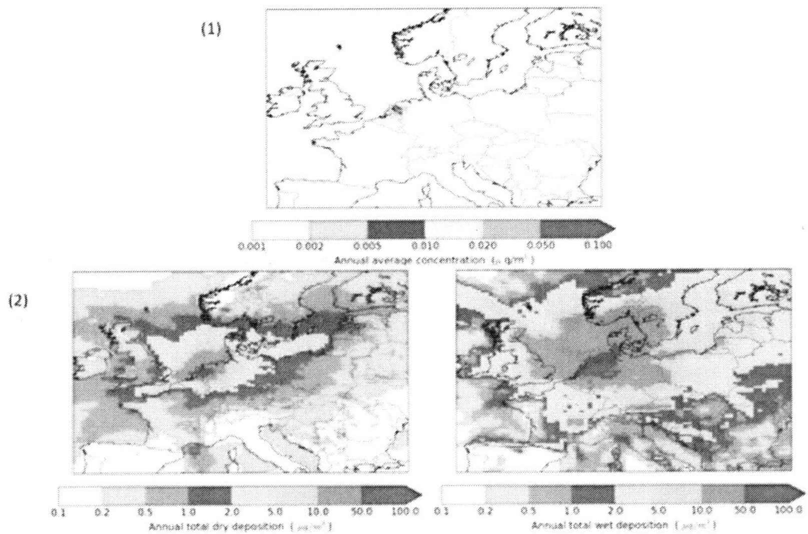

Figure 2. Annual average concentration (**1**) and total dry (**2**) and wet (**3**) deposition for a passive tracer released from a point source in The Netherlands with emission rate of 10 kg/h. Results from model simulation with LOTOS-EUROS for 2012.

6. CONCLUSIONS

Although only a small portion of ENMs is assumed to be released into ambient air compared to the other compartments, increasing use of ENMs might result in atmospheric concentrations relevant for climate, health and water, soil or biota after ENM transformation in the atmosphere. Therefore, their implications on the atmosphere have to be investigated, as well as transformations occurring in the atmospheric reactor and thus influencing ENM environmental fate.

Regarding the release of ENM, data are very scarce as measurements around production plants have not been conducted or published. Field measurements concerning the release during the life cycle are extremely important, but even more challenging due to the still small amounts of airborne ENMs and the need to differentiate between ENMs and other natural or anthropogenic (ultrafine) particles. Hence no real values of ENM emission could be given for all discussed processes (Section 2). At least some information is available from the end-of-life waste treatment.

Regarding the transformation of ENMs in the atmosphere, relevant reactions can be deduced from theoretical considerations regarding physical and chemical processes. Few experimental studies have either confirmed these assumptions or have shown new insights into the special behavior of ENMs. However, the major part of data needed to assess environmental fate of ENMs actually has to be derived from information available for UFP. These can often be used as a proxy, especially for situations when ENMs and UFP are intimately mixed together. However, differences can result e.g., from differing ENM surface properties, distinct shapes and diverging chemical composition. Problems in obtaining specific data arise inter alia due to the vast majority of different ENMs available already or in the near future, meaning that experiments are unmanageable to cover all possible reactions. Therefore, approaches are needed to compile relevant groups of ENMs for similar atmospheric implications. These groups not only need to cover pristine ENMs with and without coating, but also environmentally processed ones.

In order to reveal the fate of airborne ENMs, lab and field studies have to be complemented by regional to global modelling, for specific cases like accidental release of ENMs also by near-field local modelling.

As a first conclusion from UFP modelling, it is clear that they have residence times from minutes to several days and can be transported over substantial distances. However, more specific cases must be defined to determine the overall fate of ENMs since characteristics may vary widely from case to case, e.g., motor oil additive versus TiO_2 spraying. Another aspect is the possible difference when the same type of ENM is released during production, the use phase or at the end of life. While during the latter two, in most cases no pristine ENM will be released, this is more likely during production. Due to this we will have to deal with two different starting points for the further consideration of the ENM environmental behavior. Furthermore, transport distances for an accidental release of high ENM concentrations to protect people from exposure is a very different question than estimating transport distances for the load of ENM originating from diffuse sources received by remote ecosystems. In the first case dilution is the most important factor, in the second case deposition and rainout are more important and change of surface properties must also be taken into account.

More research is needed on in-cloud processes, which can be an important sink and determine where particles are deposited. For the fate of ENMs in terms of particle dynamics and cloud processing, clear information on their surface properties is needed to describe their interaction with ambient aerosol or cloud water, since it not only affects their atmospheric lifetime but also their availability for uptake by biota, attachment to soil and solubility in water bodies.

From the investigations carried out so far, it can be deduced that although being very interesting scientifically, the effect of ENMs on the atmosphere is not seen as an urgent research need due to their actually low concentrations and therefore lower impact compared to conventional (ultrafine) particles. On the other hand, different open questions regarding the fate of airborne ENMs in the environment have to be addressed. Prioritization of these research needs is a difficult task due to the multitude of materials, the large range of production amounts for different ENMs and the variety of release scenarios.

A major research focus regarding environmental transport should be the form of release. For ENMs that are attached to other particles or embedded in a matrix, the uncertainties are no different than for conventional particles and no nano-specific uncertainties apply. In contrast, for ENMs that are released as free particles the uncertainties are largest in all aspects, as they are most difficult to measure and deposition and coagulation are both important and difficult to separate in experiments. A second key aspect is research on the transformation of (pristine) ENMs alternating their environmental properties, information which is urgently needed for modelling. It is understood that investigations will first concentrate on the materials being (environmentally) toxic and/or being produced in large amounts.

ACKNOWLEDGMENTS

This review was prepared within the NanoFASE project that received funding from the European Union's Horizon 2020 research and innovation program under grant agreement No 646002 which also covered the costs to publish in open access.

AUTHOR CONTRIBUTIONS

Astrid John was the main author and worked on the overall manuscript, Miriam Küpper contributed to the chapters on transformation and atmospheric implications. Astrid Manders-Groot is responsible for the chapter on modelling and atmospheric lifetime. Bruno Debray and Jean-Marc Lacome are responsible for the chapter on ENM release during their lifecycle and the section on accidental modelling. Thomas Kuhlbusch triggered this review and gave valuable input to the manuscript.

CONFLICTS OF INTEREST

The authors declare no conflict of interest.

REFERENCES

1. Commission Recommendation of 18 October 2011 on the Definition of Nanomaterial (2011/696/EU). Available online:https://ec.europa.eu/research/ industrial_ technologies/pdf/policy/commission-recommendation-on-the-definition-of-nanomater-18102011_en.pdf (accessed on 2 May 2017).
2. Products of Nanotechnologies Adopted by the SCENIHR during the 28th Plenary Meeting of 19 January 2009. Available online: http://ec.europa. eu/health /archive/ph_risk/committees/04_scenihr/docs/scenihr_o_023.pdf (accessed on 14 June 2016).
3. Kreyling, W.G.; Semmler-Behnke, M.; Chaudhry, Q. A complementary definition of nanomaterial. Nano Today 2010, 5, 165–168.
4. Auffan, M.; Rose, J.; Bottero, J.Y.; Lowry, G.V.; Jolivet, J.P.; Wiesner, M.R. Towards a definition of inorganic nanoparticles from an environmental, health and safety perspective. Nat. Nanotechnol. 2009, 4, 634–641.
5. Borm, P.J.; Robbins, D.; Haubold, S.; Kuhlbusch, T.; Fissan, H.; Donaldson, K.; Schins, R.; Stone, V.; Kreyling, W.; Lademann, J.; et al. The potential risks of nanomaterials: A review carried out for ECETOC. Part. Fibre Toxicol. 2006, 3, 1.
6. Wu, J.; Pei, L.; Xuan, S.; Yan, Q.; Gong, X. Particle size dependent rheological property in magnetic fluid. J. Magn. Magn. Mater. 2016, 408, 18–25.
7. Gupta, A.; Wiggers, H. Freestanding silicon quantum dots: Origin of red and blue luminescence. Nanotechnology 2010, 22, 055707.
8. Manke, A.; Wang, L.; Rojanasakul, Y. Mechanisms of nanoparticle-induced oxidative stress and toxicity. BioMed Res. Int. 2013.
9. Garner, K.L.; Keller, A.A. Emerging patterns for engineered nanomaterials in the environment: A review of fate and toxicity studies. J. Nanopart. Res. 2014, 16, 1–28.
10. Sajid, M.; Ilyas, M.; Basheer, C.; Tariq, M.; Daud, M.; Baig, N.; Shehzad, F. Impact of nanoparticles on human and environment: Review of toxicity factors, exposures, control strategies, and future prospects. Environ. Sci. Pollut. Res. Int. 2015, 22, 4122–4143.

11. Farré, M.; Sanchís, J.; Barceló, D. Analysis and assessment of the occurrence, the fate and the behavior of nanomaterials in the environment. Trends Anal. Chem. 2011, 30, 517–527.
12. French Ministry for Ecology, Energy and Sea. Éléments Issus des Déclarations des Substances à l'État Nanoparticulaire. Direction Générale de la Prévention des Risques, Service des Risques Sanitaires Liés à l'Environnement, des Déchets et des Pollutions Diffuses, December 2015. Available online: https://www.actu-environnement.com/media/pdf/news-27025-rapport-r-nano.pdf (accessed on 2 May 2017).
13. Sun, T.Y.; Gottschalk, F.; Hungerbühler, K.; Nowack, B. Comprehensive probabilistic modelling of environmental emissions of engineered nanomaterials. Environ. Pollut. 2014, 185, 69–76.
14. Research and Markets. Available online: http://www.researchandmarkets.com/reports/2488811/the_global_market_for_metal_oxide_nanoparticles (accessed on 3 November 2016).
15. Nowack, B.; Brouwer, C.; Geertsma, R.E.; Heugens, E.H.W.; Ross, B.L.; Toufektsian, M.C.; Wijnhoven, S.W.P.; Aitken, R.J. Analysis of the occupational, consumer and environmental exposure to engineered nanomaterials used in 10 technology sectors. Nanotoxicology 2012, 7, 1152–1156.
16. Collin, B.; Auffan, M.; Johnson, A.C.; Kaur, I.; Keller, A.A.; Lazareva, A.; Lead, J.R.; Ma, X.; Merrifield, R.C.; Svendsen, C.; et al. Environmental release, fate and ecotoxicological effects of manufactured ceria nanomaterials. Environ. Sci. Nano 2014, 1, 533–548.
17. Xia, T.; Li, N.; Nel, A.E. Potential health impact of nanoparticles. Annu. Rev. Public Health 2009, 30, 137–150.
18. Montaño, M.D.; Lowry, G.V.; von der Kammer, F.; Blue, J.; Ranville, J.F. Current status and future direction for examining engineered nanoparticles in natural systems. Environ. Chem. 2014, 11, 351–366.
19. Kuhlbusch, T.A.; Quincey, P.; Fuller, G.W.; Kelly, F.; Mudway, I.; Viana, M.; Querol, X.; Alastuey, A.; Katsouyanni, K.; Weijers, E.; et al. New Directions: The future of European urban air quality monitoring. Atmos. Environ. 2014, 87, 258–260.
20. Abbott, L.C.; Maynard, A.D. Exposure assessment approaches for engineered nanomaterials. Risk Anal. 2010, 30, 1634–1644.
21. Meesters, J.A.; Veltman, K.; Hendriks, A.J.; van de Meent, D. Environmental exposure assessment of engineered nanoparticles: Why REACH needs adjustment. Integr. Environ. Assess. Manag. 2013, 9, e15–e26.
22. Nowack, B.; Ranville, J.F.; Diamond, S.; Gallego-Urrea, J.A.; Metcalfe, C.; Rose, J.; Horne, N.; Koelmans, A.A.; Klaine, S.J. Potential scenarios for nanomaterial release and subsequent alteration in the environment. Environ. Toxicol. Chem.2012, 31, 50–59.
23. Matsumoto, K.; Tanaka, H. Formation and dissociation of atmospheric particulate nitrate and chloride: An approach based on phase equilibrium. Atmos. Environ. 1996, 30, 639–648.
24. Kulmala, M.; Vehkamäki, H.; Petäjä, T.; Dal Maso, M.; Lauri, A.; Kerminen, V.M.; Birmly, W.; McMurry, P.H. Formation and growth rates of ultrafine atmospheric particles: A review of observations. J. Aerosol Sci. 2004, 35, 143–176.
25. Khoder, M.I. Atmospheric conversion of sulfur dioxide to particulate sulfate and nitrogen dioxide to particulate nitrate and gaseous nitric acid in an urban area. Chemosphere 2002, 49, 675–684.
26. Gutleb, A.C.; Cambier, S.; Fernandes, T.; Georgantzopoulou, A.; Kuhlbusch, T.A.J.; Lynch, I.; Macken, A.; Mehennaoui, K.; Mowller, R.; Nickel, C.; et al. Chapter 4: Environmental Fate and Effects of Nanomaterials in Aquatic Freshwater Environments. In Nanomaterials—A Guide to Fabrication and Applications; Krishnamoorthy, S., Ed.; CRC Press: Boca Raton, FL, USA, 2015.

27. Baalousha, M.; Cornelis, G.; Kuhlbusch, T.A.J.; Lynch, I.; Nickel, C.; Peijnenburg, W.; van den Brink, N.W. Modeling nanomaterial fate and uptake in the environment: Current knowledge and future trends. Environ. Sci. Nano 2016, 3, 323.
28. Keller, A.; McFerran, S.; Lazareva, A.; Suh, S. Global life cycle releases of engineered nanomaterials. J. Nanopart. Res.2013, 15, 1–17.
29. Gottschalk, F.; Lassen, C.; Kjoelholt, J.; Christensen, F.; Nowack, B. Modeling flows and concentrations of nine engineered nanomaterials in the Danish environment. Int. J. Environ. Res. Public Health 2015, 12, 5581–5602.
30. Mueller, N.C.; Nowack, B. Exposure modeling of engineered nanoparticles in the environment. Environ. Sci. Technol.2008, 42, 4447–4453.
31. Mahendra, S.; Zhu, H.; Colvin, V.L.; Alvarez, P.J. Quantum dot weathering results in microbial toxicity. Environ. Sci. Technol. 2008, 42, 9424–9430.
32. Fabrega, J.; Fawcett, S.R.; Renshaw, J.C.; Lead, J.R. Silver nanoparticle impact on bacterial growth: Effect of pH, concentration, and organic matter. Environ. Sci. Technol. 2009, 43, 7285–7290.
33. Li, D.; Lyon, D.Y.; Li, Q.; Alvarez, P.J. Effect of soil sorption and aquatic natural organic matter on the antibacterial activity of a fullerene water suspension. Environ. Toxicol. Chem. 2008, 27, 1888–1894.
34. Tong, Z.; Bischoff, M.; Nies, L.; Applegate, B.; Turco, R.F. Impact of fullerene (C60) on a soil microbial community. Environ. Sci. Technol. 2007, 41, 2935–2991.
35. Owen, R.; Handy, R. Viewpoint: Formulating the problems for environmental risk assessment of nanomaterials. Environ. Sci. Technol. 2007, 41, 5582–5588. [PubMed]
36. Wiesner, M.R.; Lowry, G.V.; Jones, K.L.; Hochella, M.F., Jr.; Di Giulio, R.T.; Casman, E.; Bernhardt, E.S. Decreasing Uncertainties in assessing environmental exposure, risk, and ecological implications of nanomaterials. Environ. Sci. Technol. 2009, 43, 6458–6462.
37. Tiwari, A.J.; Marr, L.C. The Role of Atmospheric Transformations in Determining Environmental Impacts on Carbonaceous Nanoparticles. J. Environ. Qual. 2010, 39, 1883–1895.
38. Bernhardt, E.S.; Colman, B.P.; Hochella, M.F.; Cardinale, B.J.; Nisbet, R.M.; Richardson, C.J.; Yin, L. An ecological perspective on nanomaterial impacts in the environment. J. Environ. Qual. 2010, 39, 1954–1965.
39. Lowry, G.V.; Gregory, K.B.; Apte, S.C.; Lead, J.R. Transformations of nanomaterials in the environment. Environ. Sci. Technol. 2012, 46, 6893–6899.
40. Ding, Y.; Kuhlbusch, T.; Tongeren, M.; Jiménez, A.S.; Tuinman, I.; Chen, R.; Larazza Alvaraz, I.; Mikolajczik, U.; Nickel, C.; Meyer, J.; et al. Airborne engineered nanomaterials in the workplace—A review of release and worker exposure during nanomaterial production and handling processes. J. Hazard. Mater. 2016, 322 Pt A, 17–28.
41. Kuhlbusch, T.; Asbach, C.; Fissan, H.; Göhler, D.; Stintz, M. Nanoparticle exposure at nanotechnology workplaces: A review. Part. Fibre Toxicol. 2011, 8.
42. Brouwer, D. Exposure to manufactured nanoparticles in different workplaces. Toxicology 2010, 269, 120–127.
43. Caballero-Guzman, A.; Nowack, B. A critical review of engineered nanomaterial release data: Are current data useful for material flow modeling? Environ. Pollut. 2016, 213, 502–517.
44. Nanoengineering—Global Approaches to Health and Safety Issues; Dolez, P.I. (Ed.) Elsevier: Amsterdam, The Netherlands, 2015.
45. Nowack, B.; Mueller, N.C.; Krug, H.F.; Wick, P. How to consider engineered nanomaterials in major accident regulations? Environ. Sci. Eur. 2014, 26, 1–10.
46. Gottschalk, F.; Nowack, B. The release of engineered nanomaterials to the environment. J. Environ. Monit. 2011, 13, 1145–1155.
47. Broomfield, M.; Foss Hansen, S.; Pelsy, F. Support for 3rd Regulatory Review on Nanomaterials: Project Report; European Commission; ENV.A.3/ETU/2015/0030; Publications Office of the European Union: Luxembourg, 2016.

48. XP CEN ISO/TS 80004-8:2013. Nanotechnologies—Vocabulary—Part 8: Nanomanufacturing Processes. Classification Index: T 16-101-8. AFNOR: Paris, June 2015. Available online: https://www.iso.org/standard/52937.html (accessed on 2 May 2017).
49. Vignes, A. Panorama des Principaux Procédés de Production, Utilisation, Transformation de Nanomatériaux; Rapport DRA-14-133557-00736A; INERIS: Verneuil-en-Halatte, France, 2013; 82 p.
50. Virji, M.A.; Stefaniak, A.B. A Review of Engineered Nanomaterial Manufacturing Processes and Associated Exposures, Comprehesive Materials Processing; Elsevier: Amsterdam, The Netherlands, 2014; Volume 8, pp. 103–125.
51. Geraci, C.; Heidel, D.; Sayes, C.; Hodson, L.; Schulte, P.; Eastlake, A.; Brenner, S. Perspectives on the design of safer nanomaterials and manufacturing processes. J. Nanopart. Res. 2015, 17, 366.
52. Gridelet, L.; Delbecq, P.; Hervé, L.; Boissolle, P.; Fleury, D.; Kowal, S.; Fayet, G. Proposal of a new risk assessment method for the handling of powders and nanomaterials. Ind. Health 2015, 53, 56–68.
53. Report for EU Project MARINA: Managing Risks of Nanomaterials; Summary Report on Processes and Activities Related to NM Lifecycle within Small and Large Industries; Grant Agreement Number 263215. Available online: http://www.marina-fp7.eu/activities/exposure/MARINA_Processes_Activities_related_NM_Lifecycle.pdf (accessed on 2 May 2017).
54. Engineered Nanomaterials: Evidence on the Effectiveness of Workplace Controls to Prevent Exposure. Safe Work Australia; ISBN 978-0-642-32884-7. 2009; 75 p. Available online: https://www.safeworkaustralia.gov.au/system/files/ documents /1702/engineerednanomaterials_evidence_effectiveness_workplacecontrolstopreventexposure_2009_pdf.pdf (accessed on 2 May 2017).
55. NF EN 1822-1:2010, High Efficiency Air Filters (EPA, HEPA and ULPA)—Part 1: Classification, Performance Testing, Marking, Classification Index: X 44-014-1, AFNOR, Paris, January 2010. Available online: https://www.boutique.afnor.org /norme/nf-en-1822-1/filtres-a-air-a-haute-efficacite-epa-hepa-et-ulpa-partie-1-classification-essais-de-performance-et-marquage/article/774500/fa157753 (accessed on 2 May 2017).
56. Lee, K.W.; Liu, B.Y.H. On the Minimum Efficiency and the Most Penetrating Particle Size for Fibrous Filters. J. Air Pollut. Control. Assoc. 1980, 30, 377–381.
57. Podgórski, A.; Bałazy, A.; Gradon, L. Application of nanofibers to improve the filtration efficiency of the most penetrating aerosol particles in fibrous filters. Chem. Eng. Sci. 2006, 61, 6804–6815.
58. Jones, A.M.; Harrison, R.M. Emission of ultrafine particles from the incineration of municipal solid waste: A review. Atmos. Environ. 2016, 140, 519–528.
59. Hleis, D.; Fernández-Olmo, I.; Ledoux, F.; Kfoury, A.; Courcot, L.; Desmonts, T.; Courcot, D. Chemical profile identification of fugitive and confined particle emissions from an integrated iron and steelmaking plant. J. Hazard. Mater. 2013, 250–251, 246–255.
60. Monfort, E.; Sanfélix, V.; Celades, I.; Gomar, S.; Martín, F.; Aceña, B.; Pascual, A. Diffuse PM10 emission factors associated with dust abatement technologies in the ceramic industry. Atmos. Environ. 2011, 45, 7286–7292.
61. Nowack, B.; David, R.M.; Fissan, H.; Morris, H.; Shatkin, J.A.; Stintz, M.; Zepp, R.; Brouwer, D. Potential release scenarios for carbon nanotubes used in composites. Environ. Int. 2013, 59, 1–11.
62. Directive 2010/75/Eu of the European Parliament and of the Council of 24 November 2010; on Industrial Emissions (Integrated Pollution Prevention and Control). Official Journal of the European Union, 17 December 2010.; p. L: 334/17. Available online: http://eur-lex.europa.eu/LexUriServ/LexUriServ.do? uri=OJ:L: 2010:334:0017:0119:en:PDF (accessed on 4 May 2017).

63. European Pollutant Release and Transfer Register. Available online: http://prtr.ec.europa.eu/ (accessed on 2 May 2017).
64. Décret no 2012-232 du 17 Février 2012 Relatif à la Déclaration Annuelle des Substances à l'Etat Nanoparticulaire Pris en Application de l'Article, L., 523-4 du Code de l'Environnement. Journal Officiel de la République Française, 19 February 2012. Available online: https://www.legifrance.gouv.fr/ affichTexte.do?cidTexte=JORFTEXT000025377246&categorieLien=id (accessed on 4 May 2017).
65. R-Nano, Eléments Issus des Déclarations des Substances à L'état Nanoparticulaire, Déclarations des Substances à l'État Nanoparticulaire—Exercice 2015: Rapport D'étude. Ministère de l'Environnement, de l'Énergie et de la Mer: Paris, France, 2015. Available online: https://www.r-nano.fr (accessed on 2 May 2017).
66. Bressot, C.; Manier, N.; Pagnoux, C.; Aguerre-Chariol, O.; Morgeneyer, M. Environmental release of engineered nanomaterials from commercial tiles under standardized abrasion conditions. J. Hazard. Mater. 2017, 322 Pt A, 276–283.
67. Wohlleben, W.; Brill, S.; Meier, M.W.; Mertler, M.; Cox, G.; Hirsh, S.; von Vacano, B.; Strauss, V.; Treumann, S.; Wiench, K.; et al. On the Lifecycle of Nanocomposites: Comparing Released Fragments and their In-Vivo Hazards from Three Release Mechanisms and Four Nanocomposites. Small 2011, 7, 2384–2395.
68. Reijnders, L. The release of TiO2 and SiO2 nanoparticles from nanocomposites. Polym. Degrad Stab. 2009, 94, 873–876.
69. Göhler, D.; Stintz, M.; Hillemann, L.; Vorbau, M. Characterization of Nanoparticle Release from Surface Coatings by the Simulation of a Sanding Process. Ann. Occup. Hyg. 2010, 54, 615–624.
70. Le Bihan, O.; Shandilya, N.; Gheerardyn, L.; Guillon, O.; Dore, E.; Morgeneyer, M. Investigation of the release of particles from a nanocoated product. Adv. Nanopart. 2013, 2, 39–44.
71. Kuhlbusch, T.; Kaminski, H. Release from Composites by Mechanical and Thermal Treatment: Test Methods. In Safety of Nanomaterials along Their Life Cycle; Wohleben, W., Kuhlbusch, T., Schnekenburger, J., Lehr, C.-M., Eds.; CRC Press: New York, NY, USA, 2015; pp. 247–276.
72. Froggett, S.J.; Clancy, S.F.; Boverhof, D.R.; Canady, R.A. A review and perspective of existing research on the release of nanomaterials from solid nanocomposites. Part. Fibre Toxicol. 2014, 11, 17.
73. Wohlleben, W.; Kuhlbusch, T.A.; Schnekenburger, J.; Lehr, C.M. Safety of Nanomaterials along Their Lifecycle: Release, Exposure, and Human Hazards; CRC Press: New York, NY, USA, 2015.
74. Holder, A.L.; Vejerano, E.P.; Zhoub, X.; Marr, L.C. Nanomaterial disposal by incineration. Environ. Sci. Process. Impacts 2013, 15, 1652–1664.
75. Ounoughene, G.; Le Bihan, O.; Chivas-Joly, C.; Motzkus, C.; Longuet, C.; Debray, B.; Joubert, A.; Le Coq, L.; Lopez-Cuesta, J.M. Behavior and fate of halloysite nanotubes (HNTs) when incinerating PA6/HNTs nanocomposite. Environ. Sci. Technol. 2015, 49, 5450–5457.
76. Ounoughene, G.; Le Bihan, O.; Chivas-Joly, C.; Longuet, C.; Motzkus, C.; Debray, B.; Joubert, A.; Lopez-Cuesta, J.M.; Le Coq, L. Aerosol of silica nanoparticles generated during the combustion of a polysiloxane nanocomposite. In Proceedings of the Aerosol Technology 2015 conference, Tampere, Finland, 15–17 June 2015; Available online: http://www.tut.fi/at2015/wp-content/uploads/abstracts_download.pdf (accessed on 2 May 2017).
77. Kiser, M.A.; Westerhoff, P.; Benn, T.; Wang, Y.; Pérez-Rivera, J.; Hristovski, K. Titanium nanomaterial removal and release from wastewater treatment plants. Environ. Sci. Technol. 2009, 43, 6757–6763.
78. Limbach, L.K.; Bereiter, R.; Müller, E.; Krebs, R.; Gälli, R.; Stark, W.J. Removal of Oxide Nanoparticles in a Model Wastewater Treatment Plant: Influence of Agglomeration and Surfactants on Clearing Efficiency. Environ. Sci. Technol.2008, 42, 5828–5833.

79. Kumar, P.; Fennell, P.; Robins, A. Comparison of the behaviour of manufactured and other airborne nanoparticles and the consequences for prioritising research and regulation activities. J. Nanopart. Res. 2010, 12, 1523–1530.
80. Hristozov, D.; Malsch, I. Hazards and Risks of Engineered Nanoparticles for the Environment and Human Health. Sustainability 2009, 1, 1161–1194.
81. Hinds, W.C. Aerosol Technology: Properties, Behavior, and Measurement of Airborne Particles, 2nd ed.; John Wiley & Sons Inc.: New York, NY, USA, 1999; 504 p.
82. Maynard, A.D.; Kuempel, E.D. Airborne nanostructured particles and occupational health. J. Nanopart. Res. 2005, 7, 587–614.
83. Quik, J.T.K.; Stuart, M.C.; Wouterse, M.; Peijnenburg, W.; Hendriks, A.J.; van de Meent, D. Natural colloids are the dominant factor in the sedimentation of nanoparticles. Environ. Toxicol. Chem. 2012, 31, 1019–1022.
84. Kulkarni, P.; Baron, P.A.; Willeke, K. Aerosol Measurement: Principles, Techniques, and Applications; John Wiley & Sons: Hoboken, NJ, USA, 2011.
85. Seipenbusch, M.; Binder, A.; Kasper, G. Temporal Evolution of Nanoparticle Aerosols in Workplace Exposure. Ann. Occup. Hyg. 2008, 52, 707–716. [PubMed]
86. Diegoli, S.; Manciulea, A.L.; Begum, S.; Jones, I.P.; Lead, J.R.; Preece, J.A. Interaction between manufactured gold nanoparticles and naturally occurring organic macromolecules. Sci. Total Environ. 2008, 402, 51–61.
87. Stankus, D.P.; Lohse, S.E.; Hutchison, J.E.; Nason, J.A. Interactions between natural organic matter and gold nanoparticles stabilized with different organic capping agents. Environ. Sci. Technol. 2010, 45, 3238–3244.
88. Graber, E.R.; Rudich, Y. Atmospheric HULIS: How humic-like are they? A comprehensive and critical review. Atmos. Chem. Phys. 2006, 6, 729–753.
89. Lee, J.; Donahue, N.M. Secondary Organic Aerosol Coating of Synthetic Metal-Oxide Nanoparticles. Environ. Sci. Technol. 2011, 45, 4689–4695.
90. Kupc, A.; Bischof, O.; Tritscher, T.; Beeston, M.; Krinke, T.; Wagner, P.E. Laboratory characterization of a new nano-water-based CPC 3788 and performance comparison to an ultrafine butanol-based CPC 3776. Aerosol Sci. Technol.2013, 47, 183–191.
91. Asbach, C.; Meyer-Plath, A.; Clavaguera, S.; Fierz, M.; Dahmann, D.; MacCalman, L.; Alexander, C.; Todea, A.M.; Iavicoli, I. nanoIndEx—Assessment of Personal Exposure to Airborne Nanomaterials—A Guidance Document. 2016., p. 47. Available online: http://www.nanoindex.eu/wp-content/uploads/2016/06/Nano_Brosch %C3%BCre.pdf (accessed on 4 October 2016).
92. Slowik, J.G.; Cross, E.S.; Han, J.H.; Kolucki, J.; Davidovits, P.; Williams, L.R.; Onasch, T.B.; Jayne, J.T.; Kolb, C.E.; Worsnop, D.R. Measurements of morphology changes of fractal soot particles using coating and denuding experiments: Implications for optical absorption and atmospheric lifetime. Aerosol Sci. Technol. 2007, 41, 734–750.
93. Weingartner, E.; Burtscher, H.; Baltensperger, U. Hygroscopic properties of carbon and diesel soot particles. Atmos. Environ. 1997, 31, 2311–2327.
94. Borm, P.; Klaessig, F.C.; Landry, T.D.; Moudgil, B.; Pauluhn, J.; Thomas, K.; Trottier, R.; Wood, S. Research strategies for safety evaluation of nanomaterials, part V: Role of dissolution in biological fate and effects of nanoscale particles. Toxicol. Sci. 2006, 90, 23–32.
95. Pruppacher, H.R.; Jaenicke, R. The processing of water vapor and aerosols by atmospheric clouds, a global estimate. Atmos. Res. 1995, 38, 283–295.
96. Desboeufs, K.V.; Losno, R.; Colin, J.L. Factors influencing aerosol solubility during cloud processes. Atmos. Environ.2001, 35, 3529–3537.
97. Spokes, L.J.; Jickells, T.D.; Lim, B. Solubilisation of aerosol trace metals by cloud processing: A laboratory study. Geochim. Cosmochim. Acta 1994, 58, 3281–3287.
98. Kaptay, G. On the size and shape dependence of the solubility of nano-particles in solutions. Int. J. Pharm. 2012, 430, 253–257.
99. Mihranyana, A.; Strømme, M. Solubility of fractal nanoparticles. Surf. Sci. 2007, 601, 315–319.

100. Quadros, M.E.; Marr, L.C. Environmental and Human Health Risks of Aerosolized Silver Nanoparticles. J. Air Waste Manag. Assoc. 2010, 60, 770–781.
101. McMahon, M.D.; Lopez, R.; Meyer, H.M., III; Feldman, L.C.; Haglund, R.F., Jr. Rapid tarnishing of silver nanoparticles in ambient laboratory air. Appl. Phys. B 2005, 80, 915–921.
102. Adeleye, A.S.; Keller, A.A.; Miller, R.J.; Lenihan, H.S. Persistence of commercial nanoscaled zero-valent iron (nZVI) and by-products. J. Nanopart. Res. 2013, 15, 1–18.
103. Xia, T.; Kovochich, M.; Liong, M.; Madler, L.; Gilbert, B.; Shi, H.; Yeh, J.I.; Zink, J.I.; Nel, A.E. Comparison of the mechanism of toxicity of zinc oxide and cerium oxide nanoparticles based on dissolution and oxidative stress properties. ACS Nano 2008, 2, 2121–2134.
104. Kadish, K.; Ruoff, R. Fullerenes: Chemistry, Physics, and Technology; John Wiley & Sons: Hoboken, NJ, USA, 2000; Available online: https://books.google.de/books?hl=de&lr=&id=SQRugQM4p9QC&oi=fnd&pg=PR7&dq=Kadish,+K.,+and+R.+Ruoff+.+2000.+Fullerenes:+Chemistry,+physics,+and+technology&ots=OwZCpaA5h_&sig=2Zy7vWWZ1iZbpTCIq4902e6uelg#v=onepage&q=solubility&f=false (accessed on 30 September 2016).
105. Geckeler, K.E.; Premkumar, T. Carbon nanotubes: Are they dispersed or dissolved in liquids? Nanoscale Res. Lett.2011, 6, X1–X3.
106. Klaine, S.J.; Alvarez, P.J.; Batley, G.E.; Fernandes, T.F.; Handy, R.D.; Lyon, D.Y.; Mahendra, S.; McLaughlin, M.J.; Lead, J.R. Nanomaterials in the environment: Behavior, fate, bioavailability, and effects. Environ. Toxicol. Chem. 2008, 27, 1825–1851.
107. Seinfeld, J.H.; Pandis, S.N. Atmospheric Chemistry and Physics: From Air Pollution to Climate Change; John Wiley & Sons: Hoboken, NJ, USA, 2012.
108. Hussein, T.; Hruška, A.; Dohányosová, P.; Džumbová, L.; Hemerka, J.; Kulmala, M.; Smolík, J. Deposition rates on smooth surfaces and coagulation of aerosol particles inside a test chamber. Atmos. Environ. 2009, 43, 905–914.
109. Zhao, Y.; Wang, F.; Zhao, J. Size-Resolved Ultrafine Particle Deposition and Brownian Coagulation from Gasoline Vehicle Exhaust in an Environmental Test Chamber. Environ. Sci. Technol. 2015, 49, 12153–12160.
110. Lowry, G.V.; Hotze, E.M.; Bernhardt, E.S.; Dionysiou, D.D.; Pedersen, J.A.; Wiesner, M.R.; Xing, B. Environmental Occurrences, Behavior, Fate, and Ecological Effects of Nanomaterials: An Introduction to the Special Series. J. Environ. Qual. 2010, 39, 1867–1874.
111. Lin, D.; Tian, X.; Wu, F.; Xing, B. Fate and transport of engineered nanomaterials in the environment. J. Environ. Qual.2010, 39, 1896–1908.
112. Zuberi, B.; Johnson, K.S.; Aleks, G.K.; Molina, L.T.; Molina, M.J.; Laskin, A. Hydrophilic properties of aged soot. Geophys. Res. Lett. 2005, 32, L01807.
113. Majestic, B.J.; Erdakos, G.B.; Lewandowsky, M.; Oliver, K.D.; Willis, R.D.; Kleindienst, T.E.; Bhave, P.V. A Review of Selected Engineered Nanoparticles in the Atmosphere: Sources, Transformations, and Techniques for Sampling and Analysis. Int. J. Occup. Environ. Health 2010, 16, 488–507.
114. Levard, C.; Hotze, E.M.; Lowry, G.V.; Brown, G.E., Jr. Environmental transformations of silver nanoparticles: Impact on stability and toxicity. Environ. Sci. Technol. 2012, 46, 6900–6914.
115. Jimenez, J.L.; Canagaratna, M.R.; Donahue, N.M.; Prevot, A.S.H.; Zhang, Q.; Kroll, J.H.; Aiken, A.C. Evolution of organic aerosols in the atmosphere. Science 2009, 326, 1525–1529.
116. Tiwari, A.J.; Morris, J.R.; Vejerano, E.P.; Hochella, M.F., Jr.; Marr, L.C. Oxidation of C60 Aerosols by Atmospherically Relevant Levels of O3. Environ. Sci. Technol. 2014, 48, 2706–2714.
117. Fortner, J.D.; Kim, D.I.; Boyd, A.M.; Falkner, J.C.; Moran, S.; Colvin, V.L.; Hughes, J.B.; Kim, J.H. Reaction of water-stable C60 aggregates with ozone. Environ. Sci. Technol. 2007, 41, 7497–7502.

118. Collins, P.G.; Bradley, K.; Ishigami, M.; Zettl, A. Extreme Oxygen Sensitivity of Electronic Properties of Carbon Nanotubes. Science 2000, 287, 1801–1804.
119. Giannozzi, P. Oxygen adsorption on graphite and nanotubes. J. Chem. Phys. 2003, 118, 1003–1006.
120. Goldoni, A.; Larciprete, R.; Petaccia, L.; Lizzit, S. Single-Wall Carbon Nanotube Interaction with Gases: Sample Contaminants and Environmental Monitoring. J. Am. Chem. Soc. 2003, 125, 11329–11333.
121. Carp, O. Photoinduced reactivity of titanium dioxide. Prog. Solid Stat. Chem. 2004, 32, 33–177.
122. Hou, W.C.; Jafvert, C.T. Photochemical transformation of aqueous C60 clusters in sunlight. Environ. Sci. Technol. 2008, 43, 362–367.
123. Nissenson, P.; Knox, C.J.; Finlayson-Pitts, B.J.; Phillips, L.F.; Dabdub, D. Enhanced photolysis in aerosols: Evidence for important surface effects. Phys. Chem. Chem. Phys. 2006, 8, 4700–4710.
124. Cheng, Y.; Yin, L.; Lin, S.; Wiesner, M.; Bernhardt, E.; Liu, J. Toxicity reduction of polymer-stabilized silver nanoparticles by sunlight. J. Phys. Chem. C Nanomater. Interfaces 2011, 115, 4425–4432.
125. Wiesner, M.; Lowry, G.; Alvarez, P.; Dionysiou, D.; Bisawas, P. Assessing the role of manufactured nanomaterials. Environ. Sci. Technol. 2006, 40, 4336–4345.
126. Egerton, T.A.; Mattinson, J.A. Comparison of photooxidation and photoreduction reactions on TiO2 nanoparticles. J. Photochem. Photobiol. A Chem. 2007, 186, 115–120.
127. Chen, L.X.; Rajh, T.; Wang, Z.; Thurnauer, M.C. XAFS studies of surface structures of TiO2 nanoparticles and photocatalytic reduction of metal ions. J. Phys. Chem. B 1997, 101, 10688–10697.
128. Linsebigler, A.L.; Lu, G.; Yates, J.T., Jr. Photocatalysis on TiO2 surfaces: Principles, mechanisms, and selected results. Chem. Rev. 1995, 95, 735–758.
129. Ndour, M.; Conchon, P.; D'Anna, B.; Ka, O.; George, C. Photochemistry of mineral dust surface as a potential atmospheric renoxification process. Geophys. Res. Lett. 2009, 36.
130. Chen, H.; Nanayakkara, C.E.; Grassian, V.H. Titanium dioxide photocatalysis in atmospheric chemistry. Chem. Rev.2012, 112, 5919–5948.
131. Henderson, M.A. A surface science perspective on photocatalysis. Surf. Sci. Rep. 2011, 66, 185–297.
132. Tatsuma, T.; Tachibana, S.I.; Miwa, T.; Tryk, D.A.; Fujishima, A. Remote bleaching of methylene blue by UV-irradiated TiO2 in the gas phase. J. Phys. Chem. B 1999, 103, 8033–8035.
133. Daniels, S.L. On the qualities of the air as affected by radiant energies (photocatalytic ionization processes for remediation of indoor environments). J. Environ. Eng. Sci. 2007, 6, 329–342.
134. Nanayakkara, C.E.; Larish, W.A.; Grassian, V.H. Titanium Dioxide Nanoparticle Surface Reactivity with Atmospheric Gases, CO2, SO2, and NO2: Roles of Surface Hydroxyl Groups and Adsorbed Water in the Formation and Stability of Adsorbed Products. J. Phys. Chem. C 2014, 118, 23011–23021.
135. Rubasinghege, G.; Elzey, S.; Baltrusaitis, J.; Jayaweera, P.M.; Grassian, V.H. Reactions on atmospheric dust particles: Surface photochemistry and size-dependent nanoscale redox chemistry. J. Phys. Chem. Lett. 2010, 1, 1729–1737.
136. Bedjanian, Y.; El Zein, A. Interaction of NO2 with TiO2 Surface under UV Irradiation: Products Study. J. Phys. Chem. A 2012, 116, 1758–1764.
137. Kebede, M.A.; Varner, M.E.; Scharko, N.K.; Gerber, R.B.; Raff, J.D. Photooxidation of Ammonia on TiO2 as a Source of NO and NO2 under Atmospheric Conditions. J. Am. Chem. Soc. 2013, 135, 8606–8615.
138. Monge, M.E.; George, C.; D'Anna, B.; Doussin, J.F.; Jammoul, A.; Wang, J.; Mellouki, A. Ozone formation from illuminated titanium dioxide surfaces. J. Am. Chem. Soc. 2010, 132, 8234–8235.

139. Chen, H.; Stanier, C.O.; Young, M.A.; Grassian, V.H. A kinetic study of ozone decomposition on illuminated oxide surfaces. J. Phys. Chem. A 2011, 115, 11979–11987.
140. Pekney, N.J.; Davidson, C.I.; Bein, K.J.; Wexler, A.S.; Johnston, M.V. Identification of sources of atmospheric PM at the Pittsburgh Supersite, Part I: Single particle analysis and filter-based positive matrix factorization. Atmos. Environ.2006, 40, 411–423.
141. Weir, A.; Westerhoff, P.; Fabricius, L.; Hristovski, K.; Von Goetz, N. Titanium dioxide nanoparticles in food and personal care products. Environ. Sci. Technol. 2012, 46, 2242–2250.
142. Li, Y.; Niu, J.; Zhang, W.; Zhang, L.; Shang, E. Influence of aqueous media on the ROS-mediated toxicity of ZnO nanoparticles toward green fluorescent protein-expressing Escherichia coli under UV-365 irradiation. Langmuir 2014, 30, 2852–2862.
143. Nakajima, N.; Nishi, C.; Li, F.M.; Ikada, Y. Photo-induced cytotoxicity of water-soluble fullerene. Fuller Sci. Technol.1996, 4, 1–19.
144. Sakai, A.; Yamakoshi, Y.; Miyata, N. Visible light irradiation of [60] fullerene causes killing and initiation of transformation in BALB/3T3 cells. Fuller Sci. Technol. 1999, 7, 743–756.
145. Misawa, M.; Takahashi, J. Generation of reactive oxygen species induced by gold nanoparticles under X-ray and UV irradiations. Nanomedicine 2011, 7, 604–614.
146. Heckert, E.G.; Karakoti, A.S.; Seal, S.; Self, W.T. The role of cerium redox state in the SOD mimetic activity of nanoceria. Biomaterials 2008, 29, 2705–2709.
147. Patakfalvi, R.; Diaz, D.; Velasco-Arias, D.; Rodriguez-Gattorno, G.; Santiago-Jacinto, P. Synthesis and direct interactions of silver colloidal nanoparticles with pollutant gases. Colloid Polym. Sci. 2008, 286, 67–77.
148. Kerminen, V.-M.; Paramonov, M.; Anttila, T.; Riipinen, I.; Fountoukis, C.; Korhonen, H.; Asmi, E.; Laakso, L.; Lihavainen, H.; Swietlicki, E.; et al. Cloud condensation nuclei production associated with atmospheric nucleation: A synthesis based on existing literature and new results. Atmos. Chem. Phys. 2012, 12, 12037–12059.
149. Tsai, Y.I. Atmospheric visibility trends in an urban area in Taiwan 1961–2003. Atmos. Environ. 2005, 39, 5555–5567.
150. Malm, W.C.; Day, D.E. Estimates of aerosol species scattering characteristics as a function of relative humidity. Atmos. Environ. 2001, 35, 2845–2860.
151. Tang, I.N.; Wong, W.T.; Munkelwitz, H.R. The relative importance of atmospheric sulfates and nitrates in visibility reduction. Atmos. Environ. 1981, 15, 2463–2471.
152. Waggoner, A.P.; Weiss, R.E.; Ahlquist, N.C.; Covert, D.S.; Will, S.; Charlson, R.J. Optical characteristics of atmospheric aerosols. Atmos. Environ. 1981, 15, 1891–1909.
153. ARIA Database, Accident N°43049, Rejet de Noir de Carbone par une Usine de Pneumatiques. Available online: http://www.aria.developpement-durable.gouv.fr (accessed on 2 May 2017).
154. Le, H.-D.; Lacome, J.-M.; Vignes, A.; Debray, B.; Truchot, B.; Fede, P.; Climent, E. A few fundamental aspects related to the modeling of an accidental massive jet release of nanoparticles. Chem. Eng. Trans. 2016, 48, 25–30.
155. Le, H.-D.; Lacome, J.-M.; Vignes, A.; Debray, B.; Truchot, B.; Fede, P.; Climent, E. Modelling of test case particle-laden jet with NEPTUNE/CFD. In Proceedings of the 14th Workshop on Two-Phase Flow Predictions, Halle, Germany, 7–10 September 2015.
156. Ketzel, M.; Berkowicz, R. Modelling the fate of ultrafine particles from exhaust pipe to rural background: An analysis of time scales for dilution, coagulation and deposition. Atmos. Environ. 2004, 38, 2639–2652.
157. Karl, M.; Kukkonen, J.; Keuken, M.P.; Lützenkirchen, S.; Pirjola, L.; Hussein, T. Modeling and measurements of urban aerosol processes on the neighborbood scale in Rotterdam, Oslo and Helsinki. Atmos. Chem. Phys. 2016, 16, 4817–4835.
158. Nikolova, I.; Janssen, S.; Vos, P.; Berghmans, P. Modelling the Mixing of Size Resolved Traffic Induced and Background Ultrafine Particles from an Urban Street Canyon to Adjacent Backyards. Aerosol Air Qual. Res. 2014, 14, 145–155.

159. Choi, W.; Paulson, S.E. Closing the UFP number concentration budget at road-to-ambient scale: Implications for particle dynamics. Aerosol Sci. Technol. 2016, 50, 448–461.
160. Keuken, M.P.; Moerman, M.; Zandveld, P.; Henzing, J.S. Total and size-resolved particle number and black carbon concentrations near an industrial area. Atmos. Environ. 2015, 122, 196–205.
161. Keuken, M.P.; Moerman, M.; Zandveld, P.; Henzing, J.S.; Hoek, G. Total and size-resolved particle number and black carbon concentrations in urban areas near Schiphol Airport (the Netherland). Atmos. Environ. 2015, 104, 132–142.
162. Petroff, A.; Zhang, L. Development and validation of a size-resolved particle dry deposition scheme for application in aerosol transport models. Geosci. Model. Dev. 2010, 3, 753–769.
163. Zhang, L.; Gong, S.; Padro, J.; Barrie, L. A size-segregated particle dry deposition scheme for an atmospheric aerosol module. Atmos. Environ. 2001, 35, 549–560.
164. Zhang, J.; Shao, Y. A new parameterization of particle dry deposition over rough surfaces. Atmos. Chem. Phys. 2014, 14, 12429–12440.
165. Manders-Groot, A.M.M.; Segers, A.J.; Jonkers, S. LOTOS-EUROS v2.0 Reference Guide; TNO Report TNO2016 R10898; TNO: Utrecht, The Netherlands, 2016.
166. Andronache, C.; Grönholm, T.; Laakso, L.; Phillips, V.; Venäläinen, A. Scavenging of ultrafine particles by rainfall at a boreal site: Observations and model estimations. Atmos Chem Phys 2006, 6, 4739–4754.
167. Laakso, L.; Grönholm, T.; Rannik, Ü.; Kosmale, M.; Fiedler, V.; Vehkamäki, H.; Kulmala, M. Ultrafine particle scavenging coefficients calculated from 6 years field measurements. Atmos. Environ. 2003, 37, 3605–3613.
168. Wang, X.; Zhang, L.; Moran, M.D. Uncertainty assessment of current size-resolved parameterizations for below-cloud particle scavenging by rain. Atmos. Chem. Phys. 2010, 10, 5685–5705.
169. Gottschalk, F.; Sonderer, T.; Scholz, R.W.; Nowack, B. Modeled environmental concentrations of engineered nanomaterials (TiO2, ZnO, Ag, CNT, fullerenes) for different regions. Environ. Sci. Technol. 2009, 43, 9216–9222.
170. Meesters, J.A.J.; Koelmands, A.A.; Quik, J.T.K.; Hendriks, A.J.; Van de Meent, D. Multimedia modeling of engineered nanoparticles with SimpleBox4nano: Model definition and evaluation. Environ. Sci. Technol. 2014, 48, 5726–5736.
171. Gottschalk, F.; Sun, T.Y.; Nowack, B. Environmental concentrations of engineered nanomaterials: Review of modeling and analytical studies. Environ. Pollut. 2013, 181, 287–300.
172. Karl, M.; Gross, A.; Pirjola, L.; Leck, C. A new flexible multicomponentmodel for the study of aerosol dynamics in the marine boundary layer. Tellus B 2011, 63, 1001–1025.
173. Korhonen, H.; Lehtinen, K.E.J.; Kulmala, M. Multicomponent aerosol dynamics model UHMA: Model development and validation. Atmos. Chem. Phys. 2004, 4, 757–771.
174. Vignati, E.; Wilson, J.; Stier, P. M7: An efficient size-resolved aerosol microphysics module for large-scale aerosol transport models. J. Geophys. Res. Atmos. 2004, 109.
175. Kokkola, H.; Korhonen, H.; Lehtinen, K.E.J.; Makkonen, R.; Asmi, A.; Järvenoja, S.; Anttila, T.; Partanen, A.-I.; Kulmala, M.; Järvinen, H.; et al. SALSA—A Sectional Aerosol module for Large Scale Applications. Atmos. Chem. Phys. 2008, 8, 2469–2483.
176. Fountoukis, C.; Riipinen, I.; Denier Van Der Gon, H.A.C.; Charalampidis, P.E.; Pilinis, C.; Wiedensohler, A.; Dowd, C.O.; Putaud, J.P.; Moerman, M.; Pandis, S.N. Simulating ultrafine particle formation in Europe using a regional CTM: Contribution of primary emissions versus secondary formation to aerosol number concentrations. Atmos. Chem. Phys. 2012, 12, 8663–8677.
177. Kukkonen, J.; Karl, M.; Keuken, M.P.; Denier van der Gon, H.A.C.; Denby, B.R.; Singh, V.; Douros, J.; Manders, A.; Samaras, Z.; Moussiopoulos, N.; et al. Modelling the dispersion of particle numbers in five European cities. Geosci. Model. Dev. 2016, 9, 451–478.

178. Riipinen, I.; Pierce, J.R.; Yli-Juuti, T.; Nieminen, T.; Häkkinen, S.; Ehn, M.; Junninen, H.; Lehtipalo, K.; Petäjä, T.; Slowik, J.; et al. Organic condensation: A vital link connecting aerosol formation to cloud condensation nuclei (CCN) concentrations. Atmos. Chem. Phys. 2011, 11, 3865–3878.

179. Pierce, J.R.; Riipinen, I.; Kulmala, M.; Ehn, M.; Petäjä, T.; Junninen, H.; Worsnop, D.R.; Donahue, N.M. Quantification of the volatility of secondary organic compounds in ultrafine particles during nucleation events. Atmos. Chem. Phys.2011, 11, 9019–9036.

180. Kranenburg, R.; Segers, A.J.; Hendriks, C.; Schaap, M. Source apportionment using LOTOS-EUROS: Module description and evaluation. Geosci. Model. Dev. 2013, 6, 721–733.

CHAPTER 8

Negative Feedback in the Polar Ice System

Peter Stallinga[1], Igor Khmelinskii[2]

[1]*FCT-DEEI, University of the Algarve, Faro, Portugal*

[2]*FCT-DQF, University of the Algarve, Faro, Portugal*

ABSTRACT

One of the ingredients of the anthropogenic global warming hypothesis is the existence of large positive feedback in the climate system. An example is polar ice that, once melted, turns into blacker water that will increase radiation absorption and this reinforces the melting. This causes a run-away scenario with a point of no return. Here it is shown that the polar ice can also have negative feedback aspects, where a melting of polar ice will cause it to reappear.

Keywords: Ice, Albedo, Negative Feedback, Fresnel

1. INTRODUCTION

The planet has been heating up in the last couple of decades and this causes large concern in society, more so since humans are allegedly responsible for the temperature rise. One of the major concerns is that we are at a point-of-no-return, caused by a positive feedback cycle of ice melting that is caused by an increased temperature and also causes further rise in temperature. The mediator of this phenomenon is the albedo—"whiteness" or reflectivity—of the Earth's surface. Since sea water has a much lower albedo compared to ice, i.e. sea water absorbs more of the radiation received from the Sun, a melting of ice will cause an increase in absorbed radiation and thus a rise in temperature and further melting of ice. We recognize a positive feedback system that can cause a catastrophic runaway situation.

Wikipedia writes it nicely: "Ice-albedo feedback (or snow-albedo feedback) is a po- sitive feedback climate process where a change in the area of snow-covered land, ice caps, glaciers or sea ice alters the albedo. This change in albedo acts to reinforce the initial alteration in ice area. Cooling tends to increase ice cover and hence the albedo, reducing the amount of solar energy absorbed and leading to more cooling. Conversely, warming tends to decrease ice cover and hence the albedo, increasing the amount of solar energy absorbed, leading to more warming." Such effects have long been known to result in positive feedback and an increased climate sensitivity [1] - [6] .

Indeed, the temperature has risen in the past decades, and the albedo of the Arctic region seems to have decreased until 2011 [7] [8] . Yet, the total amount of sea ice on this planet has not changed from 1979, when measurements started, until now. The Antarctic Sea Ice extent was actually in May 2014 at its maximum ever measured [9] . Images of icebreakers getting stuck in the ice make it to the headlines. Apparently, the ice coverage of our planet has varied dramatically, but somehow we are back on the same level as in 1979. There do not seem to be any signs of an avalanche effect—at least not in the data. This is remarkable, to say the least. There where a runaway scenario of melting ice that will lower the Earth's albedo that will raise the temperature that will melt more ice is envisaged, nature shows to be quite resilient. Somehow it does manage to escape the man-imagined scenario (if not man-made scenario). At this point, it is important to point out that talking about a "trend" can only be done if the model is linear and the measurement data are so noisy as to mask the linear behavior. In case of sea ice, the measurement noise is close to zero and talking of trends therefore does not make sense. The reasoning "I only understand a trend and that is therefore reality, with any deviations from the trend defined as noise" is a logical fallacy. Nature is not linear and there is no "trend". What you see is all there is. Deviations from the trend prove our assumptions of linearity wrong. There is something going on with the ice that we do not understand.

This made us think about what might be the underlying cause for a restoring process. A phenomenon that would cause molten ice to come back, instead of accelerating the process of melting. In this paper we present the results of our research, the things we came up with. We use finite-elements simulations to demonstrate them.

We first studied the effect of the skew angle of radiation at the polar regions com- bined with Fresnel reflectivity and see that this indeed has a negative-feedback effect. Then we studied an effect of thermal transport, both horizontal and vertical. This idea is simple and straightforward. It is based on the fact that sea water has indeed a much lower albedo, and thus absorbs more radiation, but this albedo effect is overshadowed by the effect that sea ice can be much cooler than sea water (the latter at least above melting point) because thermal transport is much lower in solid materials compared to liquid, the latter also having convection. Ice has a low thermal conductivity, in contrast to water, and the top layer of the ice—the surface that is exposed to the universe and that is radiating out—can have significantly lower temperature than water, and thus radiates much less. Heat trapped under a layer of ice cannot surface and dissipate into the universe, the ice working like a blanket. This accumulated heat will eventually melt the ice. Conversely, open seawater can be warm at the surface and emit larger amounts of radiation, efficiently cooling the planet, and enabling refreezing of the water.

In any case, what is often forgotten in the albedo argument is that anyway not much radiation is absorbed in the polar regions and the albedo effect is thus expected to be small, if not negligible. The lower temperature of ice compared to water and its resul- ting lower emission, however, is a significant contribution, since outward radiation scales with the fourth power of temperature, i.e. Stefan-Boltzmann Law of black-body radia- tion. For example, lowering the temperature by 11% from 273 K (zero °C, sea water) to 243 K ($-30°C$, ice surface) will cause a humongous 37% reduction in emission power, from 315 W/m^{2}to 198 W/m^2.

In steady state, heat to the polar regions is received not so much through inbound radiation, which is rather negligible. Firstly because of the low angle of incidence of radiation α which introduces a $\sin(\alpha)$ effect. Moreover, below certain angles, the reflection shoots up (and absorption declines) because of Fresnel's equation [10] . The refractive index of water is higher than that of air—1.33 [11] vs. 1—and at normal incidence the radiation reflected off the interface between water and air is $(1.33-1)^2$ $=10\%$. This increases to 100% for light parallel to the surface, with a significant in- crease starting at an angle of about 70°. Things like ripples and waves on the ocean modify this effect, but it has been estimated that the albedo of sea water varies with the Sun's position in the sky from zenith to horizon from about 2.5% to 42.5% [12] . So, first of all not much light is received at the polar regions to begin with, making mo- dulations of albedo insignificant, and second, the albedo does not change so drama- tically from ice (50%) to seawater (40%).

Instead, at the polar regions, most heat is received through thermal transport in the so-called Thermohaline Circulation, a moving belt of water that conveys a lot of heat to the poles. This circulation carries a tremendous amount of water and heat, estimated to be 17 sverdrup $\left(1.7\times10^{7}\ \mathrm{m}^{3}/\mathrm{s}\right)$ and 1.2 PW (petawatt) compared to the total global 120 PW radiation received from the Sun. A lot of heat is similarly received by at- mospheric heat transport. All in all, polar regions receive heat primarily from transport and no so much from direct absorption of solar radiation.

From there it radiates outward as efficiently as from any region on this planet— effectively the "albedo" of the earth is zero in the wavelength window of radiation emission, mainly infra-red. The real albedo in the visible light therefore has no impact on the outward radiation and little impact on the inward radiation.

We will now show the results of finite-elements simulations. They consist of dividing a ringlike world in many cells, each with an orientation, a temperature and an albedo and then calculate the heat dynamics.

2. IMPLEMENTATION

We use a finite-elements system as depicted in Figure 1. This system is a simplification of reality, but contains all of the necessary elements needed to make a correct phenomenological analysis of the real Earth system. In our system, a ring of water is made up of a large number of segments (elements). The ring itself is imagined to rotate about its polar axis. This rotation is much faster than other processes, e.g. 24 hours for a full rotation, compared to cooling/heating relaxation times in the order of years, and water fluxes in the order of thousands of years. As such, the received radiation is imme- diately averaged out over a full rotation of θ in the calculation. Also, no seasonal effects are taken into account (no nutation of the rotation axis).

In the simplest simulations, each element is in radiation balance. The incoming radiation is governed by the angle α of the position of the element relative to the equator; at the equator $\alpha = 0$. The Sun is far away in the plane of the equator and

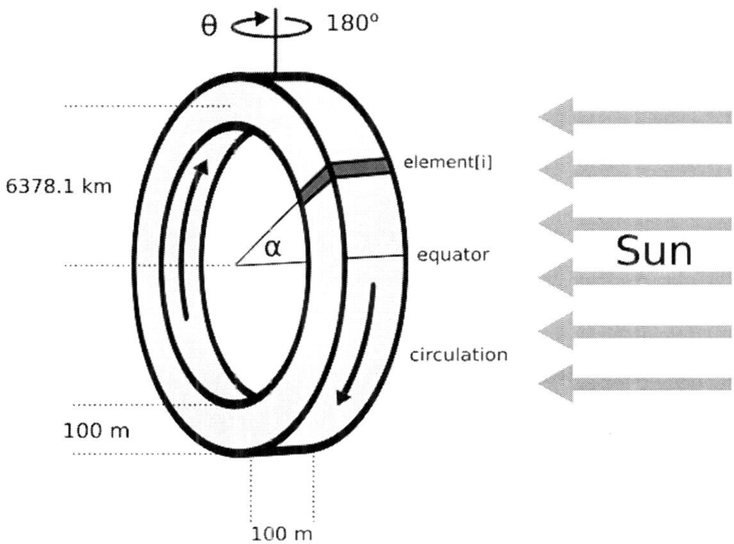

Figure 1. Schematic drawing of the system studied with the definitions of the parameters and variables.

radiates with a solar constant of $C = 1362\ \mathrm{W/m^2}$. The amount of absorbed energy per square meter p_{in} per second is then determined by this angle α and the albedo a of the element, and further averaged out by the rotation around the vertical axis (θ). The latter also suffering from the effect that half of the time the radiation received is zero because it is in the shadow:

$$p_{\mathrm{in}}(\alpha) = \frac{\int_{-\pi/2}^{+\pi/2} a(\alpha)\cos(\alpha) C \cos(\theta)\,\mathrm{d}\theta}{\int_{-\pi}^{+\pi} \mathrm{d}\theta} = a(\alpha)\cos(\alpha)\,C/\pi\ \left(\mathrm{W/m^2}\right). \quad (1)$$

The amount of energy emitted is independent of the angles α and θ and is also independent of the albedo. Where the albedo is defined as the whiteness in the visible light, the emission of an element is in the infrared window of the electromagnetic spec- trum, and it is assumed that all materials on Earth are "black" in the infrared and emit therefore like a black body,

$$p_{\mathrm{out}}(T) = \sigma T^4\ \left(\mathrm{W/m^2}\right), \quad (2)$$

with T the absolute temperature and σ the Stefan-Boltzmann constant ($5.670373 \times 10^{-8} Wm^{-2}K^{-4}$). This reasoning is close to the truth and has led for instance to the conclusion—common belief—that a change of real visible-light albedo—as in melting white snow to black seawater—will not alter the emission, but will significantly alter the absorption, and an aggravation of global warming is expected, as mentioned in the introduction.

The depth of the ring of water is chosen to be 100 m, since that is a reasonable number for the water on Earth that contributes to the climate cycle, the so-called ocean mixed layer [13] . This is important only for the time constant of the relaxation; taking deeper or shallower rings would only alter the time constants, but not alter the final temperatures. The width is taken to be 100 meters, but this parameter does not in- fluence the calculations at all (everything scales linearly with the width) and is only given for completeness' sake and to be able to visualize a 3D object.

The advanced simulations also include water circulation that, combined with a heat gradient, will cause transport of heat to the polar regions. Alternatively, we also simu- late with a heat conduction mechanism that occurs also in still water and for which no circulation is needed, even though we know that this must be of a much smaller order of magnitude.

Note that no greenhouse effects are taken into account. Basically, there is no atmos- phere. We are not simulating the greenhouse and the full climate here. We are only determining the effects of albedo and albedo changes on the temperature and, more specifically, on the amount of sea ice. To compensate for the lack of atmosphere and have anyway somewhat Earth-like temperatures in the simulations, we "crank up" the inward radiation by multiplying it with a "fudge factor".

In the first step, we make a simulation without any effects. Each element has the same steady albedo that is not a function of temperature. In the second step, we intro- duce the simple albedo effect; an element with temperature dropping below zero celsius will change the albedo from low to high, and vice versa. To better demonstrate the effect, we make, apart from a simulation with realistic changes in albedo (40% to 50%), also a simulation with exaggerated changes from 0% to 60%, as society is wont to belief. This indeed shows a non-reversible "hysteresis" effect, where the final situation de- pends on the starting conditions. In the third step, we include a more realistic Fresnel albedo, as mentioned before, and show that this mitigates the hysteresis effect. In the final simulation, we introduce the concept of heat transport. This nearly completely removes the hysteresis effect and ice, once removed, simply reappears in a time scale of about ten years.

3. RESULTS

The first simulation is a system that does not include albedo changes. All albedos are taken to be 31%. Nor does the simulation include circulation of water and accom- panying transport of heat. In this simulation all elements start at the same temperature, 273.16 K (the triple point of water, TPW), and the system is allowed to relax. For every element, the input and output radiation are calculated, a function of its position and temperature, respectively (Equations (1) and (2)). The average temperature in this way was found to be −22°C when mapped to a spherical planet (every segment enters into the average with a weight $|\cos(\alpha)|$, with α the latitude). To mimic an atmos- phere—to give values closer to real-life—a "fudge factor" of 1.75 was introduced in the inbound radiation, which then resulted in Figure 2 and an average temperature of 15.4 celsius.

As can be seen, the equator will warm up and saturate to about 35 degrees. Moderate regions saturate at about 10 degrees and the poles, because of total absence of input radiation will tend toward zero kelvin, which they will eventually reach. Figure 2(b) shows the temperature distribution over the ring from equator to poles after 500 kiloday.

The characteristic relaxation time of cooling and warming depends on the heat capacity of the layer used and thus its thickness and specific heat. The latter is well known for water, but the effective thickness is something that can be discussed. How- ever, our value of 100 m is in line with values given in the literature and we can thus expect that the relaxation time found here—in the order of some years—is close to the real one of the climate system of our planet. Note, however, that the value is much smaller than the time scale predicted by the majority of climate researchers that predict changes in CO_2 concentration in the atmosphere to become apparent on the scale of centuries.

In the next simulation, we include the change of albedo between water and ice. Common knowledge is that water has an albedo of about 0.02 and ice 0.35. (Note that albedo of ice is much lower than the albedo of snow, the latter having an albedo between 0.45 [old] and 0.85 [fresh]). The results are shown in Figure 3. A hysteresis effect now occurs in that the final result also depends on the starting conditions, the history of the planet. If we start with a marginally frozen planet (Figure 3(a) and Figure 3(b)), the system winds up with an average temperature of 6.6 degrees celsius and the planet frozen from 30.9 degrees latitude up (see inset of panel b). If, instead, we start with a fully unfrozen planet (Figure 3(c) and Figure 3(d)), the planet winds up

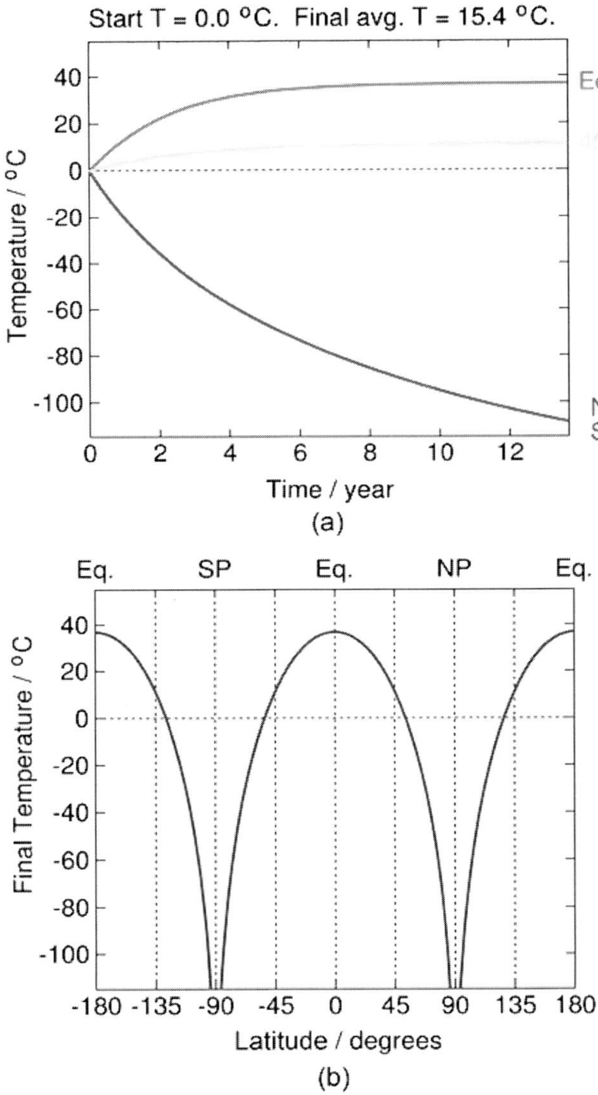

Figure 2. Simple model. No albedo changes $(a = 0.31)$ and no circulation. (a) Temperature transients when star- ting at 273 K for three key latitudes. (b) Temperature after 500 kiloday (ten times window of left panel) as a function of latitude. ("Eq." is "Equator", "NP" and "SP" are North Pole and South Pole, respectively).

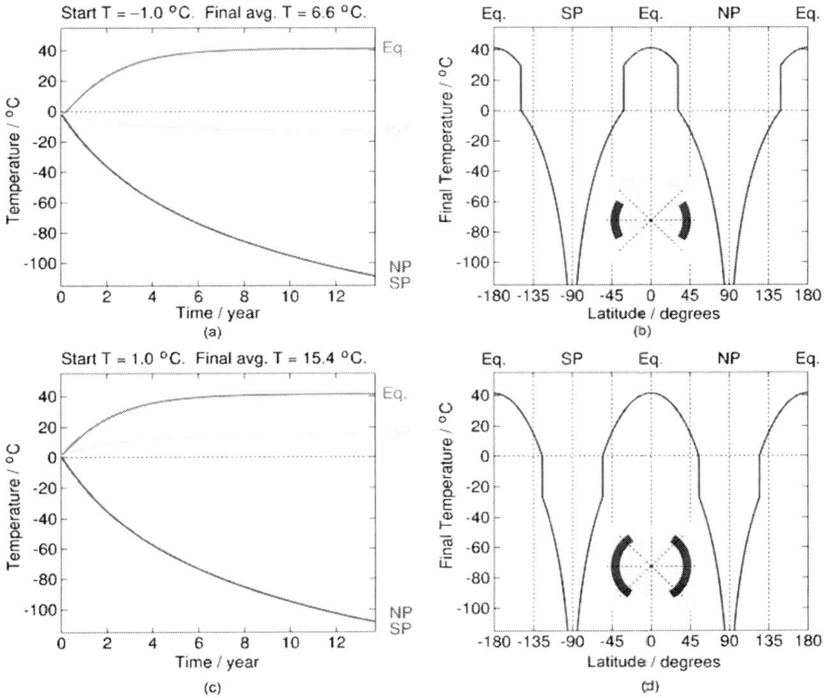

Figure 3. A system with albedo changes from sea (0.02) to ice (0.35) when the temperature drops below TPW. The final situation depends on the history; Starting with high temperature (a marginally frozen planet, (a) and (b)) less of the planet is covered with ice compared to a slightly lower starting temperature (a marginally unfrozen planet, (c) and (d); fudge factor 1.306). The insets show the distribution of ice (white) and water (dark) in the ring.

with an average temperature of 15.4 degrees celsius and covered with ice from 55.3 degrees latitude up. There is indeed a huge non-reversibility hysteresis in the climate system, both in temperature (8.8 degrees) and ice coverage.

The simulations above (Figure 3) summarize the common knowledge about positive feedback effect of the ice on our planet. We will now present our new ideas here.

First of all, the value for albedos of sea and ice are for perpendicular light. While reflection from ice is rather diffuse and the angle of incidence should not matter much, reflection off water has a more specular nature and the albedo depends heavily on the angle of incidence. At high angles of incidence, close to the poles, nearly all light is reflected. So, we expect a water albedo that is much higher close to the poles compared to the equator. Moreover, because the Earth is rotating, even at the equator, at some moments of the day light is effectively reflected off the water and even at the equator we can expect a higher albedo than 0.02.

Adequate is to use Fresnel's Equation for reflection of light at the interface of two media with different refractive indices, n_1 and n_2, respectively. Assuming no ima- ginary part in the refractive index of water (effectively seeing it as a non-absorptive, non-attenuating medium), the reflectivity (albedo) for unpolarized light is given by [10] :

$$R(\beta)=\frac{1}{2}\left[\frac{n_1\cos\beta-n_2\sqrt{1-\left(\frac{n_1}{n_2}\sin\beta\right)^2}}{n_1\cos\beta+n_2\sqrt{1-\left(\frac{n_1}{n_2}\sin\beta\right)^2}}\right]^2+\frac{1}{2}\left[\frac{n_2\cos\beta-n_1\sqrt{1-\left(\frac{n_1}{n_2}\sin\beta\right)^2}}{n_2\cos\beta+n_1\sqrt{1-\left(\frac{n_1}{n_2}\sin\beta\right)^2}}\right]^2 \quad (3)$$

For air we can take $n_1=1$ and sea water has a refractive index equal to $n_2=1.33$, which gives us a reflection as a function of angle as in Figure 4(a), where the reflec- tivity varies from 0.0201 (a value close to the 0.02 used before, justifying our as- sumption of a non-imaginary refraction index) to 1. Note that waves can change this function-in first order spread it, equalize it, and make it less pronounced—but as long as the reflection is specular, the shape is as given here. Likewise, seasonal changes (not included in these simulations) would smear out the effect in a similar way. We continue with a waveless ocean, to test the hypothesis of albedo effects.

On basis of Fresnel Equation (Equation (3) and shown in Figure 4(a)) we can cal- culate an effective albedo at every latitude. We do this by calculating the angle β of solar radiation for latitude α at every Earth rotation angle θ and calculating the reflectivity according to Equation (3). The effective albedo at latitude α is then the reflectivity thus calculated averaged over θ. The angle β is found by calculating the rotation matrices for rotations α and θ, applying it to the x-vector (the direction of the Sun) and then taking the dot product with the x-vector. This way we found that

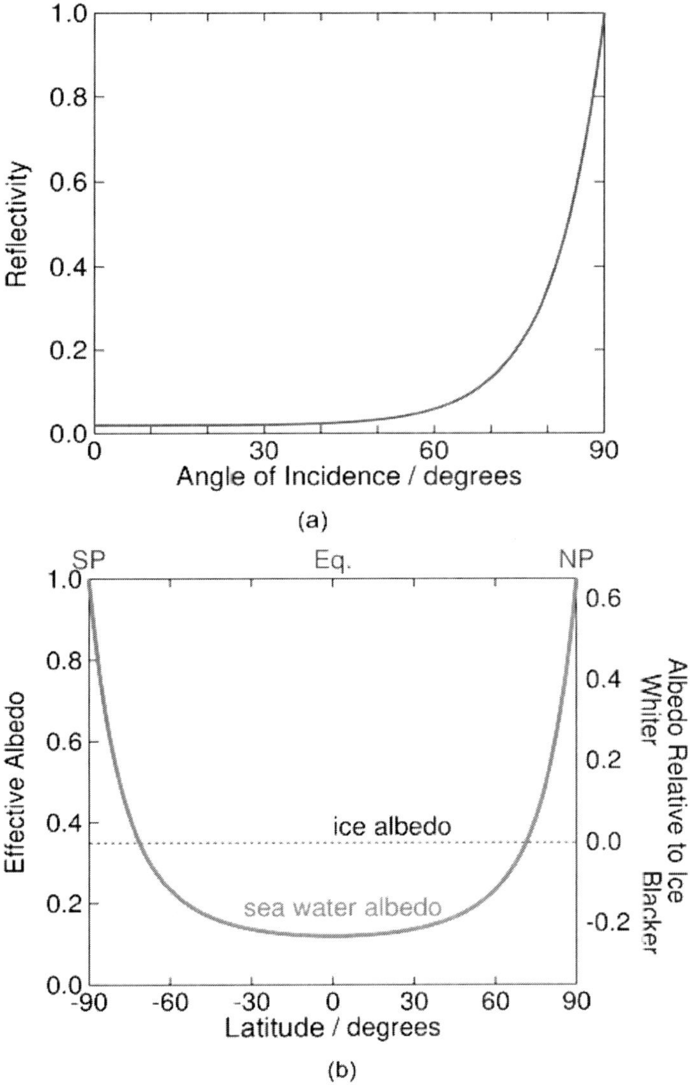

Figure 4. (a) Reflectivity as a function of incident angle of light according to Fresnel Equation (Equation (3)) with $n_1 = 1$ and $n_2 = 1.33$. (b) Effective albedo as a function of latitude, calcu- lated on basis of (a) and compared to ice albedo (0.35).

$$\beta(\alpha,\theta) = \cos^{-1}(\cos\alpha\cos\theta). \tag{4}$$

The effective albedo is then given by

$$a(\alpha) = \frac{1}{\pi}\int_{-\pi/2}^{\pi/2} R(\beta(\alpha,\theta))\,\mathrm{d}\theta. \tag{5}$$

The result of this is shown in Figure 4(b). We can see here two important effects. First of all, the albedo at the poles is unity. Meaning that polar sea water reflects sunlight even better than snow or ice does (for comparison also shown). Second, the equatorial water has, because of this specular reflection character, an albedo that is much higher than the often-assumed albedo of 0.02. That is because also these regions get the light at a very oblique angle part of the day. The effective albedo at the tropical regions is therefore about 0.12.

The implication of the high polar albedo is that the presence of ice in polar regions actually decreases the albedo and this is thus a negative-feedback "Gaia" effect; when the temperature increases ice melts and this increases the albedo and lowers the tem- perature of the planet, refreezing the water into ice again. Analyzing the figure, this happens for latitudes higher than around 70 degrees. Ice in moderate and warmer regions does have the traditional positive feedback effect. This might well be the negative feedback that we mentioned in the introduction.

On basis of this we made simulations. If the temperature of an element drops below zero °C, it will be attributed the albedo of ice $(\alpha = 0.35)$. If it melts, it gets attributed the Fresnel albedo of Figure 4. We find that, indeed, independent of the starting condition (be it a fully frozen planet, or a fully liquid planet) the polar regions wind up covered with ice. There is only a hysteresis effect for latitudes outside the polar regions, as can be seen in Figure 5. The hysteresis effect of Figure 3 of 8.6 degrees celsius has thus been reduced to 1.5 degrees. It could be absent altogether in a system without ice in the moderate regions, as is the case of our planet. We conclude that the common belief of a runaway scenario of irreversible melting of ice solely based on the albedo effects is not necessarily true.

Yet, the question arises whether this model system is adequate for the Earth radiation system? This question is more so valid considering the fact that apparently the ice cover in the North Pole region recovered 60% in a single year, from 2012 to 2013. This hints at a much larger negative feedback effect than demonstrated above. In the next simulations another idea is tested. It adds two basic phenomena to the system (see Figure 6 for all processes): 1) Heat is transported horizontally across the globe (indi- cated by arrows),

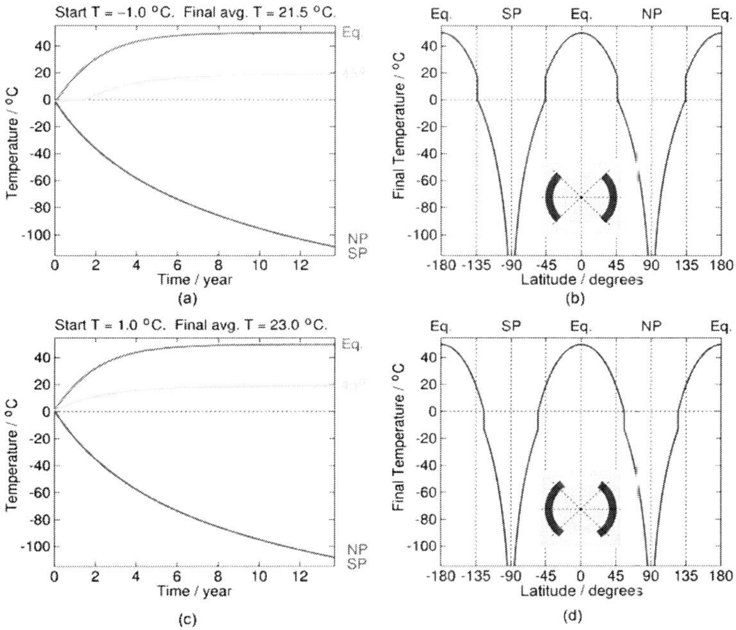

Figure 5. Model with albedo changes. Once water freezes, its albedo changes to 0.35. Liquid water has a Fresnel albedo depending on the latitude as shown in Figure 4. (Fudge factor 1.62).

2) Heat is transported vertically from the bottom to the top. The first one is something similar to the global thermohaline conveyor belt (THC), which we model with a simple exchange of heat between neighboring cells, heat flow is simply proportional to the temperature difference between neighboring cells. For the second effect of vertical transport, we have to add the possibility of a vertical gradient. We do this by duplicating the belt into two layers; every element is consisting of a top cell and a bottom cell. The effect we are trying to simulate is that ice not only has the property of a different albedo and associated change in absorbed light compared to water, but also a different heat conductivity. In this way, ice can cool down much more and ice, once formed, thus no longer emits much radiation and the planet heats up. In contrast, once the ice melts, heat is easily transported to the surface and radiated out, thus cooling down the planet.

To show what is going on, first we remove the formation of ice again and introduce the concept of heat transport (Figure 7). If we assume heat, H , to be proportional to temperature (which it isn't, but anyway, as will be seen, we are only interested in the gradient of heat transport and temperature), then the heat in an element is the temperature times the specific heat capacity of water $\left(c_m = 4.2 \text{ kJ K}^{-1}\text{kg}^{-1}\right)$, the density $\left(\rho = 1000 \text{ kg m}^{-3}\right)$ and the volume ($V = A\Delta X$, with A the cross-section and ΔX

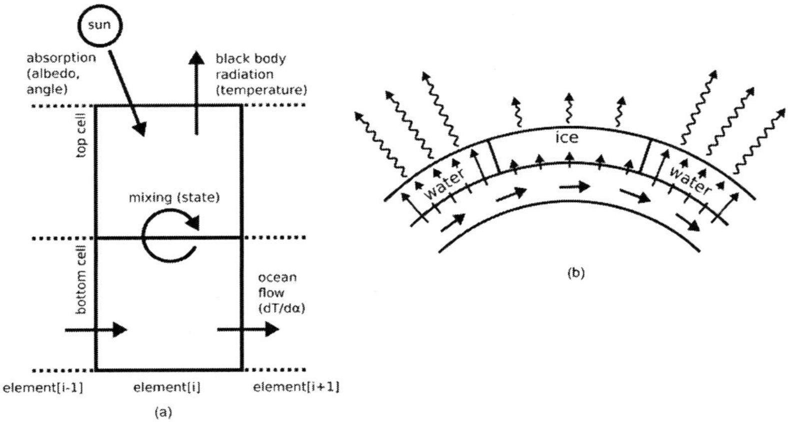

Figure 6. (a) All heat effects summarized (in brackets what they depend on). Inward solar radiation (depending on angle and albedo), outward black body radiation (depending on temperature), horizontal flow of the ocean in the deep layer (depending on temperature gradient), and mixing between the deep and shallow layer (depending on the phase state of the water, solid or liquid). (b) Schematic diagram of the effect. Ice works as a blanket and reduces IR emission.

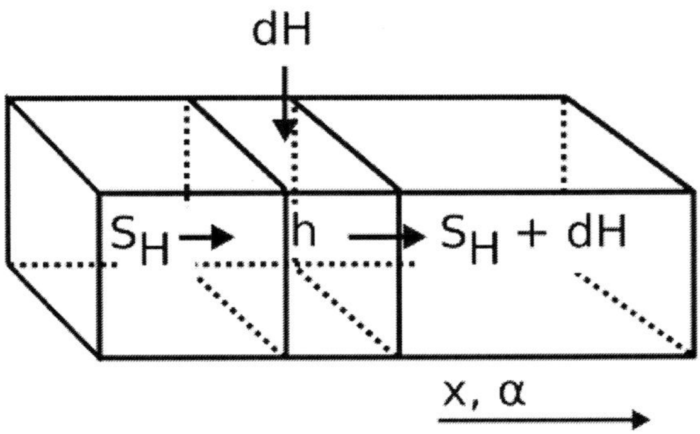

Figure 7. Heat going around in the ring. S_H is transport of heat (W), h is heat density (J/m^3, the heat in the volume being H = hV), dH is radia- tion heat balance input (W).

the thickness of an element). The density of heat is given by this heat divided by the volume of the element,

$$h(x,t) = \frac{H(x,t)}{V} = T(x,t)c_m\rho \ \left(\mathrm{J/m^3}\right). \tag{6}$$

The transport of heat, S_H, is given by the product of heat density, the velocity v of the water and the cross-section A,

$$S_H(x,t) = T(x,t)c_m\rho vA \ \left(\mathrm{J/s}\right). \tag{7}$$

To find out how much net heat enters an element in this way, we have to subtract what goes out from what comes in. Assuming that the same quantity of water enters and leaves an element, $\Delta\rho = 0$ and $\Delta v = 0$, the differential heat-flow is given by the difference of temperature of water going out and going in, $\Delta T = T_{\mathrm{out}} - T_{\mathrm{in}}$, which is proportional to the temperature gradient and the thickness of an element, ($\Delta T = \Delta x \cdot \mathrm{d}T(x)/\mathrm{d}x$, with in the limit $\Delta x = \mathrm{d}x$). The net heat accumulating in an element is thus

$$P_H(x,t) = \frac{\mathrm{d}T(x,t)}{\mathrm{d}x}\Delta X v c_m \rho A \ \left(\mathrm{J/s}\right). \tag{8}$$

The contribution of this to changes in temperature is then this heat accumulation divided by the specific heat and the volume of the element,

$$\frac{\mathrm{d}T(x,t)}{\mathrm{d}t} = \frac{P_H(x,t)}{c_m\rho A\Delta X} = \frac{\mathrm{d}T(x,t)}{\mathrm{d}x}v \ \left(\mathrm{K/s}\right), \tag{9}$$

in other words, it is the product of temperature gradient and water speed. Considering the transformation $x = R\alpha,\ \mathrm{d}x = R\mathrm{d}\alpha$ and $v = \mathrm{d}x/\mathrm{d}t = R\mathrm{d}\alpha/\mathrm{d}t = R\omega$, this can also be translated into our ringular world coordinates,

$$\frac{\mathrm{d}T(\alpha,t)}{\mathrm{d}t} = \frac{\mathrm{d}T(\alpha,t)}{\mathrm{d}\alpha}\omega \ \left(\mathrm{K/s}\right), \tag{10}$$

with ω the radial velocity, $\mathrm{d}\alpha/\mathrm{d}t$ (rad/s). In Figure 7 we can see how an area of the ring receives an external heat (equal to the radiation balance), dH (proportional to width and distance). This extra heat is added to the incoming flux of heat S_H and goes out of the cell as $S_H + dH$.

Common belief is that it can take about a thousand years for a packet of water to travel around the world. That is a speed of $\omega = 2\times10^{-10}$ rad/s. In the simulation below, the temperature difference value is added with a factor 0.6 to the adjacent cell. To give us good results we use a temperature change of $\Delta T_i = 0.3\cdot(T_{i+1} - T_i)$ in one step (i is element number). With n, the number of elements equal to 3600 and a time step of 5 days, this gives a complete cycle in 60,000 days (164 years), which is about a factor 6 faster than what is commonly assumed. One should not forget that heat transport does not only consist of slow water flow (THC), but also induces the faster atmospheric pro- cesses which are moreover of larger magnitude overall. We therefore find our value used in the simulations reasonable.

Literature is often confusing about the total heat flux of our planet, but it is in the order of 4×10^{15} W. This is confusing since in plots for heat flux of our planet, both positive as well as negative values are given, while the transport of heat is always a positive amount, according to Equation (7). On the other hand, a heat flux derivative dP_H/dx can be found by setting $\Delta X \to dx$ in Equation (8), which has then units W/m instead of W. To then find the total heat flow (W), for instance from Equator to North Pole, this function has to be integrated from Equator to North Pole. Therefore, considering this confusion, more information can be obtained from the observation that the heat flux makes the equatorial regions cool down by about 14 degrees. Using the method described above, the equatorial region cools down by some 5 degrees, which is reasonable.

Figure 8 summarizes the result of a simulation with transport of heat included. Some effects are immediately apparent. First, there are two different moderate zones, one that gets water from the Equator and is relatively warm (45° latitude), and one that receives water from a pole and is relatively cold $(135^\circ$ latitude$)$. Second, the poles themselves are much warmer than they were without circulation of heat and no longer reach absolute zero (compare with Figure 2). The final temperature plot is "averaged out". This averaging out will cause a reduction in atmospheric and oceanic temperature gradients which are the motor for the circulation in the first place. i.e., we recognize here another stabilizing negative feed-back effect.

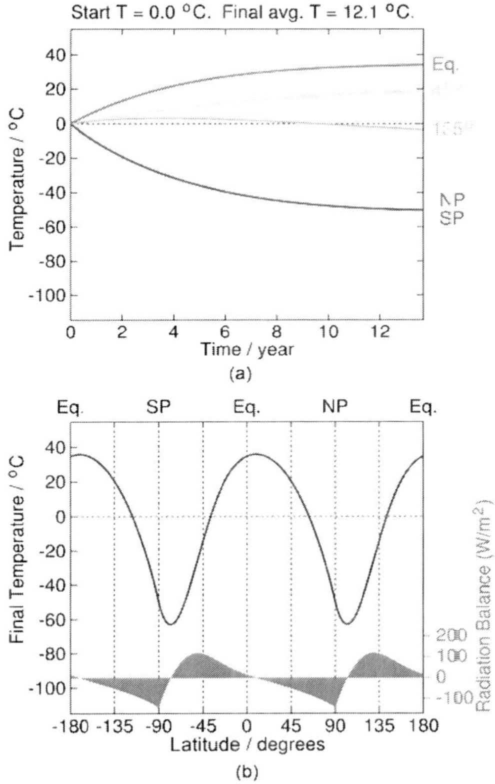

Figure 8. Model without albedo changes (all is water) but with transport of heat. Note that the poles are no longer the coldest places and that there are two distinct moderate zones, one gets a flow from the nearest pole and one from the Equa- tor. At the bottom of panel b the radiation balance is shown, the difference between inbound radiation and outbound ra- diation (Equations (1) and (2), resp.). A positive value indi- cates that the region receives more energy from the sun than it emits. (In the previous plots the radiation balance was always zero and therefore not shown).

When we reintroduce the effect of ice formation and resulting albedo changes on top of this system with horizontal and vertical heat flow, we see an interesting effect, see Figure 9. Because heat can be "trapped" under the ice the outward radiation is less efficient and the planet heats up. To compensate for that, the fudge factor has been reduced to 1.3 to give us again somewhat Earth-like temperatures. For the mixing factor we used 100% per time step (5 days) for a liquid top layer and 1% per time step for a frozen top layer. Horizontal circulation was set at 0.06 angular degrees per time step (full circulation in 87 years).

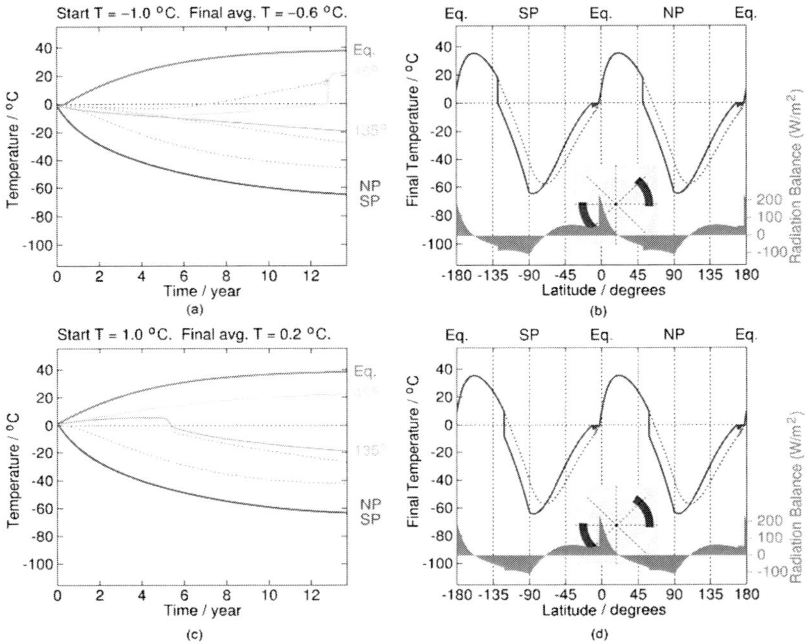

Figure 9. Model with albedo changes and transport of heat, both vertical and horizontal (as depicted in Figure 6). The solid traces are the top layer, the dashed traces the bottom layer. Starting at a fully frozen planet, or fully unfrozen planet results in nearly the same outcome.

As can be seen, the hysteresis effect has nearly completely disappeared. Starting at a fully frozen planet or fully unfrozen planet results in the same outcome, as can be seen by comparing panels b and d. Interestingly, the simulations at the edge of the ice cap are very noisy, as if smaller ice shelves are floating close to this edge. We attribute this to calculation instabilities, however, and do not claim that our system really can mimic icebergs.

The effect of the ice cap working as a blanket can be estimated from the difference between the temperature of the top layer and the bottom layer. At places where there is ice, this is easily 10 degrees, and the effect is therefore a loss of IR emission of about 50 W/m^2 (Equation (2)), which is significant.

All results are summarized in Table 1.

4. DISCUSSION

We have studied here the idea of the effect of phase changes when ice melts and water freezes into ice. The common idea is that these changes, through albedo changes, add positive feedback to the system; once ice disappears the effect is amplified by more heat absorption.

Table 1. Summary of simulations and resulting hysteresis, when starting with a two-degree dif- ference in starting conditions, TPW ±1°C (a fully frozen or unfrozen planet).

Model	Hysteresis	Results
Albedo	8.8°C	Figure 3
Fresnel	1.5°C	Figure 5
Heat transport	0.8°C	Figure 9

Contrary to this belief, we see that phase changes in fact can create negative feedback. The effects are two-fold: 1) The albedo of ice is lower than that of water at polar regions, as caused by the Fresnel effect, where water reflects all incoming radiation when the angle is at grazing incidence (with the sun close to the horizon). 2) Ice effectively blocks heat transport from lower layers of the ocean to the surface. The surface can then not radiate out the heat and the planet warms up, melting the ice.

The simulations are a proof of concept. We tried to keep here the system as simple as possible, to highlight the effects, so that everybody can easily understand what is going on. Simulations on a more realistic spherical planet with moreover an atmosphere (so as also to avoid the use of fudge factors) will have to show how much of this effect can be mapped to such a world. Things like the atmosphere [14] or non-periodic water circulation as in the ENSO (El Niño [and La Niña] Southern Oscillation) [15] have been completely ignored, along with changes in the thermohaline circulation [16] [17] [18] [19] . A more important thing that we have been ignoring is the tilt of the Earth rotation axis, which might change the game. However, we would like to point out once again that we only, and exclusively, studied the idea of water/ice phase changes and the associated albedo changes. Since we do not have access to the sophisticated equipment used in modern simulations, our contribution here is merely to point out the possibility and leave it up to others to implement them in more realistic state of the art calcu- lations.

DISCLOSURE OF INTERESTS

The authors declare not to have any conflicts of interests. This research was paid by no grant. It received no funding whatsoever, apart from our salaries at the university where we work. Nor are we members of any climate committees (political or other) or are we linked to companies or NGOs, financially or otherwise. We are not members of any political party or movement. This is an independent work that does not necessarily represent the opinion of our university or of our government.

CITE THIS PAPER

Stallinga, P. and Khmelinskii, I. (2017) Negative Feedback in the Polar Ice System. Atmospheric and Cli- mate Sciences, 7, 76-91.

REFERENCES

1. Curry, J.A., Schramm, J.L. and Ebert, E.E. (1994) Sea Ice-Albedo Climate Feedback Mechanism. Journal of Climate, 8, 240–247.
2. Flanner, M.G., et al. (2011) Radiative Forcing and Albedo Feedback from the Northern Hemisphere Cryosphere between 1979 and 2008. Nature Geoscience, 4, 151-155.
3. Hall, A. (2003) The Role of Surface Albedo Feedback in Climate. Journal of Climate, 17, 1550-1568.
4. Lian, M.S. and Cess, R.D. (1977) Energy Balance Climate Models: A Reappraisel of Ice-Albedo Feedback. Journal of the Atmospheric Sciences, 34, 1058-1062.
5. Morassutti, M.P. (1991) Climate Model Sensitivity to Sea Ice Albedo Parameterization. Theoretical and Applied Climatology, 44, 25-36.
6. Wang, W.-C. and Stone, P.H. (1979) Effect of Ice-Albedo Feedback on Global Sensitivity in a One-Dimensional Radiative-Convective Climate Model. Journal of the Atmospheric Sciences, 37, 546-552.
7. Pistone, K., Eisenman, I. and Ramanathan, V. (2013) Observational Determination of Albedo Decrease Caused by Vanishing Arctic Sea Ice. Proceedings of the National Academy of Sciences of the United States of America, 111, 3322-3326.
8. Riihela, A., Manninen, T. and Laine, V. (2013) Observed Changes in the Albedo of the Arctic Sea-Ice Zone for the Period 1982-2009. Nature Climate Change, 3, 895-898.
9. NASA http://neptune.gsfc.nasa.gov/csb/index.php?section=234
10. Hecht, E. (1975) Schaum's Outline Series: Theory and Problems of Optics. McGraw Hill.
11. Quan, X. and Fry, E.S. (1995) Empirical Equation for the Index of Refraction of Seawater. Applied Optics, 34, 3477-3480.
12. Payne, R.E. (1972) Albedo of the Sea Surface. Journal of the Atmospheric Sciences, 29, 959-970.
13. Kantha, L. and Clayson, C.A. (2014) Ocean Mixed Layer. In: North, G.R., Pyle, J.A. and Zhang, F.Q., Eds., Encyclopedia of Atmospheric Sciences, Chap. Boundary Layer (Athmospheric) and Air Pollution, 2nd Edition, Vol. 1, 290-298.

14. Jupp, T.E. and Cox, P.M. (2014) MEP and Planetary Climates: Insights from a Two-Box Climate Model Containing Atmospheric Dynamics. Philosophical Transactions of the Royal Society B, 365, 1355-1365.
15. Veettil, B.K., Maier, é.L.B., Bremer, U.F. and Souza, S.F. (2014) Combined Influence of PDO and ENSO on Northern Andean Glaciers: A Case Study on the Cotopaxi Ice-Covered Volcano, Ecuador. Climate Dynamics, 43, 3439-3448
16. Bersch, M., Yashayaev, I. and Koltermann, K.P. (2007) Recent Changes of the Thermohaline Circulation in the Subpolar North Atlantic. Ocean Dynamics, 57, 223-235.
17. Lorenzo, M.N., Taboada, J.J. and Iglesias, I. (2009) Sensitivity of Thermohaline Circulation to Decadal and Multidecadal Variability. ICES Journal Marine Science, 66, 1439-1447.
18. Marotzke, J. (2000) Abrupt Climate Change and Thermohaline Circulation: Mechanisms and Predictability. Proceedings of the National Academy of Sciences of the United States of America, 97, 1347-1350.
19. Wood, R.A., Vellinga, M. and Thorpe, R. (2003) Global Warming and Thermohaline Circulation Stability. Philosophical Transactions of the Royal Society of London A, 361, 1961-1975.

CHAPTER 9

Impact of Climate Change on the Characteristics of Indian Summer Monsoon Onset

Savita Patwardhan,[1] Ashwini Kulkarni,[1] and K. Krishna Kumar[2]

[1]*Indian Institute of Tropical Meteorology, Pune 411008, India*
[2]*Qatar Meteorology Department, Doha, Qatar*

ABSTRACT

A high resolution regional climate modeling system, known as PRECIS (Providing REgional Climate for Impact Studies), developed by Hadley Centre for Climate Prediction and Research, UK, is applied for Indian subcontinent to assess the impact of climate change on the summer monsoon onset characteristics. The present day simulation (1961–1990) with PRECIS is evaluated for the characteristics of onset over Kerala, southernmost part of India, where the monsoon sets in over Indian landmass. The meteorological parameters like precipitation, outgoing long wave radiation (OLR), and low level winds are analysed to study the monsoon onset over Kerala. The model is able to capture the sudden and sharp increase of rainfall associated with the onset. The rapid built-up of convective activity over the southeastern Arabian Sea and Bay of Bengal is well represented by the model. PRECIS simulations, under scenarios of increasing greenhouse gas concentrations and sulphate aerosols, are analysed to study the likely changes in the onset characteristics in future, towards the end of present century (2071–2100). The analysis does not indicate significant difference in the mean onset dates in A2 and B2 scenarios. However, the variability of onset date is likely to be more towards the end of the 21st century especially in A2 scenario.

1. INTRODUCTION

The onset of summer monsoon over the southern tip of Kerala is the vital meteorological phenomenon over India. It is the prime event related to the Indian agriculture and economy. Every year, summer monsoon arrives over Kerala coast (Figure 1) around the end of May/beginning of June. The delay in the onset is an issue of concern to all, from farmers to policy makers. It is very difficult to declare the onset of monsoon. Sometimes, in early May, there can be a bogus onset and the real onset will be delayed [1]. The onset of Indian summer monsoon represents significant transitions in the large scale atmospheric and oceanic circulation in the Indo-Pacific region. There is no widely accepted definition of this monsoon transition. At the surface, monsoon transitions are first revealed by variability in rainfall. Hence at the surface, monsoon onset is recognized as a rapid substantial and sustained increase in rainfall. Although no objective criterion exists for fixing the date of onset, the primary indicator from the early days of Indian Meteorology has been a sharp and sustained increase in rainfall at a group of adjacent stations [2].

Different criteria are used, by various researchers, to define the monsoon onset over Kerala. Using daily rainfall data from a dense network of rain gauges in Kerala, Ananthakrishnan and Soman [3] derived dates of monsoon onset over Kerala for south and north Kerala separately for the period 1901–1980. They defined the date of monsoon onset over Kerala as the first day of the transition from the light to heavy rainfall category with the provision that the average daily rainfall during the first 5 days after the transition should not be less than 10 mm. With this criterion, they have shown that the mean onset date over south Kerala is 30th of May and for north Kerala is 1st of June with the standard deviation of about 9 days for both. India Meteorological Department (IMD) uses a relatively qualitative method for declaring the onset, using the rainfall over stations located in Kerala, the lower tropospheric winds over Kerala, and the moisture available up to 500 hPa [4]. From the long period records of the Indian Meteorological Department, the mean onset date over Kerala is 1st of June with standard deviation of 7.28 days. From 2006, IMD is using the criterion by Joseph et al. [5] that depends on rainfall and circulation variables. Fasullo and Webster [6] used vertically integrated moisture transport instead of rainfall to define onset as well as withdrawal of monsoon. Zeng and Lu [7] defined monsoon onset and retreat using normalized precipitable water index. Xavier et al. [8] proposed an Indian summer monsoon onset index based on reversal of the large-scale meridional temperature gradient in the upper troposphere south of the Tibetan Plateau. Wang et al. [9] have used the 850-hPa zonal wind averaged over the south Arabian Sea (5°–15°N, 40°–80°E) as an onset circulation index of the Indian summer monsoon. Pai and Nair [10] have discussed various definitions of monsoon onset over Kerala (MOK) used by different researchers. Puranik et al. [11] have defined index using OLR and zonal kinetic energy to predict the monsoon onset over Kerala.

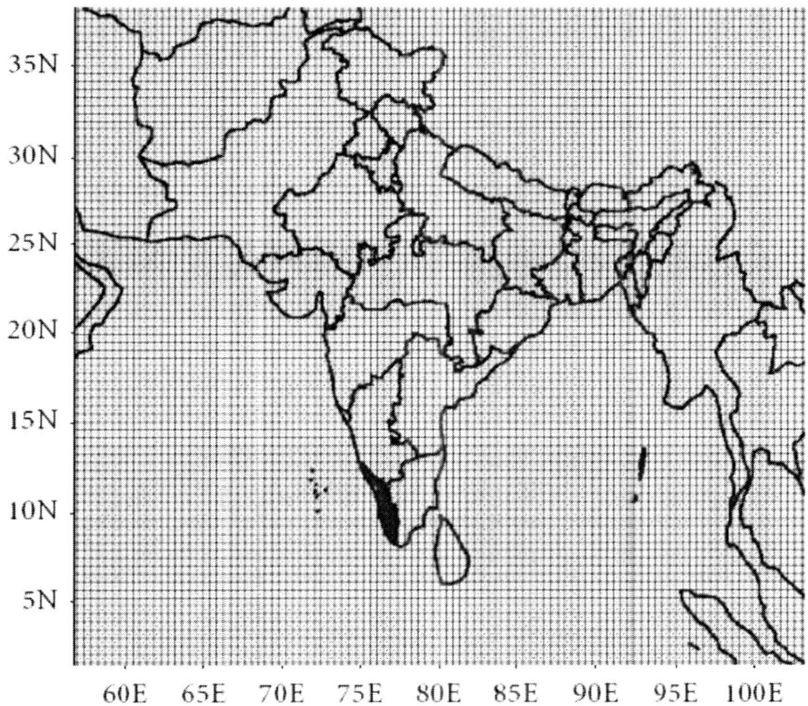

Figure 1: Indian domain used for the PRECIS simulations. Shaded region is the state of Kerala.

Global climate models, the best tool to understand and project the possible changes in climate, have large uncertainties in the simulation of monsoon precipitation compared to its simulation of temperature. All models under IPCC 4th assessment report projected a significant increase in temperature towards the end of the 21st century. However, the response of monsoon rainfall to the increased greenhouse gas concentrations varies from model to model [12], Rajeevan and Nanjundiah 2004 [13]. Bhaskaran and Mitchell [14] studied the changes in southwest monsoon precipitation using Hadley Centre global climate model, HadCM2, and showed that the monsoon onset date over the Bombay latitudes (18.975°N) delays by 10 days in the warming scenario. But the uncertainties in model projections of monsoon behavior are not clear from their study.

Here, we focused on examining the possible changes in the monsoon onset characteristics in the future under the effect of global warming, using a high resolution regional climate model, PRECIS (Providing REgional Climate for Impact Studies). The PRECIS simulation of seasonal monsoon precipitation for the present time slice (1961–1990) is quite similar to the observed, indicating that the model provides an adequate representation of present-day conditions [15]. Before analysing the future climate change scenarios using PRECIS, an evaluation of the model's capability to simulate the summer monsoon onset characteristics has been carried out using the present day (1961–1990) model simulations.

2. DATA AND METHODS

2.1. Regional Climate Model (PRECIS)

PRECIS is an atmospheric and land surface model of limited area (typically 5,000 km × 5,000 km) and high resolution (50 km × 50 km) developed by Hadley Centre for Climate Prediction and Research, UK, which can be used for simulations over any part of the globe. PRECIS has been configured for Indian domain extending from about 1.5°N to 38°N and 56°E to 103°E (Figure 1). Dynamical flow, the atmospheric sulphur cycle, clouds and precipitation, radiative processes, the land surface, and the deep soil are all formulated and boundary conditions are required at the limits of the model's domain. The model is driven at the lateral boundaries by the atmospheric general circulation model HadAM3H [16]. The basic aspects explicitly handled by the model are available in Noguer et al., 2002 [17]. The PRECIS regional model is already being used to generate scenarios for China and southern Africa (PRECIS-Update 2002). The high resolution climate change scenarios, using PRECIS simulations, are developed for the Indian region [15] and are distributed to the various impact user groups in India. Simulations using PRECIS have been performed to generate the climate for present time slice, for the period from 1961 to 1990 (taken as the baseline for model simulations) and for future period, from 2071 to 2100. The future period simulations are carried out for two different socioeconomic scenarios both characterised by regionally focused development but with priority to economic issues in one simulation (A2 scenario) and to environmental issues in the other (B2 scenario). The results for B2 scenarios are similar to A2 scenario but are on the lower side of magnitude. The detailed description of these scenarios is reported in Nakicenovic and Swart [18].

Though the results from the IPCC AR5 models are now available, we have used PRECIS simulations for A2 and B2 scenarios with lateral boundary conditions from IPCC AR4 global model due to its high resolution.

2.2. Observed Data Sets

2.2.1. Rainfall

The India Meteorological Department has developed a high resolution (lat./long.) daily gridded rainfall data set over the Indian landmass for the period 1901–2004 [19]. The data set is now updated till the year 2013 and is available on request at http://www.imdpune.gov.in. Daily rainfall data from total 1803 stations have been used in developing the gridded data set. The data for the period 1961–1990 is used in the present study for validating the model.

2.2.2. Outgoing Long Wave Radiation (OLR)

Estimates of OLR from National Oceanic and Atmospheric Administration (NOAA) polar orbiting satellites are used to distinguish areas of deep tropical convection. The National Environmental Satellite Data and Information Service (NESDIS) of NOAA archives the data onto 2.5° lat × 2.5° long grids [20]. This datum is available from June 1974 to date, except 1978. The data sets are available from the anonymous FTP site ftp://ftp.cdc.noaa.gov/.

2.2.3. Winds at 850 hPa

The NCEP-NCAR reanalysis data set [21] for the 850 hPa vector winds for the period 1961–1990 is used to examine the zonal kinetic energy in Arabian Sea during the onset phase of southwest monsoon.

2.3. Method

Daily rainfall over Kerala is represented by the simple area average rainfall over all the 16 grid points over Kerala (Figure 1). We use the following criterion for the onset of monsoon in PRECIS simulations—when the daily precipitation over Kerala is at least 5 mm and it persists for three consecutive days, then first day is taken as the onset of monsoon. Once the onset date is determined, composite analysis is performed for the parameters, namely, rainfall, low level wind, and the outgoing long wave radiation for preonset, onset, and postonset phases of the monsoon, to examine the onset characteristics simulated in the model baseline simulations (Section 3) and the likely future changes in the characteristics with reference to the above-mentioned parameters (Section 4).

3. EVALUATION OF PRECIS SIMULATIONS FOR SUMMER MONSOON ONSET

Southwest monsoon over India sets in from the southern tip of Kerala. Sudden increase in precipitation over Kerala around the end of May and beginning of June marks the onset. At the onset, rainfall increases substantially and persists for a long period. Ananthkrishnan and Soman [3] have done the superposed epoch analysis of the area weighted rainfall of south and north Kerala relative to the onset date and showed a sharp and spectacular increase of rainfall heralding the monsoon onset.

As the monsoon develops, south-westerlies are established at lower troposphere, over the Arabian Sea and the Indian subcontinent with a strong northward cross equatorial flow off the East African coast and at upper levels, a broad band of east to northeast winds gets established extending across the whole of the tropical Indian Ocean into Africa [22]. Spectacular strengthening of the westerly flow over the Arabian Sea is one of the important aspects of onset of monsoon. Krishnamurti et al. [23] analysed the MONEX data sets and showed that the kinetic energy of zonal winds at 850 hPa over Arabian Sea increases steadily after the onset.

The meteorological parameters like winds at 850 hPa, precipitation, and OLR during 1961–1990 have been analysed for preonset, onset, and postonset phases of the monsoon, to evaluate the model's ability to simulate the onset of summer monsoon over Kerala.

3.1. Rainfall

To assess the ability of model in simulating monsoon onset, the daily rainfall series simulated by PRECIS for Kerala are analysed for the present time slice (1960–1990). The criterion defined in Section 2.3 is then applied to the IMD's high resolution daily gridded rainfall data for the same period 1961–1990. It gives 30th of May as the climatological mean onset date of monsoon over Kerala with standard deviation of 12.2 days. The model simulated mean date of monsoon onset over Kerala for 1961–1990 is 25th of May, with slightly low value of standard deviation as 7.7 days.

Figure 2 shows the superposed epoch analysis of daily rainfall over Kerala. 0 day indicates the monsoon onset over Kerala. The sharp and spectacular increase in rainfall associated with the monsoon onset is reasonably well represented in the model. Prior to onset, rainfall is less than 5 mm/day. At the onset, rainfall increases substantially and persists for a long period. However, it is seen from the Figure 2(a) that the model simulated rainfall at the onset is less than the observed rainfall for the period 1961–1990.

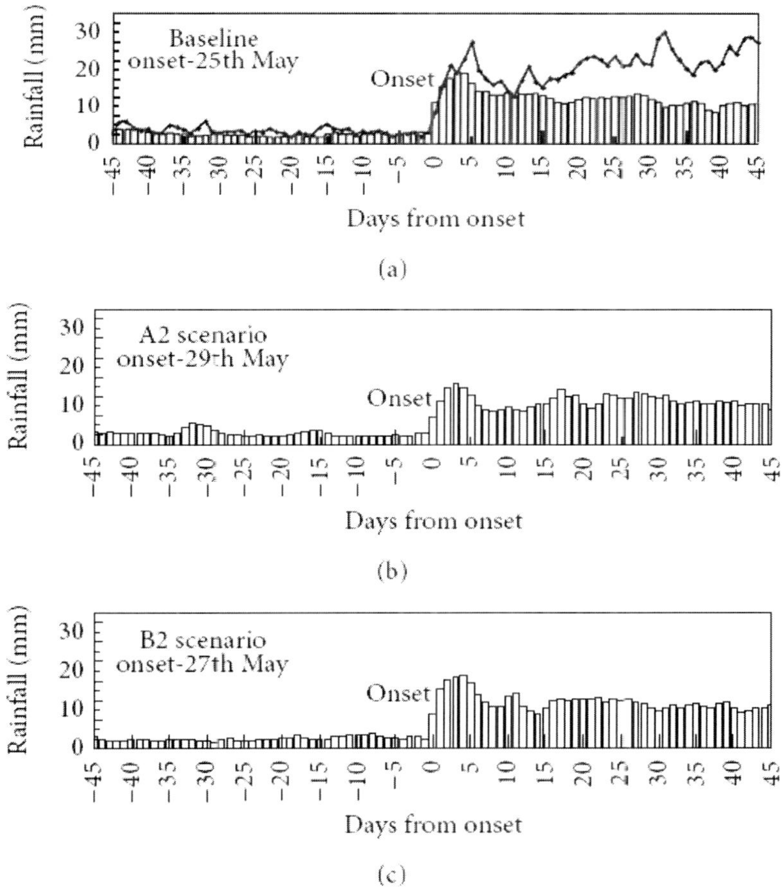

Figure 2: Daily rainfall (mm) over Kerala as simulated by PRECIS (bars) from 45 days prior to onset to 45 days after the onset for baseline (a), A2 scenario (b), and B2 scenario (c). The observed daily march of rainfall (line) is shown using IMD gridded data set.

Once the monsoon sets in over Kerala, it progresses towards northern latitudes to cover the entire country by around 15th July. The normal onset date of monsoon over west Rajasthan is 1st of July as per IMD. The model skills in simulating the northward progress of the summer monsoon are assessed using the model simulated daily rainfall over west Rajasthan. The rainfall over west Rajasthan is the average rainfall over all the 83 grids over west Rajasthan. The criterion used to define onset is similar to the Kerala onset. If the minimum daily rainfall of 1.5 mm is received for three days, then the first day is considered as the onset of monsoon over west Rajasthan. Using this criterion, the mean onset date over west Rajasthan is identified for the model baseline. The model simulated onset date over west Rajasthan is 2nd of July. Even though the model has a slight bias of early onset over Kerala, it reaches over Rajasthan on the date almost as observed; hence it can be seen that the advancement of monsoon is very well captured by the model.

3.2. Convection

Since the observed daily precipitation data is available only over the Indian land mask, we used outgoing long wave radiation (OLR) as a proxy for precipitation over the oceanic region. Higher values of OLR imply less convection and hence indicate less cloudiness and rain. According to Kripalani et al. [24] the OLR values 220–240 Wm^{-2} are associated with rainfall probability 0.7-0.8 and OLR = 180 Wm^{-2} indicates probability of rainfall greater than 0.9. Outgoing long wave radiation (OLR) serves as a good index of cloudiness and precipitation. The composite OLR pattern for the onset phase of monsoon for the period 1961–1990 is validated using observed OLR data set from NOAA. The composite OLR anomalies taken with respect to 220 Wm^{-2} are shown in Figure 3. Figure 3 shows that up to 15 days prior to the onset the OLR anomalies in the southeastern Arabian Sea are greater than 20 Wm^{-2}. Then it gradually decreases and at the onset OLR anomaly values are less than −20 Wm^{-2}. After the onset, convection shifts to east Bay of Bengal and gradually the convective activity decreases over Arabian Sea.

The spatial pattern of the OLR is well simulated by the model but the convection is more in the baseline simulation (Figure 4) as compared to the observed OLR pattern. The simulated OLR pattern reveals that in the model the convection starts building up over southeast Arabian Sea 15 days prior to the onset (OLR anomaly < -20 Wm^{-2}). Five days before the onset, the main convective area shifts to east Bay of Bengal and over southeastern Arabian Sea (OLR anomaly < -60 Wm^{-2}). On the onset day, deep convective clouds (OLR anomaly < -60 Wm^{-2} indicating probability of rain > 0.9 according to [24]) can be seen off the Kerala coast. These model results agree well with the study by Soman and Kumar [25]. After the onset, the convective activity over Arabian Sea decreases slowly and increases over northeast Bay of Bengal (OLR anomaly < −80 W m^{-2}). Northward propagation of the convection is also well reproduced in the OLR pattern (Figure 4). The colocation of model simulated heavy rainfall region with low value of OLR indicates that there is consistency among these parameters in the model (figure not shown).

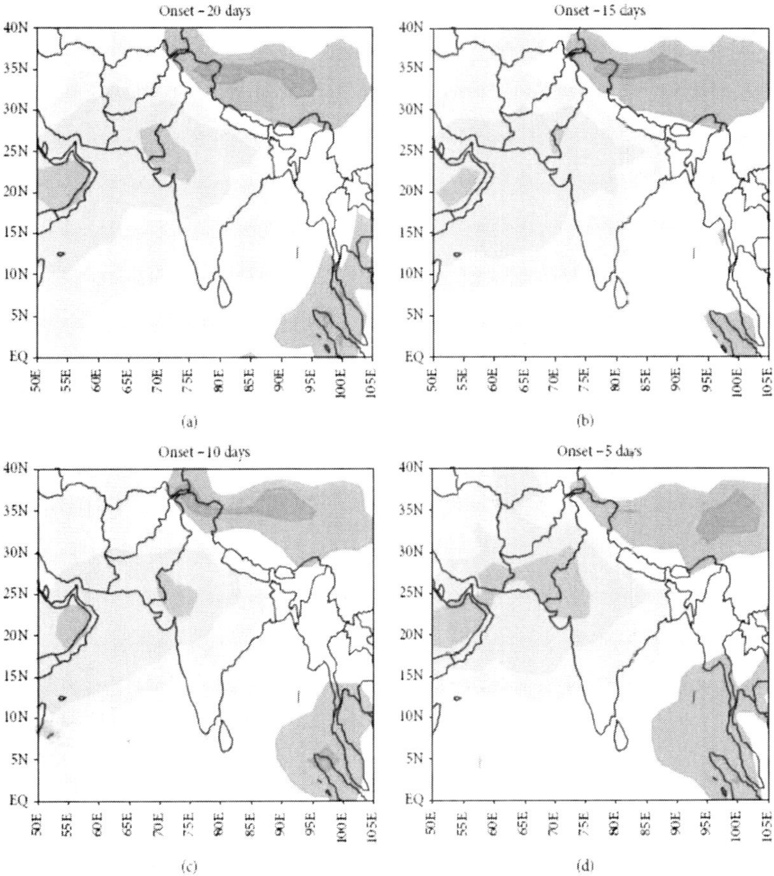

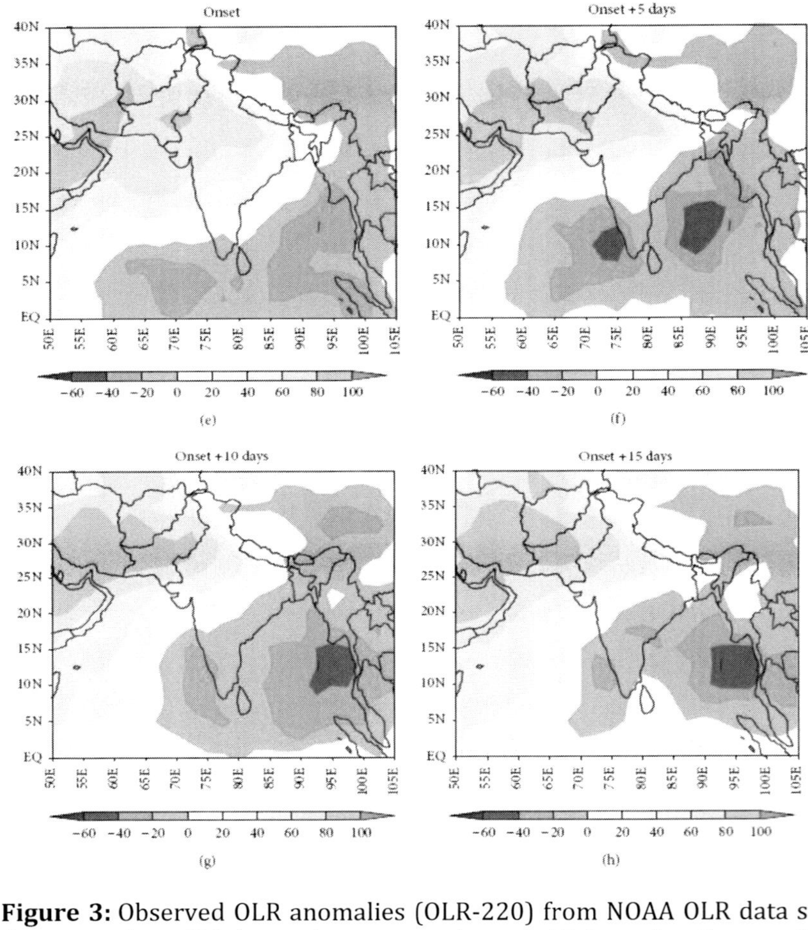

Figure 3: Observed OLR anomalies (OLR-220) from NOAA OLR data set at the onset phase (20 days prior to onset through 15 days after the onset).

3.3. Winds at 850 hPa

As the summer monsoon develops, at lower troposphere, south-westerlies are established over the Arabian Sea and the Indian subcontinent with a strong northward cross equatorial flow off the East African coast and at upper levels a broad band of east to northeast winds gets established extending across the whole of the tropical Indian Ocean into Africa [22]. In the present study, the development of cross equatorial flow could not be studied, as the model domain is restricted to the Indian region.

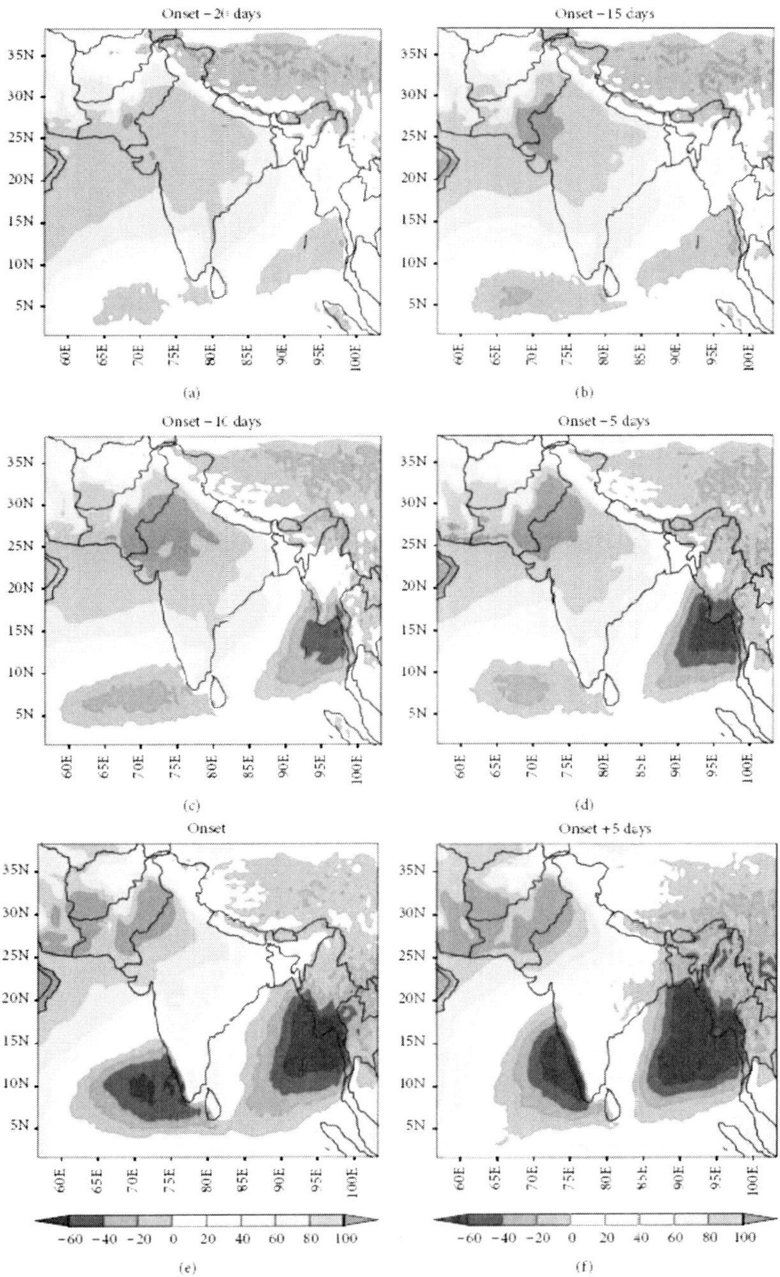
Onset − 20 days
Onset − 15 days
Onset − 10 days
Onset − 5 days
Onset
Onset + 5 days
35N
30N
25N
20N
15N
10N
5N
60E
65E
70E
75E
80E
85E
90E
95E
100E
(a)
(b)
(c)
(d)
(e)
(f)
−60 −40 −20 0 20 40 60 80 100

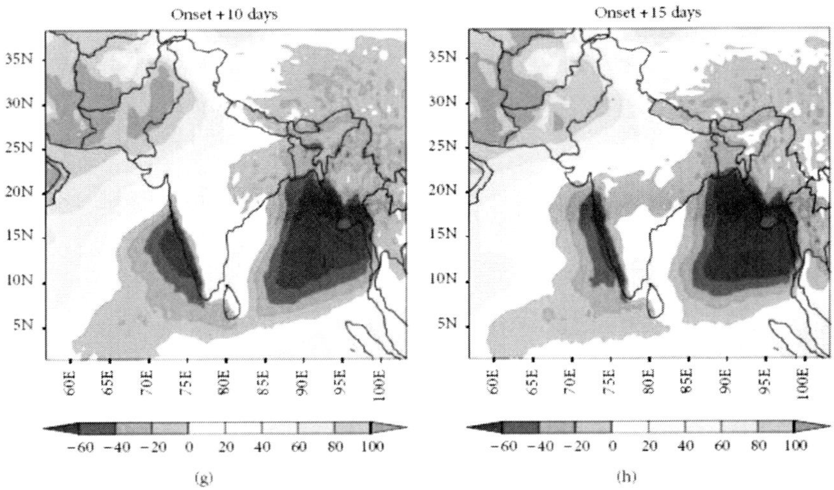

Figure 4: Same as Figure 3 but for baseline simulation.

The spectacular strengthening of the low level westerly flow over the Arabian Sea is one of the important aspects of the onset of monsoon. During the onset of monsoon over Kerala the low level westerly winds over Arabian Sea strengthen anomalously. To examine this aspect the composite analysis of 850 hPa winds is carried out from the 20th day prior to the onset to 15th day following the onset. Analysis shows that the low level westerlies over Arabian Sea and their rapid intensification after the onset are in good agreement with the observed pattern. After the onset, westerlies extend up to northern latitudes.

Krishnamurti et al. [23] analysed the MONEX data sets and showed that the kinetic energy of zonal winds (ZKE) at 850 hPa over Arabian Sea increases steadily after the onset. Raju et al. [26] analysed the ZKE in Arabian Sea and Bay of Bengal using daily NCEP-NCAR reanalysis data sets. They have shown that the sudden rise in low level kinetic energy can be a potential predictor to declare onset. The model simulated zonal winds at 850 hPa are analysed to see if this steady rise in low level ZKE is captured by the model. Figure 6 shows the ZKE at 850 hPa in the southeast Arabian Sea (56°E –70°E and 1.5° N–20°N). It increases from 20 m^2/s^2 (onset-5 days) to 50 m^2/s^2 (onset) for the baseline run (upper panel). After the onset it rises steadily up to 125 m^2/s^2. The strengthening of westerlies is well simulated by the model. However the model overestimates the ZKE compared to the observed ZKE as seen in the Figure 6.

4. FUTURE CHANGES IN THE ONSET CHARACTERISTICS

A comparison between the present time slice (1961–1990) and the future (2071–2100) emission scenarios (A2 and B2) has been made to see the possible changes in the monsoon onset characteristics in the warming scenarios. The average onset dates for A2 and B2 scenario are 29th of May and 27th of May, respectively, towards the end of the 21st century, that is, in 2080s. The change in the mean onset date is not significant in future as the variability of the onset dates in 2080s is 11.6 (9.1) days for A2 (B2) scenario. The variability of the onset dates is more in future as compared to the present (Table 1). The time required to advance the monsoon over northwest India is likely to be less in the future scenarios than the baseline. For A2 and B2 scenarios the dates are 25th of June and 27th of June, respectively. The results are comparable for A2 and B2 scenarios. However the magnitude of projections for A2 scenarios is more, being a high emission scenario.

Table 1: Onset dates of southwest monsoon over Kerala simulated by PRECIS.

	Mean onset date over Kerala	Std. dev. (days)
Observed	1st of June	7.3
Baseline	25th of May	7.7
A2 scenario	29th of May	11.6
B2 scenario	27th of May	9.1

The analysis of the future simulations of PRECIS in warming scenario also identifies the sharp rise in daily rainfall at the onset phase of monsoon towards 2080s in both emission scenarios, namely, A2 and B2. However, the rainfall at the onset phase over Kerala may be less in future as compared to the baseline, as seen from Figure 2.

Figure 5 shows the composite OLR pattern towards 2080s for A2 scenario. It can be seen that the convection at the onset phase in future may not be as deep as in baseline simulation. The OLR anomalies at the onset are $-40\ Wm^{-2}$ in southeast Arabian Sea in 2080s as compared to $-60\ Wm^{-2}$ in baseline simulation. The deeper convection in baselines simulations at the onset explains the higher rainfall over Kerala, at the onset phase in baseline simulation. The OLR anomaly pattern for A2 scenario is similar to that for B2 scenario (figure is not shown).

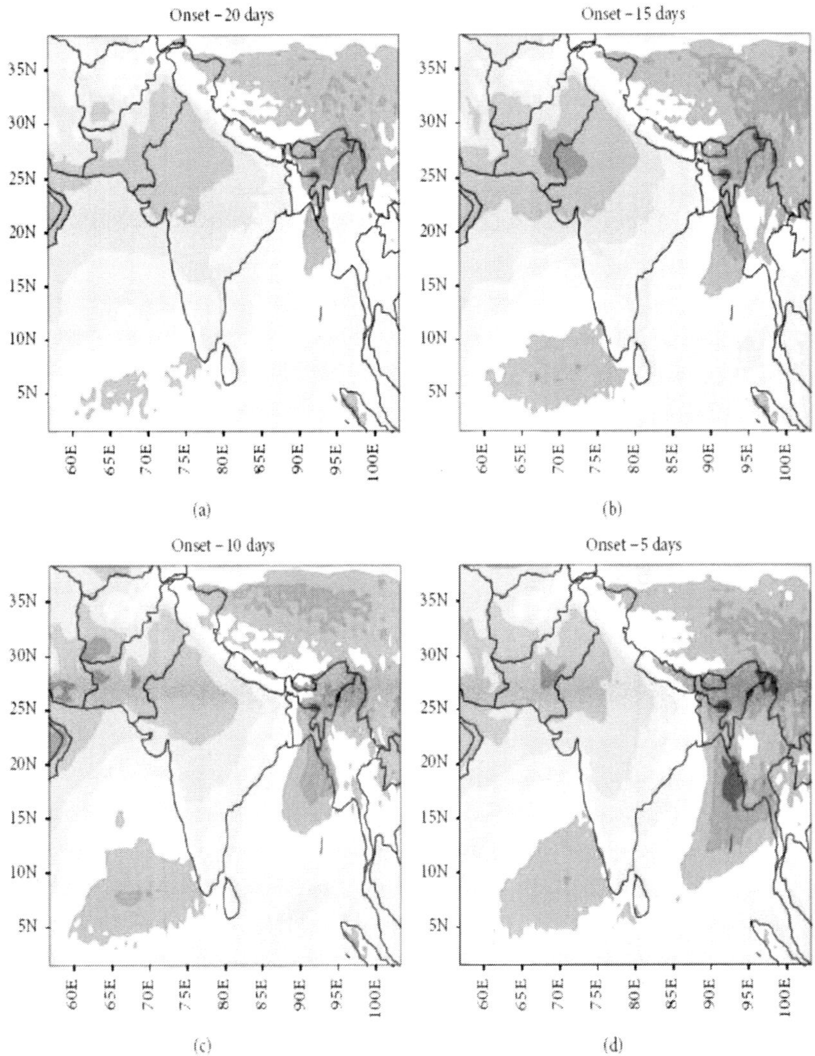

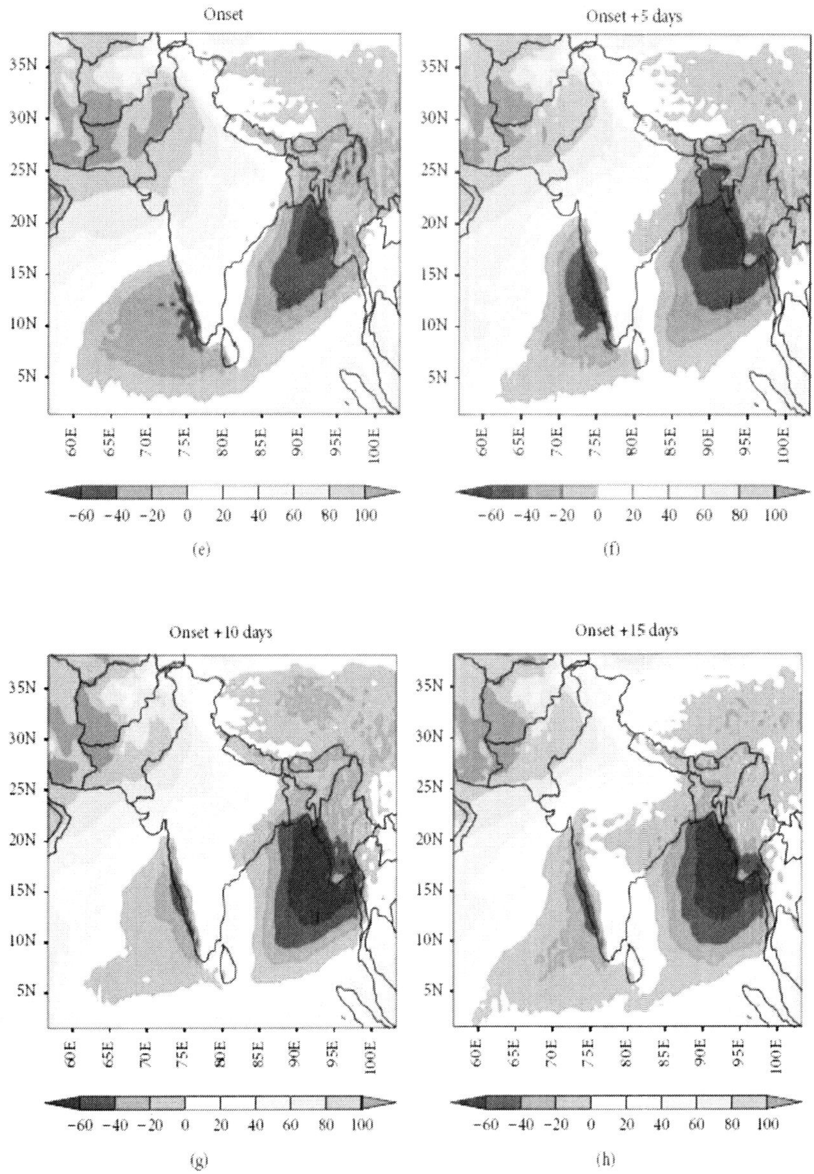

Figure 5: Same as Figure 3 but for A2 scenario.

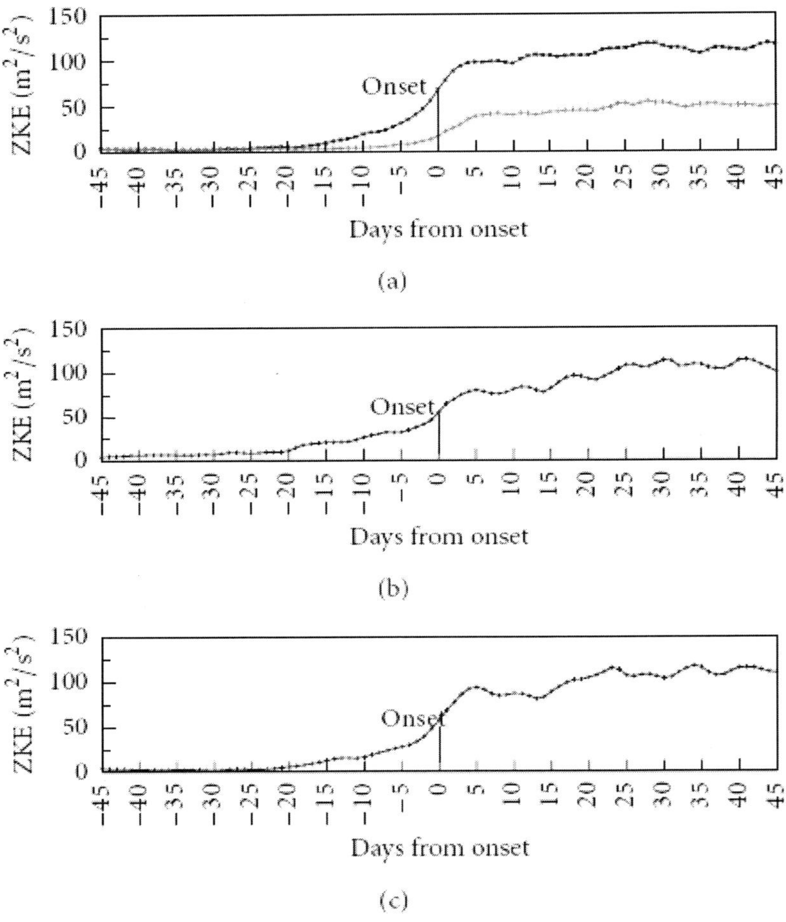

Figure 6: Zonal kinetic energy (m^2/s^2) at southeastern Arabian Sea as simulated by PRECIS for baseline (black) and observed (red) (NCEP/NCAR Reanalysis) (a) for the period 1961–1990 and in the global warming scenarios A2 (b) and B2 (c) for the period 2071–2100.

Figures 6(b) and 6(c) show the likely changes in ZKE at the end of present century. It can be seen from the figure that towards the end of the century the ZKE may not be as strong as in baseline during the monsoon onset phase. The ZKE over Arabian Sea at the onset phase is more or less the same in both emission scenarios A2 and B2 as shown in Figure 6.

5. CONCLUSIONS

The high resolution regional climate model, PRECIS, captures the onset of summer monsoon over Kerala reasonably well. The model simulates early onset as compared to the observed onset on 1st of June. The onset characteristics like the sudden and sharp increase in the daily rainfall over Kerala at the onset is well represented by the PRECIS. Model could capture the rise in the convection over southeast Arabian Sea and northeast Bay of Bengal at the time of onset. The steady rise in the zonal kinetic energy of low level winds, with the onset, at the southeast Arabian Sea, is well simulated by the model.

The model simulations under global warming scenario do not indicate significant change in the mean onset date of monsoon over Kerala in future as compared to the baseline. The onset date over Kerala does not differ much in A2 and B2 scenarios. However, the variability of onset date may be more in future especially in A2 scenario. The convection over the north Indian Ocean at the onset phase of monsoon towards the end of present century may not be as deep as the convection simulated in the baseline run.

These are the studies made using limited number of simulations using single regional climate model. To reduce the uncertainty associated with the projection, one has to have more number of simulations on more high resolution regional climate models. At present these scenario projections may be used with caution, qualitatively instead of the quantitative estimates.

Under the auspices of WCRP, a suite of high resolution regional model simulations has been made available recently to the scientific community, under the program CORDEX (Co-ordinated Regional Downscaling Experiment). The simulations over South Asian region (CORDEX-SA) are available for four models and have been archived at the data portal of Center for Climate Change Research (CCCR) of Indian Institute of Tropical Meteorology, Pune, India. These models are of latitude/longitude (~50 km × 50 km) horizontal resolution and have been derived using the lateral boundary conditions from CMIP5; however, this set does not include PRECIS. Such type of analysis has not been documented before using PRECIS simulations. Since CORDEX-SA provides the multimodel outputs for historical and RCP4.5 experiments and gives the range of uncertainty of model simulations, one can have more confidence in these projections. The present analysis will be carried out using these multimodel simulations in future.

CONFLICT OF INTERESTS

The authors declare that there is no conflict of interests regarding the publication of this paper.

ACKNOWLEDGMENTS

This work is part of a Joint Indo-UK collaborative programme on climate change impacts in India. The authors are grateful to the Department of Environment, Food and Rural Affairs (DEFRA), Government of United Kingdom, for sponsoring this project and the Ministry of Environment and Forest (MoEF), Government of India, for coordinating its implementation. The authors are thankful to the Hadley Centre for Climate Prediction and Research, UK Meteorological Office, for making available regional models and their data products required for this study. The authors are grateful to the Director, IITM, and Dr. R. Krishnan, Chief Program Scientist, CCCR, IITM, for their keen interest and the encouragement. Thanks are due to the anonymous reviewers whose suggestions helped us in improving the paper.

REFERENCES

1. M. K. Flatau, P. J. Flatau, and D. Rudnick, "The dynamics of double monsoon onsets," Journal of Climate, vol. 14, no. 21, pp. 4130–4146, 2001.
2. R. Ananthkrishnan, U. R. Acharya, and R. Krishna, "AR 1967: on the criteria for declaring the onset of the southwest monsoon over Kerala. Forecasting Manual," FMU. Report IV-18.1, India Meteorological Department, Pune, India, 1967.
3. R. Ananthakrishnan and M. K. Soman, "The onset of the southwest monsoon over Kerala: 1901–1980," Journal of Climatology, vol. 8, no. 3, pp. 283–296, 1988.
4. Y. P. Rao, Southwest Monsoon, IMD, 1976.
5. P. V. Joseph, K. P. Sooraj, and C. K. Rajan, "The summer monsoon onset process over South Asia and an objective method for the date of monsoon onset over Kerala," International Journal of Climatology, vol. 26, no. 13, pp. 1871–1893, 2006.
6. J. Fasullo and P. J. Webster, "A hydrological definition of Indian monsoon onset and withdrawal," Journal of Climate, vol. 16, pp. 3200–3211, 2003.
7. X. Zeng and E. Lu, "Globally unified Monsoon onset and retreat indexes," Journal of Climate, vol. 17, pp. 2241–2248, 2004.
8. P. K. Xavier, C. Marzin, and B. N. Goswami, "An objective definition of the Indian summer monsoon season and a new perspective on the ENSO-monsoon relationship," Quarterly Journal of the Royal Meteorological Society, vol. 133, no. 624, pp. 749–764, 2007.
9. B. Wang, Q. Ding, and P. V. Joseph, "Objective definition of the Indian summer monsoon onset," Journal of Climate, vol. 22, no. 12, pp. 3304–3316, 2009.
10. D. S. Pai and R. M. Nair, "Summer monsoon onset over Kerala: new definition and prediction," Journal of Earth System Science, vol. 118, no. 2, pp. 123–135, 2009.

11. S. S. Puranik, K. C. Sinha Ray, P. N. Sen, and P. P. Kumar, "An index for predicting the onset of monsoon over Kerala," Current Science, vol. 105, no. 7, pp. 954–961, 2013.
12. A. Kulkarni, R. H. Kripalani, and S. S. Sabade, "Examining Indian monsoon variability in coupled climate model simulations and projections," IITM Research Report RR 125, 2010.
13. M. Rajeevan and R. S. Nanjundiah, "Coupled Model Simulations of the Twentieth Century Climate of the Indian Monsoon," In Current Trends in Science. Platinum Jubilee Monograph of the Indian Academy of Sciences, Bangalore, pp. 537–557, 2009.
14. B. Bhaskaran and J. F. B. Mitchell, "Simulated changes in Southeast Asian monsoon precipitation resulting from anthropogenic emissions," International Journal of Climatology, vol. 18, no. 13, pp. 1455–1462, 1998.
15. K. Rupa Kumar, A. K. Sahai, K. Krishna Kumar et al., "High-resolution climate change scenarios for India for the 21st century," Current Science, vol. 90, no. 3, pp. 334–345, 2006.
16. R. G. Jones, M. Noguer, D. Hassell et al., "Generating high resolution climate change scenarios using PRECIS," Tech. Rep., Met Office Hadley Centre, Exeter, UK, 2004.
17. M. Noguer, R. Jones, D. Hessel et al., Workbook on Generating High Resolution Climate Change Scenarios Using PRECIS, Hadley Centre for Climate Precdiction and Research, Met Office, Bracknell, UK, 2002.
18. N. A. Nakicenovic and R. Swart, Special Report on Emission Scenarios. A special report of Working Group III of the International Panel on Climate Change, Cambridge University Press, Cambridge, U.K. Brochure on "The Hadley Centre regional climate modelling system", Update-2002, pp. 20, 2000.
19. M. Rajeevan, J. Bhate, J. D. Kale, and B. Lal, "Development of a high resolution daily gridded rainfall data for the Indian region," IMD Met Monograph Climatology 22/2005, National Climate Centre, IMD Pune, 2005.
20. B. Liebmann and C. A. Smith, "Description of a complete (interpolated) outgoing longwave radiation data set," Bulletin of the American Meteorological Society, vol. 77, pp. 1275–1277, 1996.
21. E. Kalnay, M. Kanamitsu, R. Kistler et al., "The NCEP/NCAR 40 year re-pnalysis Project," Bulletin of the American Meteorological Society, vol. 77, pp. 437–471, 1996.
22. C. S. Ramage and C. R. V. Raman, Meteorological Atlas of the International Indian Ocean Expedition. Volume 2, National Science Foundation, Washington, DC, USA, 1972.
23. T. N. Krishnamurti, P. Ardanuy, Y. Ramanathan, and R. Pasch, 'On the onset vortex of the summer monsoon," Monthly Weather Review, vol. 109, no. 2, pp. 344–363, 1981.
24. R. H. Kripalani, S. V. Singh, and P. A. Arkin, "Large-scale features of rainfall and outgoing long wave radiation over India and adjoining regions," Contributions to Atmospheric Physics, vol. 64, pp. 159–168, 1991.
25. M. K. Soman and K. K. Kumar, "Space-time evolution of meteorological features associated with the onset of Indian summer monsoon," Monthly Weather Review, vol. 121, no. 4, pp. 1177–1194, 1993.
26. P. V. S. Raju, U. C. Mohanty, and R. Bhatla, "Onset characteristics of the Southwest monsoon over India," International Journal of Climatology, vol. 25, no. 2, pp. 167–182, 2005.

CHAPTER 10

Heavy Rainfall Simulation over Sinai Peninsula Using the Weather Research and Forecasting Model

Gamal El Afandi,[1,2] Mostafa Morsy,[2] and Fathy El Hussieny[2]

[1]*Division of International Research and Development, College of Agriculture, Environment and Nutrition Sciences and College of Engineering, Tuskegee University, Tuskegee, AL 36088, USA*
[2]*Astronomy and Meteorology Department, Faculty of Science, Al Azhar University, Cairo 11884, Egypt*

ABSTRACT

Heavy rainfall is one of major severe weather over Sinai Peninsula and causes many flash floods over the region. The good forecasting of rainfall is very much necessary for providing early warning before the flash flood events to avoid or minimize disasters. In the present study using the Weather Research and Forecasting (WRF) Model, heavy rainfall events that occurred over Sinai Peninsula and caused flash flood have been investigated. The flash flood that occurred on January 18, 2010, over different parts of Sinai Peninsula has been predicted and analyzed using the Advanced Weather Research and Forecast (WRF-ARW) Model. The predicted rainfall in four dimensions (space and time) has been calibrated with the measurements recorded at rain gauge stations. The results show that the WRF model was able to capture the heavy rainfall events over different regions of Sinai. It is also observed that WRF model was able to predict rainfall in a significant consistency with real measurements. In this study, several synoptic characteristics of the depressions that developed during the course of study have been investigated. Also, several dynamic characteristics during the evolution of the depressions were studied: relative vorticity, thermal advection, and geopotential height.

1. INTRODUCTION

Heavy rainfall is one of the major severe weather in Egypt particularly in arid and semiarid regions especially if it is steep and mountainous regions such as Sinai Peninsula and the Eastern desert of Egypt. Short duration of heavy rainfall over a relatively small drainage area can lead to devastating flash flood, consequently causing a number of fatalities and tremendous damages.

It can destabilize soils along mountain slopes, resulting in landslides and mudslides that cause severe damage to nearby villages.

Heavy rainfall is usually resulting from individual mesoscale storms or mesoscale convective systems embedded in synoptic-scale disturbances [1]. High-resolution observations and numerical modeling technique are required to better predict heavy rainfall events, where the forecasting of heavy rainfall is very important for many decision makers who are sensitive to the occurrence of precipitation. An accurate quantitative precipitation forecast can identify the potential for heavy precipitation and possible associated flash flooding, as well as provide information for hydrological interests.

Since heavy rainfall and flash flood can lead to severe damage and losses for both life and infrastructure, the need to warn people in advance is, thus, an important goal, but heavy rainfall and flash floods forecasting is considered a difficult task, particularly in arid and semiarid regions, since they take place in very short time interval. Hence, rainfall forecasting is very much necessary for providing an early warning before the flash flood events to avoid or minimize disasters.

An early warning system (EWS) for flash floods has been developed for part of the Sinai Peninsula of Egypt, a hyperarid area confronted with limited availability of field data, limited understanding of the response of the valley (wadi) to rainfall, and a lack of correspondence between rainfall data and observed flash flood events [2].

Sinai Peninsula contains many complex terrains; so rainfall forecasting for complex terrains is challenging and requires numerical simulations at very high resolution.

In arid areas such as Sinai in Egypt, rainfall is mainly caused by squall line and convective cloud mechanisms and by low-intensity frontal rain, causing storm floods [3]. The complexity of terrain influences the weather in a variety of ways. Under stable atmospheric conditions, the terrain generates internal gravity waves that distribute momentum over wider areas. These processes may be related to strong winds and turbulence that influence the air traffic. Under unstable conditions, convective clouds and precipitation are generated over complex terrain which can grow into severe thunderstorms. All the mentioned processes occur on spatial length scales smaller than 100 km, usually even smaller than 10 km.

Because of the above-mentioned challenges, the mesoscale meteorological model named the Advanced Research Weather Research and Forecasting (WRF-ARW) is selected for rainfall forecasting in this research. WRF is developed mainly by the US National Centre for Atmospheric Research (NCAR) in collaboration with many other research centres and universities. WRF allows forecasting weather in complex terrains such as the one in Sinai and in the same time is considering orographic features. It is suitable for a broad spectrum of applications across scales ranging from meters to thousands of kilometres.

The main objectives of this study can be summarized as follows:(a)the understanding of the processes leading tothe flood event at Sinai using WRF?(b)the investigation of the capability of WRF to simulate successfully the rainfall amount during a flood event?

2. STUDY AREA

The Sinai Peninsula (Figure 1), study area, is located in the far northeast of Egypt. It represents about 6% of Egypt's area which is about 61,000 km^2.

The Sinai Peninsula comprises a wedge-shaped block of territory with its base along the Mediterranean Sea coast to the north and its apex bounded by the Gulfs of Suez to the west and Aqaba to the east.

Its southern portion consists of rugged, sharply serrated mountains. These reach elevations of more than 2,400 meters; among them is Mount Catherine, Egypt's highest mountain, which has an elevation of 2,642 meters. The central area of Sinai consists of two plateaus, Al-Tih and Al-Ajmah, both deeply indented and dipping northward toward Wadi (Valley) El Arish.

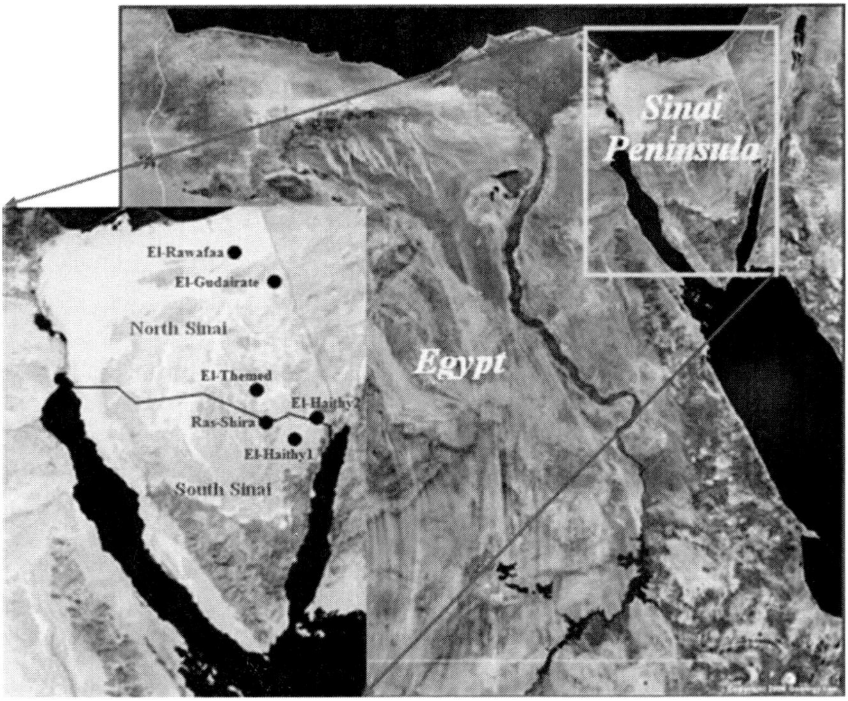

Figure 1: Study area and rain gauges location.

Toward the Mediterranean Sea, the northward plateau slope is broken by dome-shaped hills; between them and the coast are long, parallel lines of dunes, some of which are more than 100 meters high.

It is dissected by the largest wadi in Sinai, Wadi El Arish, which emerges from elevated gravelly plains and terraces in the south to a distance of about 20 km till the coast Mediterranean Sea to the north. The rainfall and floods are the only sources of renewable water resources in Sinai Peninsula.

2.1. Climate of Sinai Peninsula

It is characterized by the Mediterranean climate in the northernmost of Sinai to be close to the desert and semidesert climate to the south.

Much of the Sinai is hot, or very hot with a higher temperature inland, but there is a more temperate region near the north coast and over the mountains. During the period May/June to September/October, the mean daily maximum is 28°C to 37°C in the North, 31°C to 42°C near the south coast, and 35°C to 41°C inland. Minimum temperatures average between 20°C and 25°C in the summer.

The winter season is a little less harsh with day high's in the mid-teens and possible 20's, and evenings often falling to around 6°C–10°C, and may drop below 0°C.

The amount of rainfall in Sinai decreases from the northeast towards the southwest. The greatest amount of the annual rainfall was found at Rafah station (304 mm) in the northeast. The annual rainfall average is about 120 mm along the Mediterranean coast. It decreases in the uplands to the south to about 32 mm. The annual average of rainfall all over Sinai is about 40 millimeters, 27 millimeters from it is estimated to come from one storm that may provide 10 millimeters at a time.

Along the Mediterranean Coast, 60% of the rain occurs in the winter, while 40% falls during the transitional seasons.

The annual rainfall in the southern region is significantly less than the northern one, where it reaches 20 mm in the coastal areas over the Gulfs of Aqaba and Suez. Its amount increases to 70 millimeters over the mountain regions.

The rainfall in autumn and winter seasons is ranging from medium to heavy especially at some high terrain areas. The rainfall is completely nonsexist in summer season.

3. DATA AND METHODOLOGY

WRF model is used in this research with its nesting capability to simulate the heavy rainfall with fine grid spacing.

During the storm mesoscale, convective systems are highly interacted with synoptic-scale environment.

During this study we conducted multinested experiment for four domains with different horizontal resolutions of 81 km (DM1), 27 km (DM2), 9 km (DM3), and 3 km (DM4), respectively, as shown in Figure 2. The second domain (DM2) will be used to simulate the synoptic situation over Egypt on 18 January, 2010, while the last domain (DM4) will be used to investigate the ability of the WRF model to simulate and predict rainfall to compare it with rain gauges at Sinai.

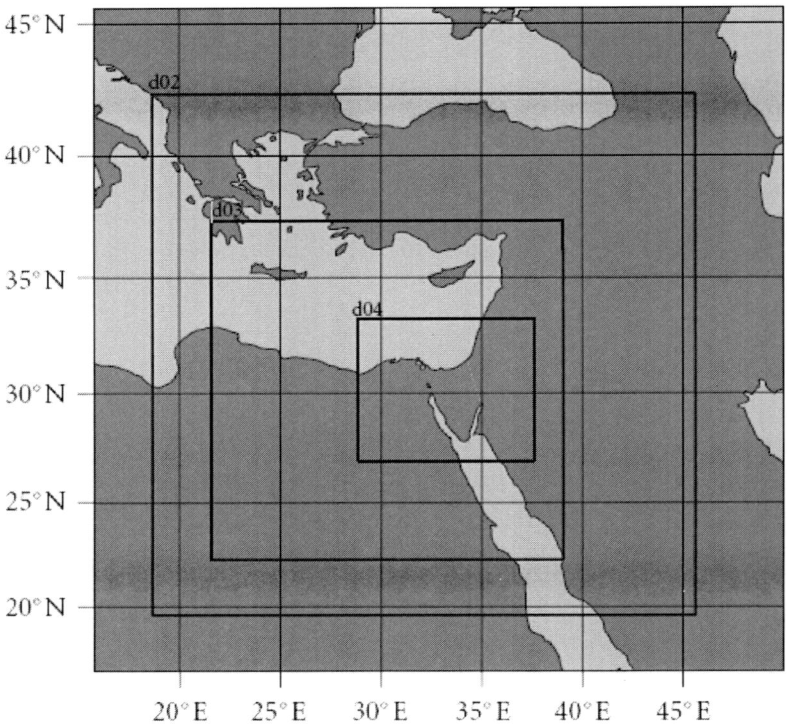

Figure 2: Model domains of 81 km, 27 km, 9 km, and 3 km horizontal resolution.

The model output is adopted to produce its output every hour to be compatible with the rain gauges observations and two-way nesting domain starting from 17 to 19 January, 2010. The 27 vertical layers with the model top of 50 hPa are used in this study.

The initial and lateral boundary meteorological data which used to run the model has been downloaded from the National Centers for Environmental Prediction (NCEP), global final analyses on degree from Global Forecasting System (GFS), and it is updated every six hours.

The hourly rainfall data, measured in millimeter (mm), was obtained from the rain gauges stations installed by the Water Resources Research Institute (WRRI); WRRI is the Egyptian governmental research institute with the mandate for flash flood management.

WRF offers multiple physics options that can be combined in any way. The options typically range from simple and efficient to sophisticated and more computationally costly and from newly developed schemes to well-tried schemes such as those in current operational models. Table 1 represents the selected physics of WRF model.

Table 1: Selected physics of the WRF model options.

Physics scheme	Type of the scheme that used in WRF run
Microphysics	Single-moment 3-class WSM3 scheme
Longwave radiation	Rapid Radiative Transfer Model (RRTM) scheme
Shortwave radiation	Dudhia scheme
Land surface	Noah Land Surface Model scheme
Surface and boundary layers	Yonsei University scheme
Cumulus parameterization	Kain-Fritsch scheme

4. RESULTS AND DISCUSSIONS

Egypt is subjected to thunderstorms and heavy rains started at the north coast, the Red Sea, and the Sinai Peninsula during 18 January, 2010. This extreme weather and intensive rainfall led to flash flood events over Sinai Peninsula.

The development of intensive weather events that invade Egypt during January 18, 2010, were characterized by "exceptional and extremely heavy rainfall," which affected a wide part of Egypt, including Sinai Peninsula and fatal to some Bedouin tribes located in its path. This heavy rainfall is accompanied by strong winds and thunderstorms.

The Mediterranean region is considered to be one of the most cyclogenetic areas in the world, usually favoring the development of weak low-pressure systems. Occasionally, these systems develop into deep cyclogenesis that cause a series of severe weather events as they cross the Mediterranean.

Generally, the winter season in Middle East is known to be associated with the development of low-pressure systems (Cyprus lows) over the eastern Mediterranean Sea [4]. Such systems have been observed at upper levels, particularly at 500 hPa. Rainfall has been observed to be heavy over Sinai and extended to southern parts of Egypt during our case study in winter season, January 18, 2010.

Kahana et al. [5] found that Mediterranean Sea cyclones deepen in general, in association with upper level troughs. Upper level troughs are regarded as the key factors in activity of midlatitudes, such as the Mediterranean Sea system, particularly in their front sides, where positive vorticity advection and enhanced convection take place [6]. Ferraris et al. [7], however, found that the polar continental air from central Asia moving toward the eastern Mediterranean, and interacting with the relatively warm sea surface temperature (SST), produces enhanced lower level instability.

Because of the importance of the Mediterranean cyclogenesis, many studies on weather systems in the region have been conducted. The synoptic situation started by the development of the subtropical jet stream in the upper troposphere. The maximum wind speed ranged between 120 and 135 knots as shown in Figure 3, with the southern extension of the polar jet to the south of the Mediterranean.

Figure 3 shows the wind speed of the subtropical branch of the jet stream at 200 hPa for 4 consecutive times per day for our case study, with the southern extension of the polar jet to the east of the Mediterranean Sea.

The synoptic situation can also be analyzed by referring to the midtroposphere. Figure 4 shows the geopotential height at 500 hPa level for the same day; the synoptic system at 500 hPa level also follows the subtropical jet development.

The axis of the trough which is located over east Europe shifts southeast. This trough line orientation indicates that the affected area has extended and engulfed the whole eastern Mediterranean area. This development agrees with Krichak et al. [8, 9] and Krichak and Alpert [10] who showed that the meridional orientation of the upper trough line represents the major synoptic process of the development of the East Mediterranean cyclone. As the subtropical jet stream shifts south and splits, the system with high geopotential heights moves eastward and allows the development of a low system in the eastern Mediterranean.

The relative vorticity advection is considered as one of the most important parameter that leads to occurrence of heavy rainfall and flash flood events in Sinai Peninsula. As shown in Figure 5, the positive relative vorticity advections leads to move the upper atmospheric trough from west to east (steering of trough) and reached to Sinai Peninsula.

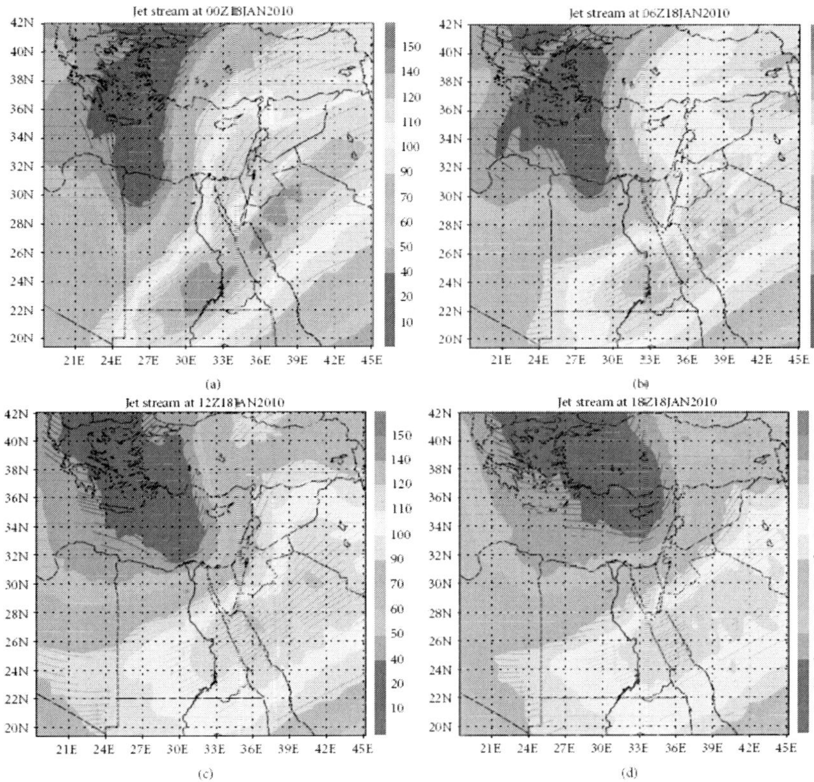

Figure 3: 200 hPa subtropical Jet stream (knots) at over northeastern Africa with the southern extension of the polar jet to the east of the Mediterranean at January 18, 2010.

From Figure 6, we can also follow the temperature advection that calculated using centered finite difference scheme of second order. In the stage of rapid development, there is a cooperative interaction between the upper level and surface flows; strong cold advection is seen to occur west of the trough at the surface, with warm advection to the east. This pattern of thermal advection is a direct consequence of the fact that the trough at 500 hPa lags (lies to the west of) the surface trough. These are consistent with the results of Charney [11], Holton [6, 12], and Bluestein [13], where the growth of midlatitude cyclone is associated with a meridional gradient of temperature. We can expect that the cold advection on the west of the trough is responsible for the decreasing of temperature over Egypt in that period.

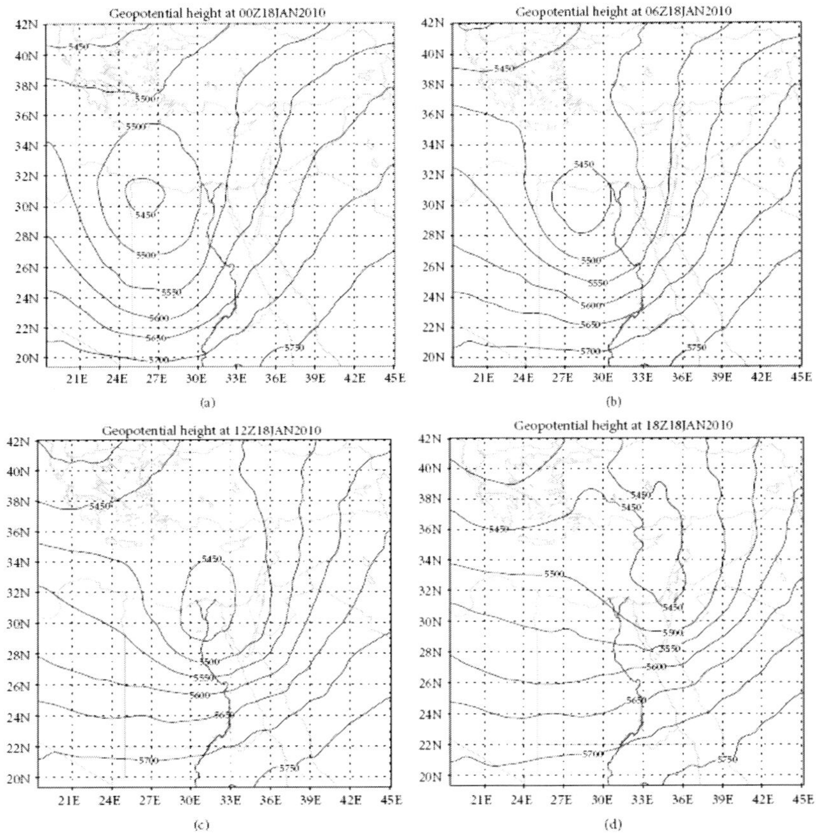

Figure 4: Geopotential height (m), at 500 hPa level every six hours at January 18, 2010.

The mean sea level pressure (MSLP) of the case study is shown in Figure 7 and follows the system development at the 850 hPa level as shown by the heat advection (Figure 6). The intensification of Azores high, to the west of the region, leads the low system to be developed in the eastern of the Mediterranean region. In addition, this produces a larger air mass temperature difference between the east and west of the low trough line and allows the Red Sea trough to extend northward. The Red Sea trough extension and the meridional orientation of the upper trough permit favorable conditions for the formation and development of the eastern Mediterranean cyclone.

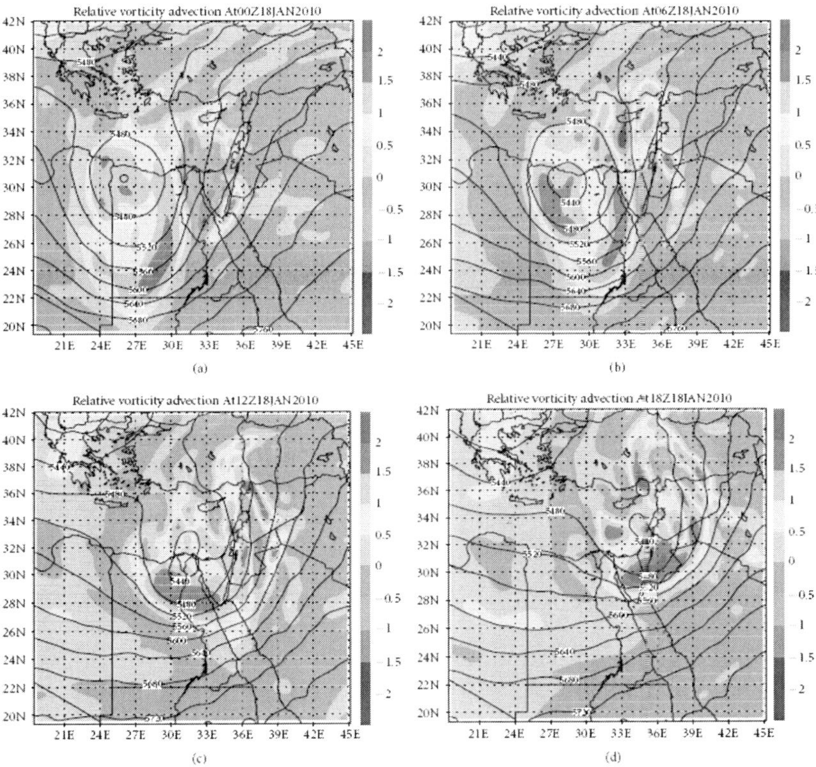

Figure 5: Relative vorticity advections ($10^{-5}\,S^{-1}$) with geopotential height contour at 500 hPa level every six hours at January 18, 2010.

Figure 8 shows the total precipitation associated with the east Mediterranean storm. The rainfall amount increases as the low-pressure system deepens and the precipitation area extends eastward following the surface low.

Evaluation of the performance of WRF model for prediction of heavy rainfall events over Sinai Peninsula will be done by comparison of the rainfall predicted by the WRF model and the rain gauges measurements; the complete information of the rain gauges is shown in Table 2.

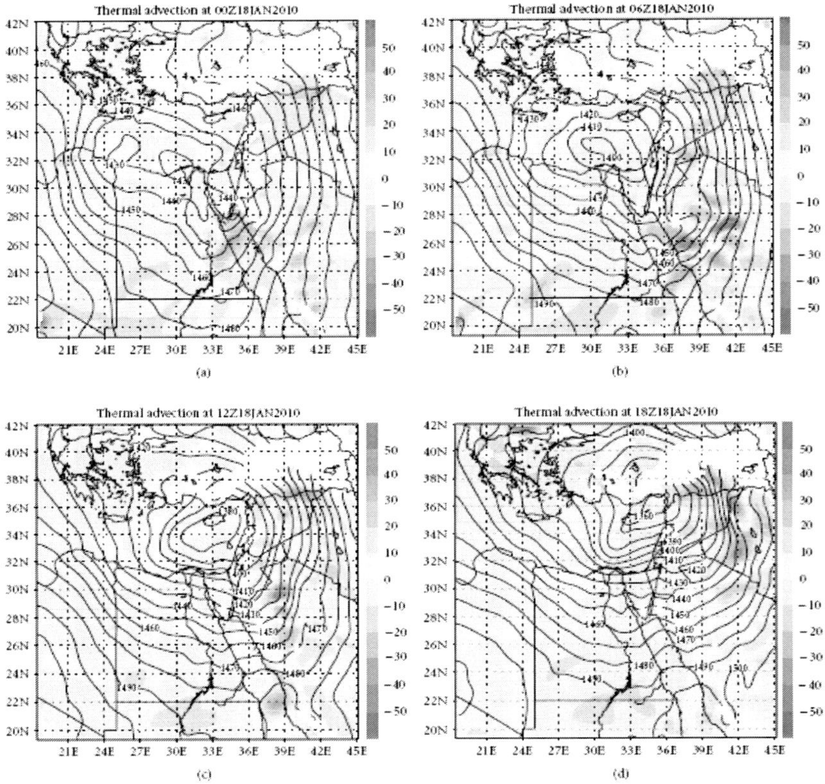

Figure 6: Thermal advection (°C) with geopotential height contour at 850 hPa level every six hours at January 18, 2010.

Table 2: Rain gauges name and location which used in the comparison.

No.	Station name	Lon.	Lat.
1	El-Rawafaa	34° 08′ 51″	30° 49′ 37″
2	El-Gudairate	34° 24′ 35″	30° 38′ 28″
3	El-Themed	34° 18′ 19″	29° 40′ 37″
4	El-Haithy1	34° 36′ 36″	29° 29′ 11″
5	El-Haithy2	34° 42′ 24″	29° 28′ 04″
6	Ras-Shira	34° 27′ 57″	29° 31′ 02″

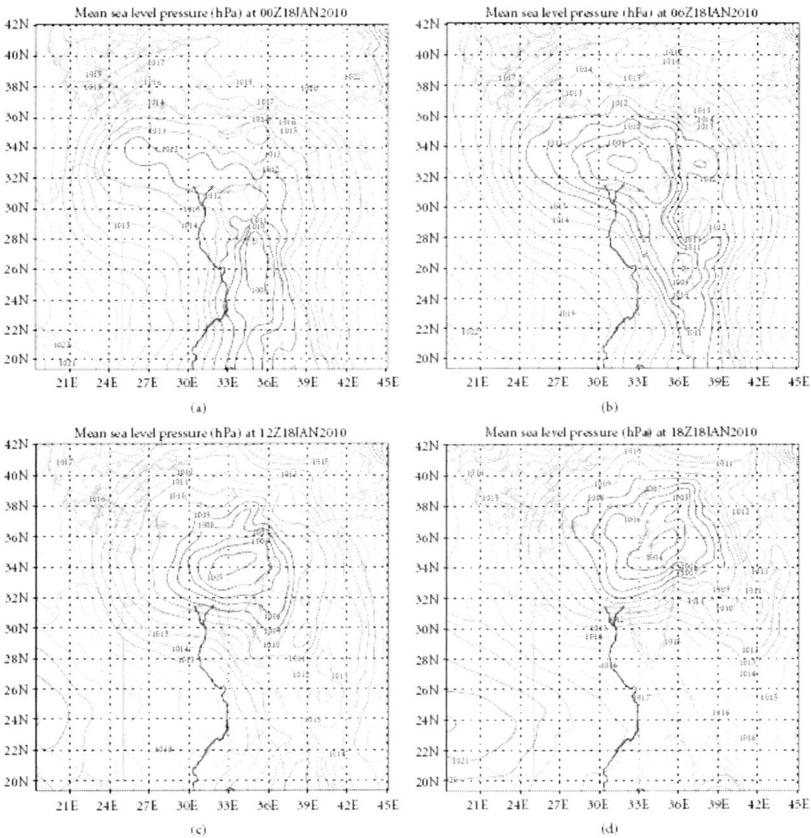

Figure 7: Mean sea level pressure (hPa) every six hours at January 18, 2010.

One of the very important statistical tools is the root mean square error (RMSE) which gives a good overall measure of model performance. The weighting of (prediction-observation) by its square tends to inflate RMSE, particularly when extreme values are present. With respect to a perfect model, the root mean square error should approach zero.

On the other hand, the mean bias (MB) is considered also to evaluate the model performance which represents the degree of correspondence between the mean prediction and the mean observation. Lower numbers are best and values less than 0 indicate underprediction.

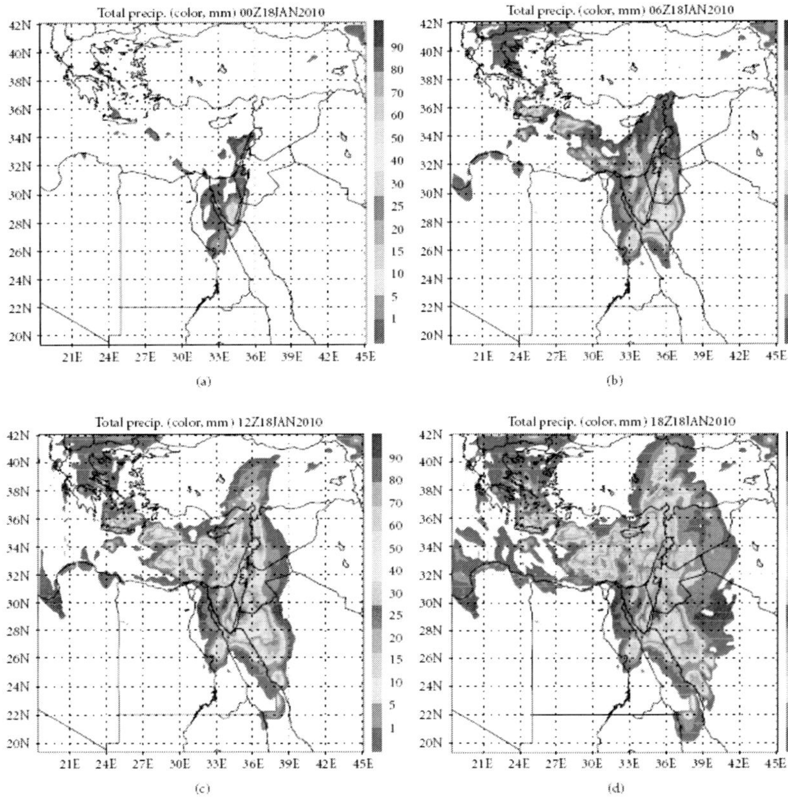

Figure 8: Accumulated rainfall (mm) during January 18, 2010 plotted every six hours with the maximum volume of rainfall over Sinai Peninsula.

The equations for RMSE, MB and its percentages are given as follows:

$$\text{RMSE} = \left(\sum_{i=1}^{n} \frac{(X_p - X_o)^2}{n} \right)^{0.5}, \qquad \text{MB} = \sum_{i=1}^{n} X_p - \frac{X_o}{n},$$

$$\text{RMSE}\ \% = \left(\frac{\text{RMSE}}{\overline{X_o}} \right) * 100, \qquad \text{MB}\ \% = \left(\frac{\text{MB}}{\overline{X_o}} \right) * 100, \tag{1}$$

where is number of observations, and are the predicted and observed values, respectively, and is the average of observed values.

These statistical measures are used extensively for evaluation of model forecasts of precipitation (e.g., [14–16]). The quantitative validation of the simulated precipitation is done by calculating their differences with the corresponding rain gauges.

Figure 9 shows the comparisons between the predicted hourly rainfall and the measured ones in different regions over Sinai Peninsula during January 18, 2010.

From Figure 9, it can be noticed that the predicted rainfall by WRF model is in a good consistency and harmony with observed rainfall for all rain gauge stations.

In El-Gudairate, El-Rawafaa, El-Haithy2, and El-Themed Stations rainfall prediction by WRF is very close to the measured ones. But in the other stations such as El-Haithy1 and Ras-Shira, it gave small differences with the measured rainfall during January 18, 2010.

Table 3 shows the percentages of root mean square error and mean bias for predicted rainfall and the measured ones at different stations.

Table 3: The percentages of root mean square error and mean bias for rainfall from WRF and the measurements at different stations.

Station	RMSE %	MB %
El-Gudairate	4.5	−0.3
El-Rawafaa	8.6	0.7
El-Themed	11.8	1.2
El-Haithy1	30.7	20.4
El-Haithy2	4.7	1.3
Ras-Shira	15.8	−4.2

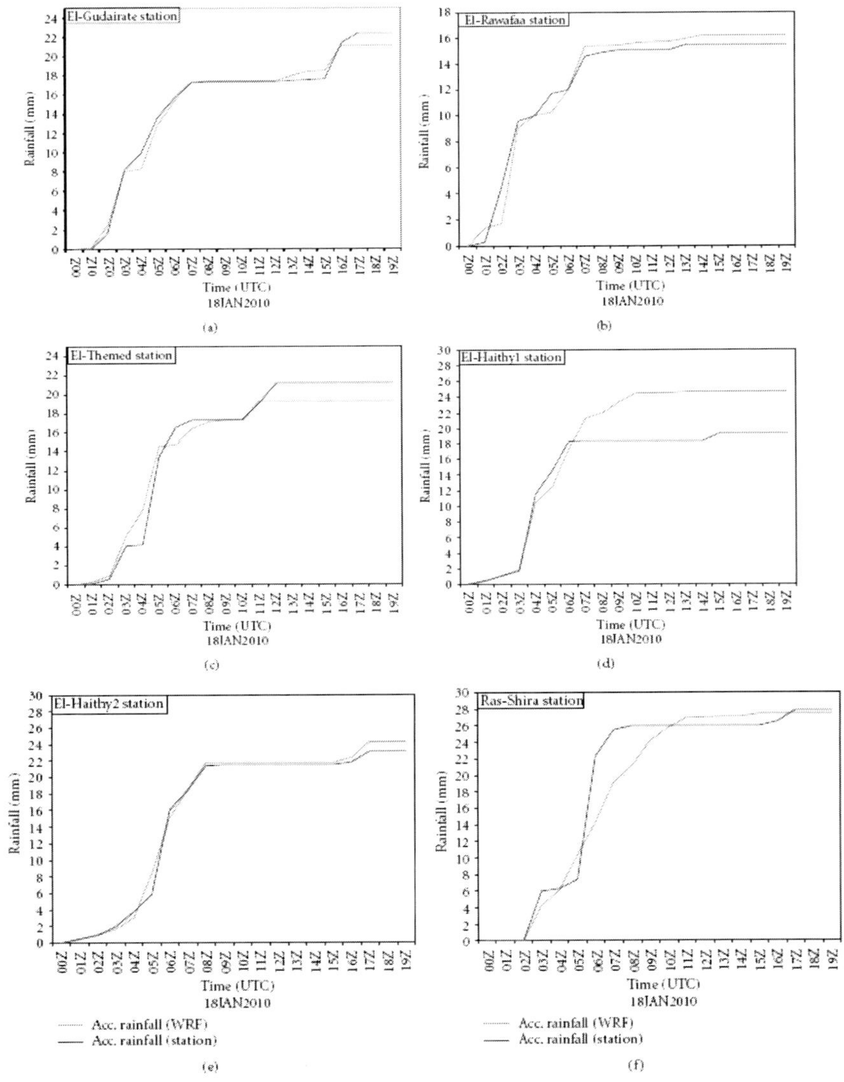

Figure 9: Comparison between accumulated observed and predicted precipitations (mm) for El-Gudairate, El-Rawafaa, El-Themed, El-Haithy1, El-Haithy2, and Ras-Shira stations through January 18, 2010.

It can be deduced from this table that the maximum values of root mean square error and mean bias are for El-Haithy1 station, while the minimum values of root mean square error and mean bias are for El-Gudairate, El-Rawafaa, El-Haithy2, El-Themed, and Ras-Shira Stations.

This means the WRF model has high-performance rainfall prediction for all stations except El-Haithy1 station.

The WRF model was able to simulate the synoptic situation which invaded Egypt and accompanied by heavy rains started at the north coast, the Red Sea, and the Sinai Peninsula during 18 January, 2010.

In the same time, the model simulation of the synoptic situation was in a good compatibility with the previous studies.

It may conclude that the performance of WRF model approximately gave a good and reasonable rainfall prediction compared to measurements at all stations over Sinai Peninsula.

5. CONCLUSIONS

The WRF model was able to simulate the synoptic situation which invaded Egypt and accompanied by heavy rains started at the north coast, the Red Sea, and the Sinai Peninsula during January 18, 2010.

In the same time, the model simulation of the synoptic situation was in a good compatibility and agreement with the previous studies over the Mediterranean region.

It may be concluded that the performance of WRF model approximately gave a good and reasonable rainfall prediction compared to measurements at all stations over Sinai Peninsula.

REFERENCES

1. D.-K. Lee, H.-R. Kim, and S.-Y. Hong, "Heavy rainfall over Korea during 1980–1990," Korean Journal of the Atmospheric Sciences, vol. 1, no. 1, pp. 32–50, 1998.
2. J. Cools, P. Vanderkimpen, G. El Afandi et al., "An early warning system for flash floods in hyper-arid Egypt," Natural Hazards and Earth System Sciences, vol. 12, no. 2, pp. 443–457, 2012.
3. G. El Afandi, "Developing flash-flood guidance in Egypt's Sinai Peninsula with the Weather Research forecast (WRF) model," International Journal of Meteorology, vol. 35, no. 347, pp. 75–82, 2010.
4. C. A. Babu, A. A. Samah, and H. Varikoden, "Rainfall climatology over Middle East region and its variability," International Journal of Water Resources & Arid Environments, vol. 1, no. 3, pp. 180–192, 2011.

5. R. Kahana, B. Ziv, U. Dayan, and Y. Enzel, "Atmospheric predictors for major floods in the Negev Desert, Israel," International Journal of Climatology, vol. 24, no. 9, pp. 1137–1147, 2004.
6. J. Holton, An Introduction to Dynamic Meteorology, Academic Press, New York, NY, USA, 2nd edition, 1992.
7. L. Ferraris, R. Rudari, and F. Siccardi, "The uncertainty in the prediction of flash floods in the northern Mediterranean environment," Journal of Hydrometeorology, vol. 3, pp. 714–727, 2002.
8. S. O. Krichak, P. Alpert, and T. N. Krishnamurti, "Interaction of topography and tropospheric flow—a possible generator for the Red Sea trough?" Meteorology and Atmospheric Physics, vol. 63, no. 3-4, pp. 149–158, 1997.
9. S. O. Krichak, P. Alpert, and T. N. Krishnamurti, "Red Sea Trough/cyclone development–numerical investigation," Meteorology and Atmospheric Physics, vol. 63, no. 3-4, pp. 159–169, 1997.
10. S. O. Krichak and P. Alpert, "Role of large scale moist dynamics in November 1-5, 1994, hazardous Mediterranean weather," Journal of Geophysical Research D, vol. 103, no. 16, pp. 19453–19468, 1998.
11. J.G. Charney, "The dynamics of long waves in a baroclinic westerly current," Journal of Atmospheric Sciences, vol. 4, no. 5, pp. 136–162, 1947.
12. J. Holton, An Introduction to Dynamic Meteorology, Academic Press, New York, NY, USA, 3rd edition, 1992.
13. H. B. Bluestein, Synoptic-Dynamic Meteorology in Midlatitudes, Vol. II: Observations and Theory of Weather Systems, Oxford University Press, Oxford, UK, 1992.
14. F. Mesinger, T. L. Black, D. W. Plummer, and J. H. Ward, "Eta model precipitation forecasts for a period including tropical storm Allison, Wea," Forecasting, vol. 5, pp. 483–493, 1990.
15. F. Mesinger, "Improvements in quantitative precipitation forecasts with the eta regional model at the national centers for environmental prediction: the 48-km upgrade," Bulletin of the American Meteorological Society, vol. 77, no. 11, pp. 2637–2649, 1996.
16. K. Lagouvardos, V. Kotroni, A. Koussis, C. Feidas, A. Buzzi, and P. Malguzzi, "The meteorological model BOLAM at the National Observatory of Athens: assessment of two-year operational use," Journal of Applied Meteorology, vol. 42, pp. 1667–1678, 2003.

CHAPTER 11

Concentrations and Size Distributions of Bacteria-Containing Particles over Oceans from China to the Arctic Ocean

Ming Li [1,2], Xiawei Yu [1], Hui Kang [1], Zhouqing Xie [1,*] and Pengfei Zhang [3]

[1]*Anhui Province Key Laboratory of Polar Environment and Global Change, School of Earth and Space Sciences, University of Science and Technology of China, Hefei 230026, China*

[2]*State Key Laboratory of Loess and Quaternary Geology, Institute of Earth Environment, Chinese Academy of Sciences, Xi'an 710075, China*

[3]*Department of Earth and Atmospheric Sciences, City College of New York, New York, NY 10031, USA*

ABSTRACT

During the third China Arctic Research Expedition (July–September 2008), size-resolved measurements of bacteria-containing particles (BCPs) in the marine boundary layer (MBL) air were conducted during a cruise through the East China Sea, the Yellow Sea, the Japan Sea, the Okhotsk Sea, the Bering Sea, the Chukchi Sea, and the Arctic Ocean. The concentrations of total airborne BCPs (TBCPs), non-salt tolerant airborne BCPs (NSBCPs), and salt tolerant airborne BCPs (SBCPs) varied from 29 to 955 CFU m^{-3} (CFU = Colony Forming Unit), 16 to 919 CFU m^{-3}, and 4 to

276 CFU m^{-3}, with an average value of 275, 182, and 92 CFU m^{-3}, respectively. Although the SBCP concentrations were less than the NSBCP concentrations when averaged over all measurements, there are several cases where the reverse is true (e.g., in the high Arctic Ocean). During the cruise, the TBCP sizes were dominated by the diameter >4.7 μm fraction (accounted for 46.3% on average), while the fine fraction (diameter <2.1 μm) accounted for 27.8%. For NSBCPs and SBCPs, the coarse fraction also was the dominant fraction over most regions. The influence of local meteorological conditions on the abundance, size distributions, and species of airborne bacteria is discussed. Notably, in the atmosphere over the Arctic Ocean the abundance of airborne bacteria was apparently related to the distribution of sea ice. As cultivation based methodologies may underestimate the environmental bacterial communities, it is expected that the abundance of bacteria in the ambient air would be higher than that observed in this study. In order to distinguish different species of bacteria, molecular biological techniques (e.g., 16S rDNA analysis) are required for identification in future investigations.

Keywords: airborne bacteria; abundance; size distribution; marine boundary layer; the Arctic Ocean; sea ice

1. INTRODUCTION

Airborne bacteria are ubiquitous and play an important role in public health [1,2]. Moreover, airborne bacteria may potentially influence the formation of clouds and precipitation by serving as biogenic nuclei of water droplets and ice crystals [2,3,4,5]. At present, however, the concentrations and size distributions of airborne bacteria in different regions of the Earth are not well characterized [2]. Most studies on the concentrations of airborne bacteria have been carried out in continental areas, with a handful in coastal regions [6,7,8,9,10,11]. There are only a few studies measuring bacterial concentrations in marine air, and even fewer over polar waters [12,13].

In recent decades, several studies [8,14] have been carried out to characterize the abundance of airborne bacteria at continental sites of various landscape types (mostly in short measurement campaigns of up to a few weeks in duration). For example, mean spring concentrations in aerosols were found to be 1.1 × 10^{4} CFU m^{-3} (CFU = Colony Forming Unit) [7] and 1.6 × 10^{4} CFU m^{-3} [15] at rural sites in Austria and England, respectively. Lighthart and Shaffer [16] reported a mean cultivable bacteria concentration of 609 CFU m^{-3} at a forest site. In some urban sites, the concentrations of airborne bacteria ranged from 0 to 7220 CFU m^{-3} [17,18].

Background concentrations of airborne bacteria in clean continental air observed at Mt. Rex, Austria were about 1.1×10^4CFU m^{-3} [7]. However, the concentrations of atmospheric bacteria of some coastal regions were lower than the concentrations over land atmosphere. For example, concentrations of atmospheric bacteria collected from marine air arriving at a coastal, cliff-top site in Barbados were about 2–3 orders of magnitude lower than those over land [19]. Concentrations of total bioaerosols ranged from 8.5×10^4 to 1.7×10^5 cells m^{-3} in the Qingdao coastal region [10] and from 580 to 778 CFU m^{-3}in an embayment adjacent to the East River in western Long Island Sound [11]. Bacterial concentrations in the marine atmosphere are even lower, e.g., a few to ~100 CFU m^{-3} over the open Baltic Sea [20].

As to the sources of bacteria in marine air, it has been suggested that bacteria and other primary biological particles can be released into the atmosphere from the sea-surface micro-layer (SML) by the bursting of bubbles in breaking waves, a process that occurs both during wave-breaking at the coasts, and in whitecaps on wind-driven waves in the open ocean [3,21]. Bacterial concentrations in sea spray aerosols can be significantly enriched relative to concentrations in the SML, which in turn are enriched relative to concentrations in the bulk water. Such enrichment has been observed both in laboratory-generated aerosols [3] and in marine aerosols relative to underlying waters [22]. Some studies have shown indirect evidence that some bacteria collected in marine air may originate from a marine source, for instance by demonstrating genetic similarity between some strains of marine bacteria and airborne bacteria [23,24,25]. In addition, atmospheric turbulence can lead to the transmission of biological aerosols between different regions [26,27], so airborne bacteria over oceans may be derived from other sources and carried to the environment through long-range transport processes. Burrows et al. [2,28] simulated large-scale atmospheric transport of 1μm particles emitted from different ecosystems including coastal, deserts, forests, urban, seas, grasslands, and so on. The results demonstrated that bacteria from other ecosystems could also contribute significantly to the concentrations of bacteria over oceans.

The measurement of airborne bacterial concentrations is in general conducted in two steps: aerosol collection in the field and analysis in the laboratory. Filtration and impaction are the two most widely used collection methods, which are followed by cultivation techniques or epi-fluorescence microscopy to determine the concentrations of bacteria-containing particles [2].

During the third China Arctic Research Expedition (July–September 2008), bacteria-containing particles (BCPs) were investigated using a six-stage air sampler with cultivation. The cruise path covered many regions including the Eastern China Sea, the Yellow Sea, the Japan Sea, the Okhotsk Sea, the Bering Sea, and the Arctic Ocean (38°36′ N to 85°25′ N). The project provides an opportunity to understand the sources and changes in concentrations of airborne BCPs in the marine boundary layer from low to high latitudes. Here we report the abundance and size distributions of BCPs measured during the expedition. The potential impacts of meteorological variables on the spatial distribution of airborne bacteria are also discussed.

2. EXPERIMENTAL METHODS

2.1. Sampling Data

Samples were collected in the marine boundary layer along a cruise path from Shanghai, China to the Arctic Ocean during the third China Arctic Research Expedition (11 July–24 September 2008). The sampling details are given in Table 1. The procedure for sample collection and bacteria cultivation was based on the method described by a previous report [10]. Briefly, the bioaerosol samples were collected using a six-stage cultivable microorganism FA-1 multi-orifice cascade impactor (Applied Technical Institute of Liaoyang, Liaoyang, China) with a flow rate of 28.3 L min^{-1}. No cyclone inlet was installed, thus total suspended particles were collected and separated according to their individual sizes on the different stages of the sampler. The particle sizes were fractionated into six size ranges: >7.0 μm (stage 1), 4.7–7.0 μm (stage 2), 3.3–4.7 μm (stage 3), 2.1–3.3 μm (stage 4), 1.1–2.1 μm (stage 5), 0.65–1.1 μm (stage 6). Two samplers with a 50-cm Teflon inlet each were installed on the front deck of the ship 0.5 m away from each other and about 30 m above sea level to avoid ship contamination. The cultivable microorganism samples were collected on 9.0-cmPetri dishes containing an agar medium. The collection duration for each sample ranged from 10 min to 20 min (Table 1). Two groups of samples were collected at each of the 20 sites. As the sampler has six stages, the total number of samples in each group was 120. To control the collection process in both sampling devices blanks were taken routinely at every site. Agar plates were mounted into the sampler and removed directly without turning the pump on. Blanks were made and treated as the real samples. Since the FA-1 cascade impactor collects the airborne particles, the individual colonies formed on the culture medium may be formed by the presence of multiple bacteria contained in the particle, so in this study we use the number of bacteria-containing particles instead of cultivable bacteria.

Table 1. Sampling location, date, and meteorological conditions during the third China Arctic Research Expedition, 2008.

Site	Date	UTC Time	Sampling Duration (min)	Coordinate	Location	Wind Speed (m/s)	Wind Direction (°)	RH (%)	T (°C)
1	14 July	1:40	10	133° E 38° N	East China Sea	13	36	89	22.7
2	17 July	23:30	10	139° E 45° N	Sea of Japan	11.3	66	99	17.5
3	22 July	22:30	15	161° E 51° N	North Pacific Ocean	4	195	96	13.2
4	25 July	22:30	15	179° E 59° N	North Pacific Ocean	14	282	90	7
5	31 July	1:00	15	169° W 62° N	Bering Sea	15.1	131	86	5.5
6	2 August	23:00	15	165° W 64° N	Bering Sea	5.6	237	100	1.2
7	5 August	21:30	18	168° W 72° N	Chukchi Sea	7.8	128	99	0
8	8 August	21:30	18	157° W 73° N	Chukchi Sea	3.3	245	94	3.5
9	11 August	21:30	20	152° W 75° N	Arctic Ocean	10.1	7	89	0.2
10	14 August	21:30	20	147° W 76° N	Beaufort Sea	18.4	298	100	0.8
11	17 August	21:30	20	147° W 81° N	Arctic Ocean	9	329	92	0.2
12	21 August	22:00	20	144° W 84° N	Arctic Ocean	1	176	99	0.6
13	26 August	22:00	20	147° W 85° N	Arctic Ocean	14.4	28	98	−0.5
14	27 August	22:40	20	102° W 88° N	Arctic Ocean	11.6	125	100	−1.7
15	31 August	21:40	20	148° W 84° N	Arctic Ocean	0.9	203	100	−3.6
16	2 September	21:40	20	170° W 80° N	Arctic Ocean	2	47	100	−4.2
17	6 September	21:50	20	169° W 73° N	Chukchi Sea	15.5	131	97	1.7
18	9 September	22:10	20	165° W 64° N	Bering Sea	6.3	47	63	13
19	12 September	22:00	20	170° E 59° N	North Pacific Ocean	9.9	216	91	12.5
20	15 September	23:10	20	151° E 44° N	Sea of Okhotsk	4	222	92	12.8

UTC: Coordinated Universal Time; RH: Relative humidity.

2.2. Bacterial Cultivation

As soon as the sample was collected on the culture medium, it was immediately carried to the laboratory on the ship and cultivated. As in the previous research [10], two different kinds of agar were used for the cultivation of bacteria: beef extract-peptone medium (3 g beef extract, 10 g peptone, 5 g NaCl, 16 g agar, 1 L distilled H_2O, pH 7.2–7.6) and 2216E medium (5 g peptone, 1 g yeast extract, 0.1 g FePO4, 15 g agar, 400 mL distilled H_2O, and 600 mL aging seawater, pH 7.2–7.6). Both agars had 50 mg/L cycloheximide to suppress fungal growth. In addition, 2216E medium yielded 3.2% NaCl to reflect the salty environment in oceans [29,30,31]. The difference between these two media is that the bacteria cultured on the 2216E medium are more resistant to salt, as the 2216E medium is usually used to culture marine bacteria [10,32]. The 2216E medium was used here because the marine atmosphere may contain more salt-tolerant bacteria. Here, the bacteria incubated on the beef extract-peptone medium are named non-salt tolerant airborne BCPs (NSBCPs), and the bacteria cultivate on the 2216E medium are named salt tolerant airborne BCPs (SBCPs). The incubation time for both media was 48 h and was chosen in a way that the bacterial colonies could grow large enough to be visible. The incubation temperature, however, was different between the two media. The NSBCP samples were incubated at 37 °C, as suggested by the literature, whereas the SBCP samples were incubated at RT (Room Temperature, approximately 28 °C) [10]. As the room had air conditioning, the temperature did not vary much during the cruise.

2.3. Ancillary Data

The meteorological data were recorded simultaneously, including information on the temperature, relative humidity, atmospheric pressure, wind direction, and wind speed. Ice data were obtained from the NASA (National Aeronautics and Space Administration) Earth Observations website (https://neo.sci.gsfc.nasa.gov/ view.php? datasetId =NISE_D).

3. RESULTS AND DISCUSSION

3.1. Concentrations of Bacteria-Containing Particles

The atmospheric concentrations of bacteria-containing particles (BCPs) can be calculated from the colony counts and the total volume of air sampled. The concentrations of BCPs for each sampling site are shown in Figure 1. All controls were negative.

During the entire sampling period, the total BCP (TBCP) concentrations (including all six stages, non-salt tolerance and salt tolerance) ranged from 29 CFU m^{-3} (site 7) to 955 CFU m^{-3} (site 1), with an average concentration of 275 CFU m^{-3} and coefficient of variation of 76.7%. Most of the high TBCP concentrations were found near coast areas (e.g., site 1, 2, 4, 18). The concentrations of NSBCPs of all six stages ranged from 16 CFU m^{-3} (site 7) to 919 CFU m^{-3} (site 1), with a mean of 182 CFU m^{-3} and coefficient of variation of 112.0%. Many of the high concentrations were found in the coast regions, e.g., site 1 (919 CFU m^{-3}, 133° E 38° N), site 4 (445 CFU m^{-3}, 179° E 59° N) and site 18 (346 CFU m^{-3}, 165° W 64° N), but some high concentrations were also observed in the Arctic Ocean (site 9, 10, and 15). The large number of NSBCPs in the marine boundary layer may be influenced by the continental bacteria. As latitude increases, the concentrations of NSBCPs decreased ($R = -0.473, p < 0.05$). The concentrations of SBCPs varied considerably, ranging from 4 CFU m^{-3} (site 5, 165° W 64° N) to 276 CFU m^{-3} (site 15, 148° W 84° N) with an average value of 92 CFU m^{-3} and coefficient of variation of 80.3%. Although the SBCP concentration on average was lower than that of NSBCPs (92 vs. 182 CFU m^{-3}), the former could be occasionally higher than the latter, e.g., in the high Arctic Ocean. This was consistent with earlier observations [33,34] and model predictions [2] that bacteria are abundant in marine air in this region, where continental outflow contributes significantly to the atmospheric aerosols.

Figure 1. Bacterial concentrations along the cruise path for the R/V Xuelong during the third China Arctic Research Expedition. The concentrations of airborne non-salt tolerant bacteria-containing particles (NSBCPs) and airborne salt tolerant bacteria-containing particles (SBCPs) are shown by red and blue columns, respectively (unit: CFU m^{-3}). The blue shaded area represents sea ice.

Sampling locations were separated into three geographic groups for further analysis: Group I: the East China Sea, the Japan Sea, and the Northwest Pacific (sites 1–4, 19, 20); Group II: the Bering Sea and the Chukchi Sea (sites 5–8, 17, 18); Group III: the western Arctic Ocean (sites 9–16). Geographically, the concentrations of TBCPs ranged from 89 to 955 CFU m^{-3} with an average of 401 ± 317 CFU m^{-3}, 29–412 with an average of 177 ± 132 CFU m^{-3}, and 104–478 CFU m^{-3} with an average of 253 ± 117 CFU m^{-3} in Group I, Group II, and Group III, respectively. The average concentrations of NSBCPs varied from 298 CFU m^{-3} in Group I (coefficient of variation 112.6%), 133 CFU m^{-3} in Group III (coefficient of variation 86.1%), to 133 CFU m^{-3} in Group II (coefficient of variation 63.3%). The spatial distribution of SBCPs appeared to differ from that of NSBCPs: observed concentrations of SBCPs were 103 CFU m^{-3} in Group I (coefficient of variation 66.4%), 44 CFU m^{-3} in Group II (coefficient of variation 95.3%), and 120 CFU m^{-3} in Group III (coefficient of variation 71.1%). In the Group I area, the ratios of NSBCPs and SBCPs to TBCPs were 61% and 39%, respectively. In Group II and Group III areas, NSBCPs and SBCPs accounted for 72% and 28%, and 51% and 49% of TBCPs, respectively. The results showed that there are more salt tolerant bacteria in the Arctic Ocean.

3.2. Size Distributions

The size-resolved bacteria fraction at different locations is shown in Figure 2. Figure 2a presents the percentages of each stage's BCP to TBCPs of each sampling site. The average concentrations of TBCPs over different regions by size were 108 CFU m^{-3} (stage 1), 49 CFU m^{-3} (stage 2), 36 CFU m^{-3} (stage 3), 52 CFU m^{-3} (stage 4), 43 CFU m^{-3} (stage 5), and 52 CFU m^{-3}(stage 6), respectively. The coarse fraction (>4.7 μm, the sum of stage 1 and stage 2) was the dominant size fraction for most sampling locations. In contrast, over the Japan Sea and the Bering Sea, the dominant sizes were 2.1–4.7 μm and 0.65–3.3 μm (62% and 76%, respectively). In the Arctic Sea, the BCP concentrations of each stage did not show big differences.

Figure 2b shows the percentages of each stage's NSBCPs to the total NSBCPs at each site, and Figure 2c presents the percentages of each stage's SBCPs to the total SBCPs at each site. The average concentrations of NSBCPs by size were 55 CFU m^{-3} (stage 1), 33 CFU m^{-3} (stage 2), 20 CFU m^{-3} (stage 3), 24 CFU m^{-3} (stage 4), 21 CFU m^{-3} (stage 5), and 30 CFU m^{-3}(stage 6), respectively. For most of the samples, NSBCPs in the coarse fraction (sum of stage 1 and stage 2) show comparable levels to the fine fraction (sum of stage 5 and stage 6). On average, the amount of coarse fraction and fine fraction accounted for 42% and 31% of the total NSBCPs during the whole cruise, respectively.

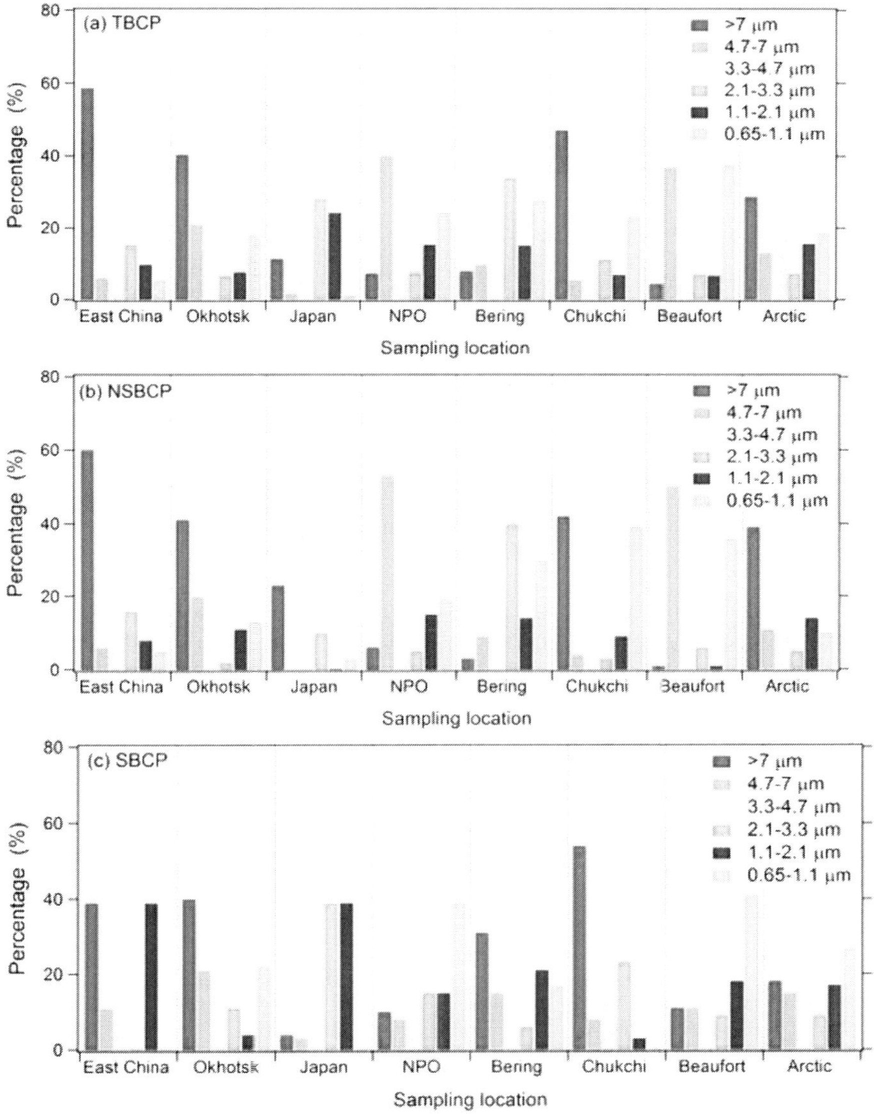

Figure 2. Distribution plot for (**a**) total airborne bacteria-containing particles (TBCPs); (**b**) non-salt tolerant airborne bacteria-containing particles (NSBCPs); and (**c**) salt tolerant bacteria-containing particles (SBCPs) fraction over each region.

On average, the amount of SBCPs with respect to the particle size was 19 CFU m^{-3} (stage 1), 11 CFU m^{-3} (stage 2), 12 CFU m^{-3} (stage 3), 13 CFU m^{-3} (stage 4), 17 CFU m^{-3} (stage 5), and 21 CFU m^{-3} (stage 6) (Figure 2b). The coarse and fine fractions contributed to about 33% and 36% of the SBCP concentrations, respectively. There were occasionally higher proportions of SBCPs in the fine fraction in the Arctic region. For example, at sites 10, 11, 15, and 16, fine SBCPs were several times higher than coarse SBCPs.

3.3. Comparison with Literature Data

From previously reported values in different ecosystems, lower concentrations of bacteria were found in marine air than in continental air. A model simulation by Burrows et al. [28] predicted the presence of continental bacteria, exported from the continents, in air over the Northern Hemisphere Ocean. However, no observations of bacterial concentrations in high-latitude marine air were available to constrain the model.

To date, there are few reports of airborne bacterial abundance over the Arctic Ocean and Northwest Pacific Ocean in one cruise. This study thus provides the first direct observation over these regions. Overall, the concentrations of TBCPs in this study ranged from 29 to 955 CFU m^{-3}. At one measurement site in the remote ocean (a drilling station near the Mid-Atlantic ridge), the concentration of cultivable bacteria was 16 CFU m^{-3} [35]. Higher concentrations were observed at coastal sites, e.g., up to about 8×10^4 CFU m^{-3} [15]. This is likely due to the strong influence of continental sources. Prospero et al. [19] found that NSBCP concentrations were 2–3 orders of magnitude lower in selectively sampled marine air than those sampled at an island only a short distance inland. Montero et al. [11] reported that the offshore and onshore total culturable microbial counts were 778 CFU m^{-3} and 580 CFU m^{-3}, respectively. Because different sources can cause microbial composition to be different and marine bacteria grow selectively on saline growth media, these previous reports may not meaningfully account for bacteria from ocean waters.

3.4. Potential Meteorological Factors Influencing the Variation of Concentrations of Airborne Bacteria

3.4.1. Temperature

Some previous studies have observed a correlation between ambient air temperature and the concentrations of continental cultivable airborne bacteria [36]. Likewise, in observations over the Northwestern Pacific Ocean, airborne microorganism concentrations were also found to be correlated with air temperature [37]. Temperature directly affects the rate of bacterial metabolism and reproduction, as well as cultivability [2]. Temperature is also correlated with a number of other important meteorological and climatological variables that may affect bacterial concentrations in air, such as boundary layer turbulence, time of day, and season. In this study, the sampling cruise covered the regions from Shanghai through the mid-latitudes to the Arctic Ocean, and the temperature ranged from 22.7 °C to −4.7 °C. The average temperature of Group I, Group II, and Group III was 13.95 °C, 4.48 °C, and −1.09 °C, respectively. As latitude increases, the temperature decreases ($R = -0.95$, $p < 0.01$). The abundance of TBCPs was significantly correlated with ambient air temperature at the time of sampling ($R = 0.516$, $p < 0.05$). The correlation was even stronger for the subset of six samples collected over the Group II area (Chukchi Sea and Bering Sea, $R = 0.878$, $p < 0.05$). Over this region, the temperature ranged from 0 to 5.5 °C during the sampling period. The same correlation with temperature was also found for NSBCPs during the whole cruise ($R = 0.564$, $p < 0.01$) and over the Group II area ($R = 0.980$, $p < 0.01$). There was no correlation found between SBCP concentrations and temperature.

3.4.2. Relative Humidity and Precipitation

Although some studies have shown positive correlations between relative humidity (RH) and ambient bacterial concentrations [38], this study found negative correlations between TBCPs and RH as well as NSBCPs with RH over the Group II area (Table 2). No correlation was found between SBCP concentrations and RH.

Table 2. Correlation of bacteria-containing particles (BCPs) with wind speed (WS), relative humidity (RH), and temperature (T).

Species		Whole Cruise			Group I			Group II			Group III		
		WS	RH	T	WS	RH	T	WS	RH	T	WS	RH	T
TBCP	Pearson Correlation	296	−291	516 *	−063	−171	−365	101	−858 *	878 *	028	236	−242
	Sig. (2-tailed)	205	214	020	906	746	477	848	029	022	947	574	563
	N	20	20	20	6	6	6	6	6	6	8	8	8
NSBCP	Pearson Correlation	312	−414	564 **	112	−322	−212	006	−965 **	976 **	216	−240	−099
	Sig. (2-tailed)	180	069	010	832	534	686	991	002	001	607	568	816
	N	20	20	20	6	6	6	6	6	6	8	8	8
SBCP	Pearson Correlation	−019	316	−087	−180	−653	−660	302	−083	115	−174	559	−234
	Sig. (2-tailed)	937	175	714	732	160	153	561	875	829	680	150	577
	N	20	20	20	6	6	6	6	6	6	8	8	8

** Correlation is significant at the 0.01 level (2-tailed); * Correlation is significant at the 0.05 level (2-tailed).TBCP: total airborne bacteria-containing particles; NSBCP: non-salt tolerant airborne bacteria-containing particles; SBCP: salt tolerant airborne bacteria-containing particles.

Bauer et al. [7] and Morris et al. [39] reported that the presence of clouds and precipitation, e.g., clear skies, rain, snow or fog, may affect the deposition of airborne bacteria. During the sampling period, weather was rainy at nine sampling sites including sampling sites 2, 3, 6, 7, 10, 12, 14, 15, and 16. As shown in Figure 3, the average concentration of NSBCPs during rainy days (101 ± 70 CFU m^{-3}) was lower than that during the dry days (249 ± 254 CFU m^{-3}). In contrast, the average concentration of SBCPs during rainy days (115 ± 88 CFU m^{-3}) was slightly higher than that during the dry days (73 ± 58 CFU m^{-3}), albeit the difference is not statistically different. It might be that some airborne bacteria deposit as cloud condensation nucleus by rain in rainy days, and the precipitation caused the splashing of more sea water, bringing the bacteria into the atmosphere. As such, the concentration of NSBCPs looks higher in rainy days more than in dry days but there is no difference for SBCPs.

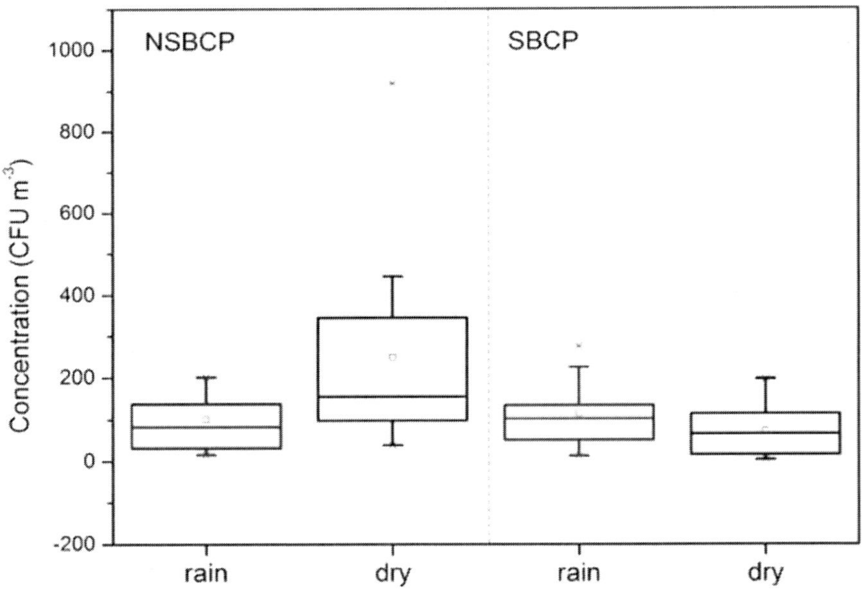

Figure 3. The concentration of NSBCPs and SBCPs during rainy days and dry days.

3.4.3. Wind Speed

Many previous studies have found relationships between the concentrations of airborne cultivable or microscopically observed bacteria and wind speed [40,41,42,43]. While particle sources from the surface are typically stronger under windier conditions, stronger winds are also associated with greater boundary-layer turbulence and dilution, leading to a reduction in boundary-layer concentrations. For these reasons, we measured wind speed (Table 1) concurrently with particle sampling, however, no relationship between wind speed and bacteria concentrations was found in the data collected in this study (as shown Table 2).

3.5. The Role of Sea Ice in the Arctic Ocean

Bacteria have been found to live and metabolize in sea ice and snow [44]. At the South Pole, Carpenter et al. [45] found a population of about 200 to 5000 cells per mL of melted surface snow. Elevated concentrations of SBCPs were observed in the Arctic Ocean in the presence of seasonal ice. During the sampling time, ice concentrations ranged from 0 to 20% and 0 to 91% over Group II and Group III, respectively. In the Bering Sea and the Chukchi Sea where average sea ice concentration was 3.3% during the cruise (almost open water during the cruise), the average concentrations of TBCPs, SBCPs, and NSBCPs were 176, 44, and 133 CFU m^{-3}, respectively. In the area of 75° N to 85° N, the so-called "floating sea ice region" covered by seasonal ice [46] with an average ice concentration of 57.4% during the cruise, the average TBCP, SBCP, and NSBCP concentrations of the Arctic Ocean were 253, 120, and 133 CFU m^{-3}, respectively. The result showed that the concentrations of TBCPs and SBCPs over the floating sea ice region were higher than those found over the open sea areas. Moreover, there is a positive correlation between ice concentration and SBCPs ($R = 0.612, p < 0.05$),indicating a potential relationship between the presence of seasonal sea ice and the source of SBCPs [47]. In contrast, the NSBCPs showed no difference between these two areas.

It is known that precipitation in the form of snow results in the accumulation of airborne bacteria in the seasonal snowpack and sea ice (50 to over 40,000 cells mL^{-1}) [48,49]. Resuspension of snow containing accumulated bacteria may thus be one potential source. According to the data of the National Snow and Ice Data Center (NSIDC), the monthly ice extent in August 2008 was the second-lowest recorded during the period from 1979 to 2008.

Another possible explanation is that bacteria could collect on the ice in "frost flowers", and then be aerosolized [50]. Also, there area lot of biological activities around the sea ice, which contains nutrients. During summer, ice melt lead increases can cause increases in biological activities [12]. Leck and Bigg [51] suggested that spray production from leads may be an important source of bacteria to the Arctic atmosphere. In summer, marine gel also plays an important role in the cloud condensation nuclei in the high Arctic (north of 80° N) [13]. Therefore, in the high Arctic region, marine source bacteria probably are a big contribution to atmospheric aerosols.

4. CONCLUSIONS

The abundance and size distributions of airborne bacteria in the marine boundary layer were investigated during the third China Arctic Research Expedition (July–September, 2008). To our knowledge, this is the first quantitative study of airborne bacterial concentrations in the Arctic. The concentrations of NSBCPs and SBCPs varied considerably over the whole sampling period (NSBCPs 16–919 CFU m^{-3}; SBCPs 4–276 CFU m^{-3}). These concentrations are much higher than those reported earlier using non-saline agar at a mid-Atlantic marine site and in marine background air at Barbados. NSBCPs were also observed in the high Arctic Ocean. The NSBCPs and SBCPs presented in both the coarse and fine aerosols. Concentrations of TBCPs and NSBCPs correlated with air temperature. Changes in sea ice in the Arctic Ocean were also found to influence the distribution of bacteria. Higher concentrations of SBCPs were observed in the presence of seasonal Arctic sea ice, suggesting a possible sea-ice related source, probably from biological activities in the leads, re-blowing of snow, or frost flower formation on the seasonal ice pack.

Since culture methods may underestimate the environmental bacterial communities, it is expected that the abundance in the ambient air would be even higher than those observed in this study. Future investigations should apply molecular biological techniques (e.g., 16S rDNA analysis) to distinguish species between different bacteria.

ACKNOWLEDGMENTS

This research was supported by grants from the National Natural Science Foundation of China (Project No. 41676173), the Chinese Arctic and Antarctic Administration (Projects No. CHINARE2012-2017 and IC201202), and the External Cooperation Program of BIC, CAS (Project No.211134KYSB20130012). We thank Hang Su and Susannah Burrows for their valuable contributions to study related discussions.

AUTHOR CONTRIBUTIONS

Z.Q.X. conceived and designed the experiments; L.M. and H.K. performed the experiments; L.M., X.W.Y., Z.Q.X. and P. Z. analyzed the data; L.M., X.W.Y. and Z.Q.X. wrote the paper. All of the authors contributed to the discussion and revise the manuscript.

CONFLICTS OF INTEREST

The authors declare no conflict of interest.

REFERENCES

1. Després, V.R.; Huffman, J.A.; Burrows, S.M.; Hoose, C.; Safatov, A.S.; Buryak, G.; Fröhlich-Nowoisky, J.; Elbert, W.; Andreae, M.O.; Pöschl, U.; et al. Primary biological aerosol particles in the atmosphere: A review. Tullus B 2012, 64, 15598.
2. Burrows, S.M.; Butler, T.; Jöckel, P.; Tost, H.; Kerkweg, A.; Pöschl, U.; Lawrence, M.G. Bacteria in the global atmosphere—Part 2: Modeling of emissions and transport between different ecosystems. Atmos. Chem. Phys. 2009, 9, 9281–9297.
3. Blanchard, D.C. The ejection of drops from the sea and their enrichment with bacteria and other materials: A review. Estuaries 1989, 12, 127–137.
4. Christner, B.C.; Morris, C.E.; Foreman, C.M.; Cai, R.; Sands, D.C. Ubiquity of biological ice nucleators in snowfall. Science 2008, 319, 1214.
5. Pöschl, U.; Martin, S.; Sinha, B.; Chen, Q.; Gunthe, S.; Huffman, J.; Borrmann, S.; Farmer, D.; Garland, R.; Helas, G. Rainforest aerosols as biogenic nuclei of clouds and precipitation in the amazon. Science 2010, 329, 1513–1516.
6. Lighthart, B. The ecology of bacteria in the alfresco atmosphere. FEMS Microbiol. Ecol. 1997, 23, 263–274.
7. Bauer, H.; Kasper-Giebl, A.; Löflund, M.; Giebl, H.; Hitzenberger, R.; Zibuschka, F.; Puxbaum, H. The contribution of bacteria and fungal spores to the organic carbon content of cloud water, precipitation and aerosols. Atmos. Res. 2002, 64, 109–119.
8. Amato, P.; Ménager, M.; Sancelme, M.; Laj, P.; Mailhot, G.; Delort, A.-M. Microbial population in cloud water at the puy de dôme: Implications for the chemistry of clouds. Atmos. Environ. 2005, 39, 4143–4153.
9. Dueker, M.E.; Weathers, K.C.; O'Mullan, G.D.; Juhl, A.R.; Uriarte, M. Environmental controls on coastal coarse aerosols: Implications for microbial content and deposition in the near-shore environment. Environ. Sci. Technol. 2011, 45, 3386–3392.
10. Li, M.; Qi, J.; Zhang, H.; Huang, S.; Li, L.; Gao, D. Concentration and size distribution of bioaerosols in an outdoor environment in the qingdao coastal region. Sci. Total Environ. 2011, 409, 3812–3819.
11. Montero, A.; Dueker, M.E.; O'Mullan, G.D. Culturable bioaerosols along an urban waterfront are primarily associated with coarse particles. PeerJ 2016, 4, e2827.
12. Matrai, P.A.; Tranvik, L.; Leck, C.; Knulst, J.C. Are high arctic surface microlayers a potential source of aerosol organic precursors? Mar. Chem. 2008, 108, 109–122.

13. Orellana, M.V.; Matrai, P.A.; Leck, C.; Rauschenberg, C.D.; Lee, A.M.; Coz, E. Marine microgels as a source of cloud condensation nuclei in the high arctic. PNAS 2011, 108, 13612–13617.
14. Zuraimi, M.S.; Fang, L.; Tan, T.K.; Chew, F.T.; Tham, K.W. Airborne fungi in low and high allergic prevalence child care centers. Atmos. Environ. 2009, 43, 2391–2400.
15. Harrison, R.M.; Jones, A.M.; Biggins, P.D.E.; Pomeroy, N.; Cox, C.S.; Kidd, S.P.; Hobman, J.L.; Brown, N.L.; Beswick, A. Climate factors influencing bacterial count in background air samples. Int. J. Biometeorol. 2005, 49, 167–178.
16. Shaffer, B.T.; Lighthart, B. Survey of culturable airborne bacteria at four diverse locations in oregon: Urban, rural, forest, and coastal. Microb. Ecol. 1997, 34, 167–177.
17. Vladavets, V.; Mats, L. The influence of meteorological factors on the microflora. Microbiologya 1958, 59, 539–544.
18. Kelly, C.D.; Pady, S.M. Microbiological studies of air masses over montreal during 1950 and 1951. Can. J. Bot. 1954, 32, 591–600.
19. Prospero, J.M.; Blades, E.; Mathison, G.; Naidu, R. Interhemispheric transport of viable fungi and bacteria from africa to the caribbean with soil dust. Aerobiologia 2005, 21, 1–19.
20. Marks, R.; Kruczalak, K.; Jankowska, K.; Michalska, M. Bacteria and fungi in air over the gulf of gdańsk and baltic sea. J. Aerosol Sci. 2001, 32, 237–250.
21. Baylor, E.; Baylor, M.; Blanchard, D.; Syzdek, L.; Appel, C. Virus transfer from surf to wind. Science 1977, 198, 575–580.
22. Aller, J.Y.; Kuznetsova, M.R.; Jahns, C.J.; Kemp, P.F. The sea surface microlayer as a source of viral and bacterial enrichment in marine aerosols. J. Aerosol Sci. 2005, 36, 801–812.
23. Cho, B.C.; Hwang, C.Y. Prokaryotic abundance and 16s rrna gene sequences detected in marine aerosols on the east sea (Korea). FEMS Microbiol. Ecol. 2011, 76, 327–341.
24. Dueker, M.E.; O'Mullan, G.D.; Weathers, K.C.; Juhl, A.R.; Uriarte, M. Coupling of fog and marine microbial content in the near-shore coastal environment. Biogeosciences 2012, 9, 803–813.
25. Urbano, R.; Palenik, B.; Gaston, C.; Prather, K. Detection and phylogenetic analysis of coastal bioaerosols using culture dependent and independent techniques. Biogeosciences 2011, 8, 301.
26. Griffin, D.W. Atmospheric movement of microorganisms in clouds of desert dust and implications for human health. Clin. Microbiol. Rev. 2007, 20, 459–477.
27. Hallar, A.; Chirokova, G.; McCubbin, I.; Painter, T.H.; Wiedinmyer, C.; Dodson, C. Atmospheric bioaerosols transported via dust storms in the western united states. Geophys. Res. Lett. 2011.
28. Burrows, S.M.; Elbert, W.; Lawrence, M.G.; Pöschl, U. Bacteria in the global atmosphere—Part 1: Review and synthesis of literature data for different ecosystems. Atmos. Chem. Phys. 2009, 9, 9263–9280.
29. Clark, S.; Rylander, R.; Larsson, L. Airborne bacteria, endotoxin and fungi in dust in poultry and swine confinement buildings. Am. Ind. Hyg. Assoc. J. 1983, 44, 537–541.
30. Liu, Y.; Han, J.-X.; Huang, H.-Y.; Zhu, B. Development and evaluation of 16s rdna microarray for detecting bacterial pathogens in cerebrospinal fluid. Exp. Biol. Med. 2005, 230, 587–591.
31. Slabova, O.I.; Nikitin, D.I. Influence of the incubation temperature on the reaction of oligotrophic bacteria to stress. Microbiology 2004, 73, 650–653.
32. Li, E.; Wu, J.; Wang, P.; Zhang, D. D-Phenylalanine inhibits biofilm development of a marine microbe, Pseudoalteromonas sp. SC2014. FEMS Microbiol. Lett. 2016, 363, fnw198.
33. Boreson, J.; Dillner, A.M.; Peccia, J. Correlating bioaerosol load with PM2.5 and PM10cf concentrations: A comparison between natural desert and urban-fringe aerosols. Atmos. Environ. 2004, 38, 6029–6041.

34. Kuske, C.R.; Barns, S.M.; Grow, C.C.; Merrill, L.; Dunbar, J. Environmental survey for four pathogenic bacteria and closely related species using phylogenetic and functional genes. J. Forensic Sci. 2006, 51, 548–558.
35. Griffin, D.W.; Westphal, D.L.; Gray, M.A. Airborne microorganisms in the african desert dust corridor over the mid-atlantic ridge, Ocean Drilling Program, Leg 209. Aerobiologia 2006, 22, 211–226.
36. Sobek, A.; Olli, K.; Gustafsson, Ö. On the relative significance of bacteria for the distribution of polychlorinated biphenyls in arctic ocean surface waters. Environ. Sci. Technol. 2006, 40, 2586–2593.
37. Deguillaume, L.; Leriche, M.; Amato, P.; Ariya, P.A.; Delort, A.M.; Pöschl, U.; Chaumerliac, N.; Bauer, H.; Flossmann, A.I.; Morris, C.E. Microbiology and atmospheric processes: Chemical interactions of primary biological aerosols. Biogeosciences 2008, 5, 1073–1084.
38. Lighthart, B.; Shaffer, B.T. Bacterial flux from chaparral into the atmosphere in mid-summer at a high desert location. Atmos. Environ. 1994, 28, 1267–1274.
39. Morris, C.; Georgakopoulos, D.; Sands, D. Ice nucleation active bacteria and their potential role in precipitation. J. Phys. 2004, 121, 87–103.
40. Jones, A.M.; Harrison, R.M. The effects of meteorological factors on atmospheric bioaerosol concentrations—A review. Sci. Total Environ. 2004, 326, 151–180.
41. Lindemann, J.; Upper, C.D. Aerial dispersal of epiphytic bacteria over bean plants. Appl. Environ. Microbiol. 1985, 50, 1229–1232. [PubMed]
42. Mouli, P.; Mohan, S.; Reddy, S. Assessment of microbial(bacteria) concentrations of ambient air at semi-arid urban region: Influence of meteorological factors. Appl. Ecol. Environ. Res. 2005, 3, 139–149.
43. Tong, Y.; Lighthart, B. Diurnal distribution of total and culturable atmospheric bacteria at a rural site. Aerosol Sci. Technol. 1999, 30, 246–254.
44. Arrigo, K.R.; Thomas, D.N. Large scale importance of sea ice biology in the southern ocean. Antarct. Sci. 2004, 16, 471–486.
45. Carpenter, E.J.; Lin, S.; Capone, D.G. Bacterial activity in south pole snow. Appl. Environ. Microbiol. 2000, 66, 4514–4517.
46. Lu, P.; Li, Z.; Cheng, B.; Lei, R.; Zhang, R. Sea ice surface features in arctic summer 2008: Aerial observations. Remote Sens. Environ. 2010, 114, 693–699.
47. Harding, T.; Jungblut, A.D.; Lovejoy, C.; Vincent, W.F. Microbes in high arctic snow and implications for the cold biosphere. Appl. Environ. Microbiol. 2011, 77, 3234–3243.
48. Ariya, P.A.; Domine, F.; Kos, G.; Amyot, M.; Côté, V.; Vali, H.; Lauzier, T.; Kuhs, W.F.; Techmer, K.; Heinrichs, T.; et al. Snow–A photobiochemical exchange platform for volatile and semi-volatile organic compounds with the atmosphere. Environ. Chem. 2011, 8, 62–73.
49. Møller, A.K.; Barkay, T.; Al-Soud, W.A.; Sorensen, S.J.; Skov, H.; Kroer, N. Diversity and characterization of mercury-resistant bacteria in snow, freshwater and sea-ice brine from the high arctic. FEMS Microbiol. Ecol. 2011, 75, 390–401.
50. Bowman, J.; Deming, J. Elevated bacterial abundance and exopolymers in saline frost flowers and implications for atmospheric chemistry and microbial dispersal. Geophys. Res. Lett. 2010.
51. Leck, C.; Bigg, E.K. Biogenic particles in the surface microlayer and overlaying atmosphere in the central arctic ocean during summer. Tellus B 2005, 57, 305–316.

CHAPTER 12

Extended-Range Ensemble Predictions of Convection in the North Australian Monsoon Region

Wasyl Drosdowsky and Matthew C. Wheeler*

Research and Development, Bureau of Meteorology, Melbourne, VIC, Australia

ABSTRACT

Extended-range (<35 day) predictions of area-averaged convection over northern Australia are investigated with the Bureau of Meteorology's Predictive Ocean-Atmosphere Model for Australia (POAMA). Hindcasts from 1980-2011 are used, initialized on the 1st, 11th, and 21st of each month, with a 33-member ensemble. The measure of convection is outgoing longwave radiation (OLR) averaged over the box 120°E-150°E, 5°S-17.5°S. This averaging serves to focus on the intraseasonal and longer time scales, and is an area of interest to users. The raw hindcasts of daily OLR show a strong systematic adjustment away from their initial values during the first week, and then converge to a mean seasonal cycle of similar amplitude and phase to observations. Hence, forecast OLR anomalies are formed by removing the model's own seasonal cycle of OLR, which is a function of start time and lead time, a usual practice for dynamical seasonal prediction. Over all hindcasts, the model forecast root-mean-square (RMS) error is smaller than the RMS error of persistence and climatological reference forecasts for leads 3–35 days. Ensemble spread is less than the forecast RMS error (i.e., under-spread) for days 1–12, but slightly greater than the RMS error for longer leads. Binning the individual forecasts based on ensemble spread shows a generally positive relationship between spread and error. Therefore, greater certainty can be given for forecasts with smaller spread.

Keywords: Australian monsoon, monsoon prediction, tropical prediction, intraseasonal, extended-range, subseasonal, POAMA, dynamical prediction system

INTRODUCTION

As defined by the World Meteorological Organization, extended-range weather forecasts cover the lead-time range of 10–30 days. Stakeholders in agriculture, industry, and the resources sector have continually called out for forecasts on this intermediate range, but few operational products exist. At the Bureau of Meteorology, for example, the only operational product that currently focuses on this range is the Weekly Tropical Climate Note (WTCN: http://www.bom.gov.au/climate/tropnote/tropnote.shtml), which currently provides non-quantitative text-based outlooks of likely large-scale tropical conditions for the coming few weeks.

In recognition of the demand for quantitative extended-range forecast products, the Bureau of Meteorology has sought to provide such quantitative guidance through further development of its dynamical coupled ocean-atmosphere prediction system (Hudson et al., 2011, 2013). Testing of the evolving model/system for skill on the extended range (also known as the intraseasonal or multi-week range) has mostly concentrated on weekly or longer averages of grid-point fields (Hudson et al., 2011, 2013; Marshall et al., 2014b) or on daily indices of large-scale climate phenomena such as the Madden–Julian Oscillation (MJO; Marshall et al., 2011), Southern Annular Mode (SAM; Marshall et al., 2012), and blocking (Marshall et al., 2014a). Zhu et al. (2014) took the approach of examining the skill of a seamless range of time scales, including the extended range, by using time averages equal in length to the forecast lead time. Recently, Marshall and Hendon (2015) examined the skill of predicting Australian monsoon indices of rain and wind. Together, this research provides encouraging signs of useful extended-range skill in many locations provided there is suitable time-averaging or selection of the intraseasonal climate signals.

For northern Australia, strong intraseasonal variability of tropical convection and rainfall has been long appreciated and documented (see review by Wheeler and McBride, 2011). A frequently-used measure of convection in the Australian monsoon region, and one that is of relevance to the WTCN, is the area-averaged outgoing longwave radiation (OLR) over

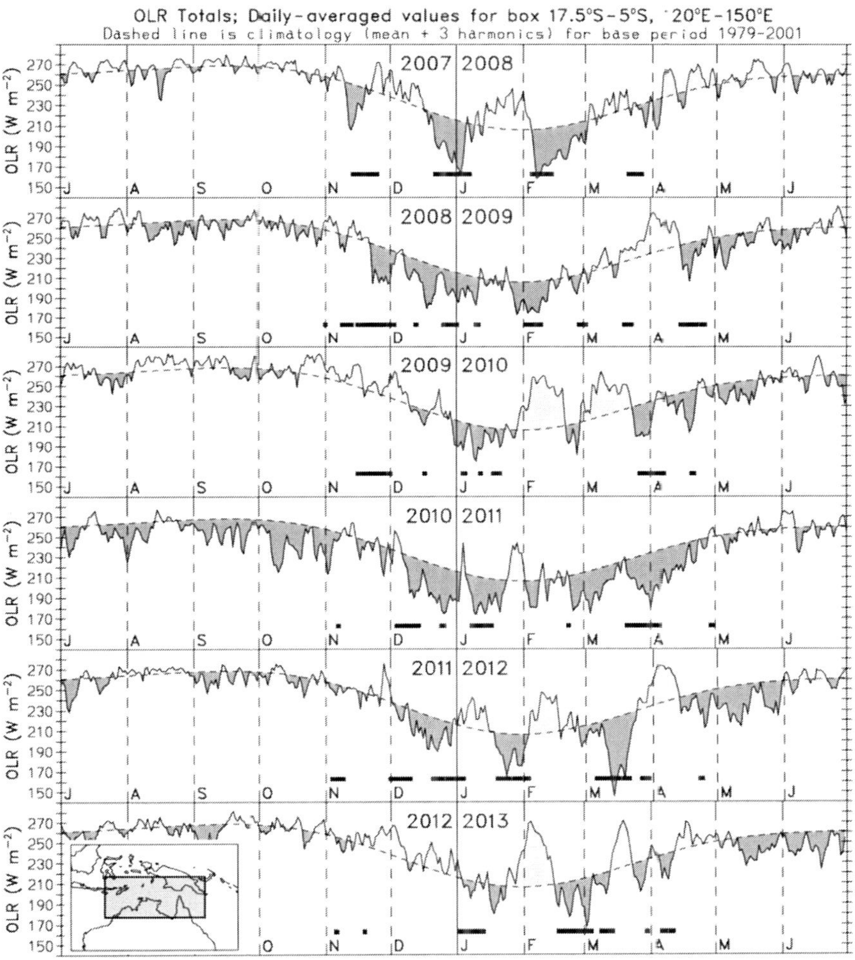

FIGURE 1. TIME SERIES OF DAILY NOAA SATELLITE-OBSERVED OUTGOING LONGWAVE RADIATION (OLR), AVERAGED FOR THE BOX 17.5°S–5°S, 120°E–150°E, FOR JULY 2007-JUNE 2013. Dashed line shows the smoothed climatological seasonal cycle with dark and light shading to indicate negative and positive OLR anomalies respectively. Thick horizontal lines indicate when the Wheeler-Hendon RMM index of the MJO was in phases 4, 5, or 6, for the months of November through April only.

the box 120°E-150°E, 5°S-17.5°S, as displayed in Figure 1. This figure uses dark shading to indicate a negative OLR anomaly, which is indicative of a greater number of cold clouds and/or colder cloud tops than normal in the region, i.e., enhanced convection. Strong intraseasonal variability of convection can be seen in most years. For example, the 2007/08 wet season is made up of about three complete intraseasonal cycles, with monsoon "bursts" occurring in mid-November 2007, late December 2007 to early January 2008, and most of February 2008 (see also Wheeler, 2008). The intraseasonal variability in 2007/08 can be seen to be well correlated to the MJO, as indicated by the times of MJO Phases 4-6 (horizontal thick lines in Figure 1, using the definition of MJO phases of Wheeler and Hendon, 2004). It is this empirical relationship that is one of the main inputs to the WTCN. However, in some other years (e.g., 2009/10, 2010/11) any relationship with the MJO is less apparent, and other variability plays an important role (examples provided in Wheeler and McBride, 2011). Noting that the WTCN is currently heavily reliant on the MJO for its outlooks, and that the Bureau's dynamical prediction system attempts to model all of the important sources of variability, it is of interest to see how well the dynamical prediction system performs for this region.

In this work we therefore investigate the quantitative extended-range prediction of the aforementioned area-averaged OLR using an ensemble of hindcasts from the Bureau's operational coupled modeling system, the Predictive Ocean Atmosphere Model for Australia (POAMA) version 2 M (hereafter POAMA-2M). We investigate the model forecast bias, the removal of this bias, the resulting prediction skill, the ensemble spread versus error, and a real-time forecast display.

DATA AND MODEL FORECAST SYSTEM

Observational OLR and MJO Data

The observed OLR data is the NOAA satellite interpolated OLR (Liebmann and Smith, 1996) available from 1974 to the present. Daily MJO index data are the Real-time Multivariate MJO (RMM) index of Wheeler and Hendon (2004) obtained from http://www.bom.gov.au/climate/mjo.

POAMA-2M Forecast System

We analyse POAMA-2M (Hudson et al., 2013) which currently (early 2017) produces the Bureau of Meteorology's operational monthly and seasonal forecasts. This version of POAMA was developed specifically to provide more skilful output on the extended-range time scale (hence the letter "M" for multi-week). Improvements included in this version of POAMA are the use of perturbed atmosphere and ocean initial conditions and a burst ensemble (an ensemble starting from a single initial time as opposed to a lagged ensemble), as well as the use of three different model configurations (using different convective parameterizations or flux correction at the ocean surface) to form a multi-model ensemble (Hudson et al., 2013).

The atmospheric component of POAMA-2M is a spectral model with resolution T47 (~250 km grid) and 17 vertical levels. The ocean component has a zonal resolution of 2° and a varying meridional resolution of 0.5–1.5° with 25 vertical levels. The unperturbed initial conditions are provided by separate data assimilation schemes for the ocean versus the atmosphere and land. For the atmospheric and land initial states, they are generated by nudging of wind, temperature, and humidity toward one of two analysis products that come from different global models (i.e., different to the model used in POAMA-2M). The hindcasts and forecasts are therefore more likely to suffer from "initial shock" than a model that has its own atmospheric data assimilation (Hudson et al., 2011). Perturbations to the initial conditions of the central member are generated using a coupled breeding scheme. Ten perturbed states are produced, providing 11 different initial states that are input to three different configurations of the model, providing a 33-member ensemble (Hudson et al., 2013). This description applies to both the hindcasts and real-time forecasts.

HINDCASTS AND BIAS REMOVAL

We analyse hindcasts from POAMA-2M that have been initialized on the 1st, 11th, and 21st of each month for the period 1980 to 2011. Observations of OLR are not used as part of the model initialization. Instead, the model OLR is computed by the model's radiation scheme and depends critically on the production of convection and clouds by the model's convective parameterization. Therefore, the model OLR is not necessarily the same as observed at the initial condition. Further, since the atmospheric initial conditions are produced by nudging toward a different model (see above), there is high potential for initial shock of the model OLR as the model shifts toward its own attractor. This is explored in Figure 2, which shows the annual cycle of observed OLR (black curve) together with the annual cycle of all the day 1 hindcasts (blue) and day 20 hindcasts (green), for the region of interest. Interestingly, the initial (day 1) OLR is close to observed during the wetter months of December-April, but is systematically about 15 Wm^{-2} higher than observed during the drier months of June-October, whereas the day 20 OLR is close to observed during the drier months and systematically too high during the wetter months. Therefore, the initial shock in the OLR field is to increasing OLR (i.e., less convection) in the climatologically-wet months but decreasing OLR (i.e., more convection or colder surface temperatures) in the dry months. Further, the individual hindcast climatologies from each start date (thin colored lines) demonstrate that most of this initial shock occurs in the first few days of the hindcasts. This figure is a strong demonstration for the need to compute anomalies for the model with respect to the lead-time dependent hindcast climatology, a procedure that is now common for dynamical seasonal prediction (Stockdale, 1997; Hudson et al., 2011). This is a first-order linear correction for the initial shock, model drift, and model mean bias.

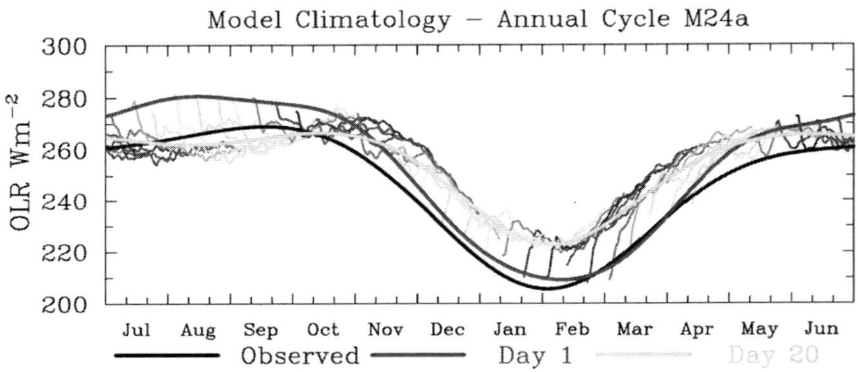

FIGURE 2. POAMA-2M OLR (AREA-AVERAGED FOR SAME BOX AS FIGURE 1) HINDCAST CLIMATOLOGY (THIN RAINBOW-COLORED LINES) FORMED BY AVERAGING ALL AVAILABLE HINDCASTS FOR EACH DIFFERENT START DATE, E.G., FOR 1ST JULY, 11TH JULY, 21ST JULY, 1ST AUGUST, ETC. Thick lines are interpolated and smoothed versions of all the day 1 hindcasts (blue line) day 20 hindcasts (green line) and observations (black line). The smoothed seasonal cycle is obtained by retaining only the first 3 annual harmonics.

Separate lead-dependent hindcast climatologies are calculated for each of the three model configurations making up the multi-model ensemble. Resulting individual ensemble anomalies, and the multi-model ensemble mean anomaly, may then be plotted relative to the observed climatology, as shown for an example hindcast in Figure 3. In this example the forecast ensemble mean (thick pink line) is shown to track the verifying observations quite well in the first couple of weeks, with the ensemble members gradually spreading around it. Note that the same color is used for ensemble members using the same initial condition, as input to the three different model configurations, resulting in three lines of each color. Interestingly, there is a grouping of these ensemble members for short lead times, showing that for the first few days it is the initial condition that most determines the forecast trajectory, rather than the model physics. We will return to this issue in the next section.

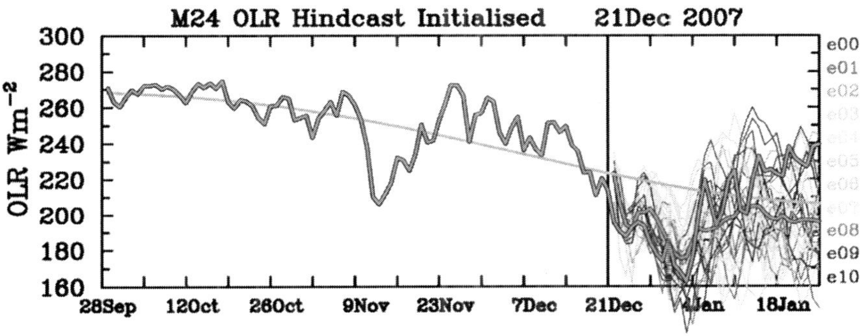

FIGURE 3. EXAMPLE BIAS-CORRECTED HINDCAST FOR START DATE 21 DECEMBER 2007, SHOWING ENSEMBLE MEAN (THICK PINK LINE) AND ALL 33 ENSEMBLE MEMBERS (3 MEMBERS PER RAINBOW COLOR). Thick gray line shows the observed OLR from both before and after the model start date, and thinner gray line is the observed climatological seasonal cycle.

HINDCAST PERFORMANCE AND SKILL-SPREAD RELATIONSHIP

Model OLR vs. Observed OLR

Model performance is evaluated for the bias-corrected multi-model ensemble mean OLR anomalies in comparison with the verifying observed OLR anomalies. Two metrics have been calculated; the correlation (shown later) and the root mean square (RMS) error for lead times from 1 to 35 days. These measures are calculated over all the hindcasts, and over various subsets. Figure 4 shows the RMS error for the ensemble mean from all hindcasts (upper panel) and for the summer monsoon months of December through March (DJFM; lower panel). Also shown are the RMS error for forecasts computed using climatology (i.e., a zero anomaly), the RMS error for persistence of the initial daily anomaly, and the hindcast "skill" computed as a percentage improvement of POAMA over climatology (using the same numerical scale as the error, i.e., from 0 to 40%).

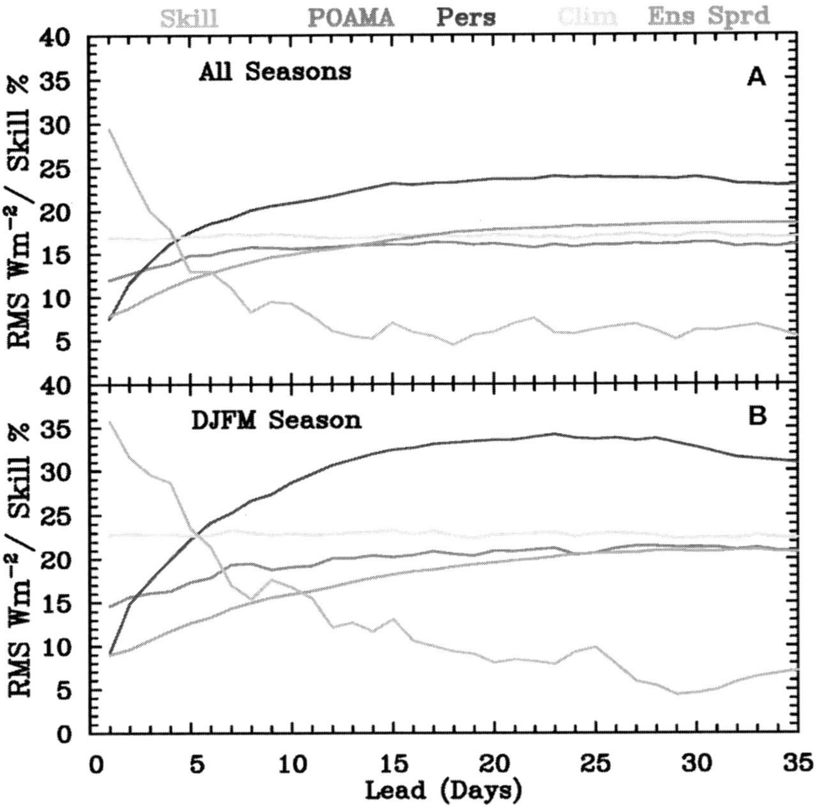

FIGURE 4. ROOT-MEAN-SQUARE (RMS) ERROR OF AREA-AVERAGED OLR (IN W M^{-2}, ON A SCALE FROM 0 TO 40 W M^{-2}) FOR POAMA ENSEMBLE-MEAN HINDCASTS (RED CURVE), REFERENCE FORECASTS USING A FORECAST OF ZERO ANOMALY (I.E., CLIMATOLOGY; GREEN), AND PERSISTENCE OF THE INITIAL DAILY ANOMALY (BLUE). Also shown is the ensemble spread (magenta) in W m^{-2}, and skill score as a percentage improvement of POAMA over climatology (orange) on a scale from 0 to 40%. **(A)** Is for all seasons, and **(B)** for summer monsoon (DJFM) season only.

The RMS error of POAMA is smaller than that of both persistence and climatology for the range 2–35 days for both the all season and the summer monsoon (i.e., DJFM) season. Although the POAMA hindcast RMS error is greater in DJFM, the hindcasts are more skilful relative to climatology during this season, with a percentage improvement of about 36% at a lead of 1 day compared to 30% for the all-seasons case, and a percentage improvement of about 16% compared to 9% at a lead of 10 days. We contend that these percentage improvements represent a useful level of skill.

The spread (pink lines in Figure 4) for the all-season case is less than the model RMS error for days 1 to 12, i.e., indicating that the ensemble is under-spread. Between days 12 and 16 the spread is very close to the RMS error for both the model forecasts and climatological forecasts. Beyond day 16 the spread of the ensemble exceeds the RMS error of a climatological forecast, which implies that the POAMA forecasts have slightly greater variance than observed. For the summer months only, the ensemble appears to be under-spread for lead times up to about 20 days, and the spread stays below the RMS error of a climatological forecast for all leads.

Impact of MJO on Forecast Skill

As noted in the introduction, some, but not all, seasons show a strong relationship between the OLR over northern Australia and the phase of the MJO. It has also been previously shown that prediction of the RMM index of the MJO is somewhat more skilful when the MJO is strong at the initial time (Rashid et al., 2011). It therefore seems a reasonable hypothesis that the prediction of OLR over northern Australia should be more skilful when the MJO is strong in the initial conditions. We test this hypothesis by examining the impact of the MJO on the forecast skill when stratifying the hindcasts according to the presence or absence of a strong MJO signal (Figure 5A), and also by the phase of the MJO (Figure 5B) at the initial time. However, there is no evident impact of the amplitude or phase of the MJO on the hindcast RMS error. For the correlation skill (not shown), there appears a weak increase in correlation between days 2 and 9 for hindcasts initialized with a strong MJO, but it is not statistically significant. Thus we cannot confirm our hypothesis above. This appears consistent with the result of Marshall et al. (2011) who found that although POAMA was able to correctly simulate and predict the relationship between the MJO and rainfall (i.e., convection) over most of the tropical Indo-Pacific, it was not able to do this over the Maritime Continent and northern Australia, indicating that there is still room for improvement in these extended-range forecasts for northern Australia.

Ensemble Spread vs. Observed Uncertainty and Error

The relationship between the spread and the underlying observed variability is further highlighted by the rank histogram or "Talagrand diagram" (Talagrand et al., 1997; Hamill, 2001), shown in Figure 6 using all hindcasts. This histogram is constructed by counting where the verifying

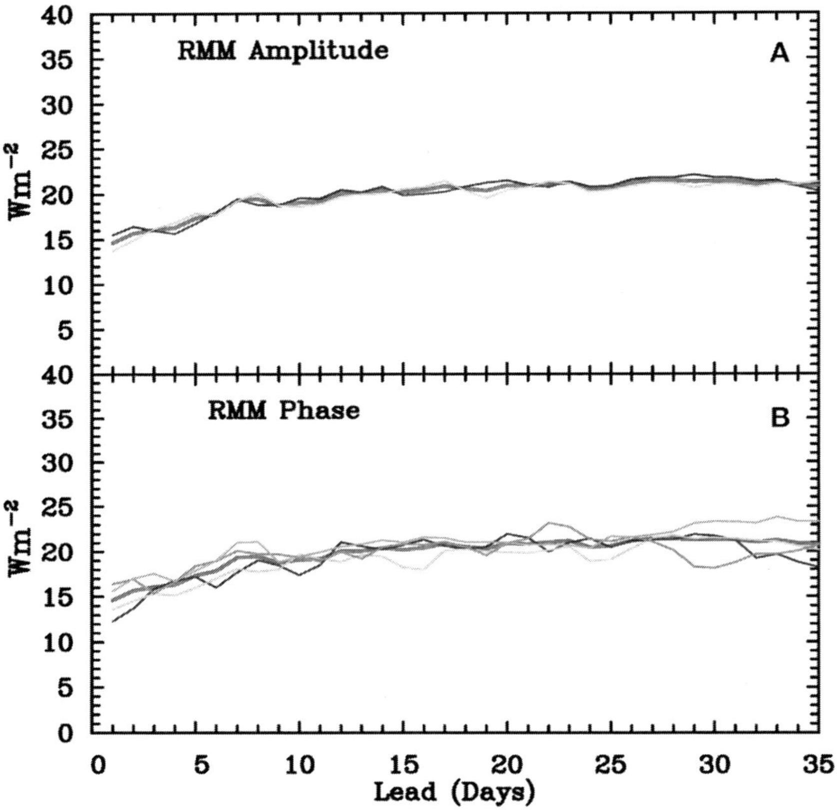

FIGURE 5. RMS ERRORS OF ALL POAMA HINDCASTS (A) stratified by RMM amplitude, blue for RMM > 1.3, green RMM ≤ 1.3, and red for all cases, and **(B)** stratified by RMM phase, with red for all cases, blue for phases 1 and 8, green for phases 2 and 3, magenta for phases 4 and 5, and orange for phases 6 and 7.

observation lies amongst all the ensemble members for each hindcast. Ideally the distribution should be flat, which occurs when the ensemble spread matches the observed variability, whereas a U-shaped distribution indicates insufficient spread with many verifying observations falling near the extremes or outside the range of the ensemble members, and a domed distribution indicates too much ensemble spread. All three types of distributions can be seen in Figure 6, which shows too little spread at 3 days lead, too much spread at 20 days lead, and about the right level of spread at 10 days lead. This result is consistent with Figure 4A which suggested that ensemble spread was insufficient at short leads (up to ~12 days) and over spread beyond about 16 days. The lack of ensemble spread at short lead times is exacerbated by the use of the same initial conditions for each of the three model configurations. This shows up in Figure 6 in the strong peak found at every third bin, due to the very small separation between the three ensemble members with identical initial conditions. As discussed by Hudson et al. (2013), this clearly signifies the need for using different initial conditions for each of the three model configurations.

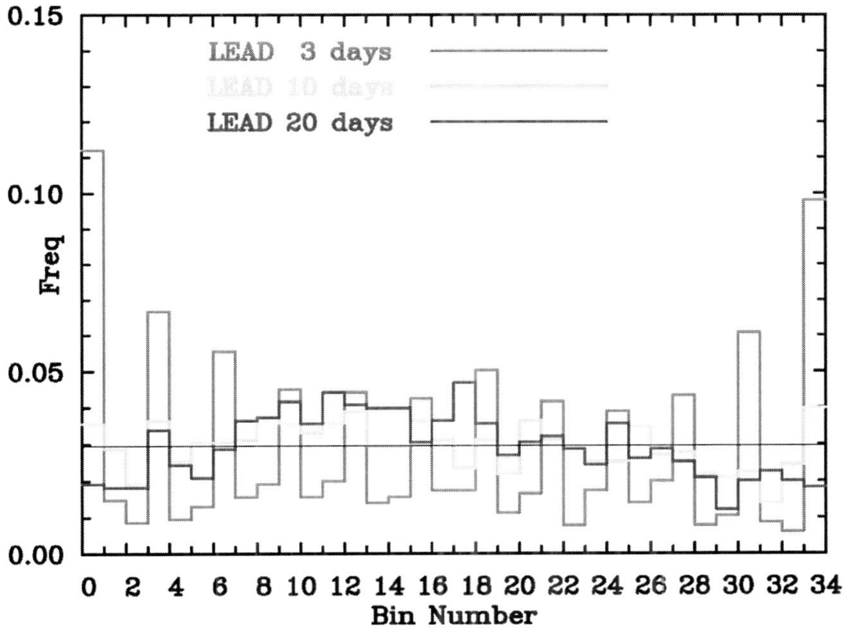

FIGURE 6. RANK HISTOGRAM FOR HINDCASTS AT 3-DAY (RED), 10-DAY (GREEN), AND 20-DAY (BLUE) LEAD FOR ALL HINDCASTS. Thin black line is expected distribution assuming hindcasts fully represent observed uncertainty.

Binning the individual forecasts based on ensemble spread shows a generally positive relationship between spread and error (Figure 7). The spread-error correlations are 0.55 for the 3-day lead, 0.57 for the 10-day lead, and 0.90 for the 20-day lead. These correlations are more encouraging of a positive relationship between spread and error than others that have been listed in the literature (as reviewed by Grimit and Mass, 2007), especially for the 20-day lead. Therefore, especially for the 20-day lead, greater certainty can be given to forecasts with smaller spread.

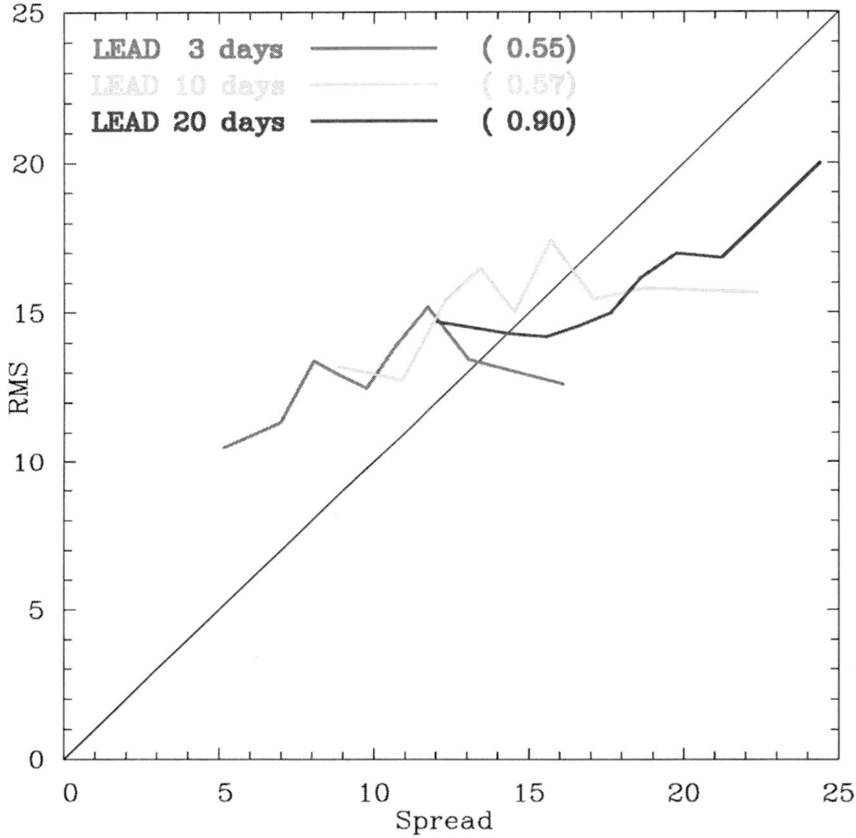

FIGURE 7. SPREAD VS. RMS ERROR FOR HINDCASTS AT A LEAD OF 3 DAYS (RED) 10 DAYS (GREEN) AND 20 DAYS (BLUE)**.** Binning of the hindcasts is based on the ensemble spread, using 9 bins containing 128 hindcasts each. Numbers in brackets are correlation between spread and RMS error for each lead time.

REAL-TIME FORECASTS

Real-time forecasts have been performed once a week since August 2011 and twice a week since February 2013, and run out to at least 120 days. For most forecasts, the start date will not coincide with the dates for which hindcast climatologies are available. There are a number of possible strategies to obtain climatological values for any arbitrary start date. These include simply using the nearest available climatology or linearly interpolating between the two closest dates. The approach adopted here is to fit a smoothed annual cycle to each lead time, similar to the annual cycles for the observed and initialized data, and then obtain a climatological value for any start date and lead time by interpolation from this smoothed annual cycle. Smoothed annual cycles for lead times of day 1 (blue) and day 20 (green) are shown in Figure 2.

Real-time anomalies are then calculated in a similar manner to that used for the hindcasts. The consistency of the forecasts over a run of forecasts is demonstrated in Figure 8, which shows the ensemble mean of a sequence of 11 forecasts initialized between 04 Feb 2013 and 11 Mar 2013. This example shows a general agreement between one start time and the next, although with a notable exception to this for forecasts initialized before 1st March vs. those initialized after 1st March, showing that the model had a "change of heart" at that time. Comparison with the observed OLR anomalies shows that the forecast of positive OLR anomalies for 12–25 March by those forecasts initialized after 1st March, indicating suppressed convection, verified well. Overall, verification of all the available real-time forecasts (Figure 9) shows that the hindcast skill is generally maintained by independent forecasts. The limited real-time POAMA forecasts have lower RMS errors than persistence beyond the first day of the forecast, which is better than was the case for the hindcasts (Figure 4), which took 2–3 days to beat persistence. However, compared to climatological forecasts, the POAMA real-time forecasts lose skill more swiftly than the hindcasts, with the POAMA RMS error slightly exceeding that of climatology at lead times greater than about 15 days. We cannot think of a difference in the model prediction system between the hindcasts and real-time forecasts that could explain these skill differences (e.g., the model physics and initialisation remain the same), so we assume that the skill differences are a result of the rather short period (<5 years) of real-time data, and the associated statistical uncertainty this causes.

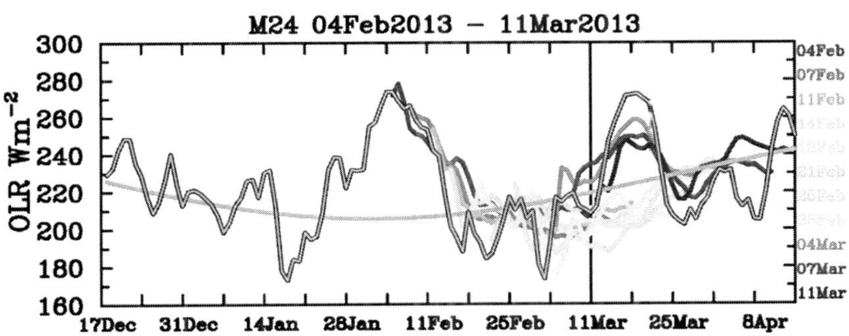

FIGURE 8. REAL-TIME FORECASTS (AREA-AVERAGED ENSEMBLE MEAN OLR) FOR ALL START DATES FROM 4 FEBRUARY TO 11 MARCH 2013.

Thick gray line shows the observed OLR from both before and after the model start dates, and thinner gray line is the observed climatological seasonal cycle.

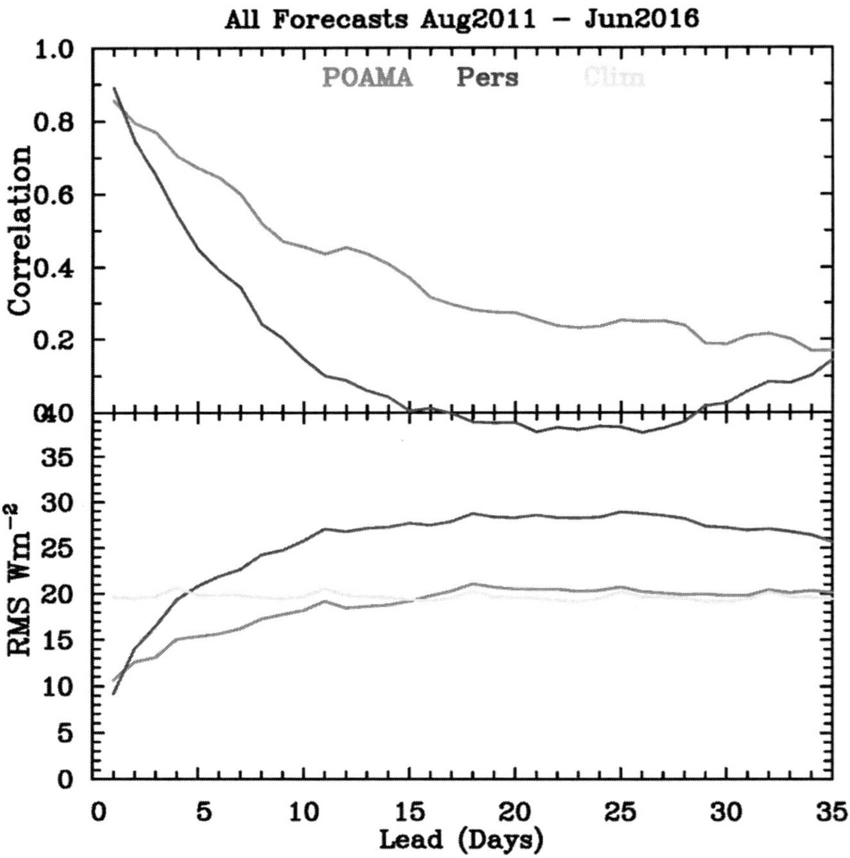

FIGURE 9. VERIFICATION OF ALL AVAILABLE POAMA FORECASTS FROM AUGUST 2011 TO JUNE 2016 (RED CURVES) COMPARED WITH PERSISTENCE (BLUE CURVES) AND CLIMATOLOGICAL FORECASTS (GREEN CURVE) FOR THE SAME PERIOD.

CONCLUSIONS

We examine the ability of the latest version of the POAMA system (POAMA-2M) to simulate the seasonal cycle of OLR over the north Australian monsoon region, and to predict its variability on timescales out to 35 days. After the period of initial shock is over (i.e., after the first few days), the model's seasonal cycle shows that it does not have enough convection during the wet season and consequently has smaller amplitude than observed. After correcting for this bias, model hindcasts show skill, over persistence and climatological forecasts, at lead times beyond 2 days. These hindcasts show that the model has greater skill during the summer monsoon season from December to March, but this skill is largely independent of the state of the MJO. Examination of the spread of hindcasts suggests that the model has insufficient spread at short lead times (<12 days), and slightly too much spread at longer lead times (>16 days). Real-time forecasts are constructed in a similar manner to the hindcasts, with the required bias correction obtained by interpolating the model seasonal cycle to give values at each real-time forecast start time. Verification of all available forecasts from August 2011 through to June 2016 suggests that the hindcast skill relative to persistence and climatology is mostly maintained in completely independent forecasts.

AUTHOR CONTRIBUTIONS

WD and MW jointly designed the research, interpreted the results, and wrote the paper. WD analyzed the prediction system output and made Figures 2–9. MW made Figure 1, undertook the milestone reporting to funding bodies and managers, formatted the manuscript, and submitted the manuscript.

FUNDING

This work was partially supported during 2011–13 by the Northern Australia/Monsoon Prediction project of the Managing Climate Variability Program managed by the Grains Research and Development Corporation.

CONFLICT OF INTEREST STATEMENT

The authors declare that the research was conducted in the absence of any commercial or financial relationships that could be construed as a potential conflict of interest.

The reviewer SW and the handling Editor declared their shared affiliation, and the handling Editor states that the process nevertheless met the standards of a fair and objective review.

ACKNOWLEDGMENTS

We thank the POAMA and related teams for their dedication to producing and maintaining POAMA, and for supporting its products. In particular, we thank Griffith Young for the POAMA web pages and computing support. Greg Browning and Andrew Marshall kindly provided internal reviews.

REFERENCES

1. Grimit, E. P., and Mass, C. F. (2007). Measuring the ensemble spread–error relationship with a probabilistic approach: stochastic ensemble results. *Mon. Weather Rev.* 135, 203–221.
2. Hamill, T. M. (2001). Interpretation of rank histograms for verifying ensemble forecasts. *Mon. Weather Rev.* 129, 550–560.
3. Hudson, D., Alves, O., Hendon, H. H., and Wang, G. (2011). The impact of atmospheric initialisation on seasonal prediction of tropical Pacific SST. *Clim. Dyn.* 36, 1155–1171.
4. Hudson, D., Marshall, A. G., Yin, Y. H., Alves, O., and Hendon, H. H. (2013). Improving intraseasonal prediction with a new ensemble generation strategy. *Mon. Weather Rev.* 141, 4429–4449.
5. Liebmann, B., and Smith, C. A. (1996). Description of a complete (interpolated) outgoing longwave radiation dataset. *Bull. Am. Met. Soc.* 77, 1275–1277.
6. Marshall, A. G., and Hendon, H. H. (2015). Subseasonal prediction of Australian summer monsoon anomalies. *Geophys. Res. Lett.* 42, 10913–10919.
7. Marshall, A. G., Hudson, D., Hendon, H. H., Pook, M. J., Alves, O., and Wheeler, M. C. (2014a). Simulation and prediction of blocking in the Australian region and its influence on intra-seasonal rainfall in POAMA-2. *Clim. Dyn.* 42, 3271–3288.
8. Marshall, A. G., Hudson, D., Wheeler, M. C., Alves, O., Hendon, H. H., Pook, M. J., et al. (2014b). Intra-seasonal drivers of extreme heat over Australia in observations and POAMA-2. *Clim. Dyn.* 43, 1915–1937.
9. Marshall, A. G., Hudson, D., Wheeler, M. C., Hendon, H. H., and Alves, O. (2011). Assessing the simulation and prediction of rainfall associated with the MJO in the POAMA seasonal forecast system. *Clim. Dyn.* 37, 2129–2141.
10. Marshall, A. G., Hudson, D., Wheeler, M. C., Hendon, H. H., and Alves, O. (2012). Simulation and prediction of the Southern Annular Mode and its influence on Australian intra-seasonal climate in POAMA. *Clim. Dyn.* 38, 2483–2502.
11. Rashid, H. A., Hendon, H. H., Wheeler, M. C., and Alves, O. (2011). Prediction of the Madden-Julian oscillation with the POAMA dynamical prediction system. *Clim. Dyn.* 36, 649–661.
12. Stockdale, T. N. (1997). Coupled ocean–atmosphere forecasts in the presence of climate drift. *Mon. Weather Rev.* 125, 809–818.
13. Talagrand, O., Vautard, R., and Strauss, B. (1997). "Evaluation of probabilistic prediction systems," in *Proceedings, ECMWF Workshop on Predictability.* Reading: ECMWF, 1–25.

14. Wheeler, M. C. (2008). Seasonal climate summary southern hemisphere (summer 2007-08): mature La Nina, an active MJO, strongly positive SAM, and highly anomalous sea-ice. *Aust. Meteorol. Mag.* 57, 379–393.
15. Wheeler, M. C., and Hendon, H. H. (2004). An all-season real-time multivariate MJO index: development of an index for monitoring and prediction. *Mon. Weather Rev.* 132, 1917–1932.
16. Wheeler, M. C., and McBride, J. L. (2011). "Australasian monsoon," in *Intraseasonal Variability in the Atmosphere-Ocean Climate System, 2nd Edn*, eds W. K. M. Lau and D. E. Waliser (Berlin: Springer), 147–198.
17. Zhu, H., Wheeler, M. C., Sobel, A. H., and Hudson, D. (2014). Seamless precipitation prediction skill in the tropics and extratropics from a global model. *Mon. Weather Rev.* 142, 1556–1569.

CHAPTER 13

Understanding Drought Dynamics during Dry Season in Eastern Northeast Brazil

Denis D. Costa[1*], Thiago A. da Silva Pereira[2], Carlos R. Fragoso Jr.[3], Kaveh Madani[1,4] and Cintia B. Uvo[1]

[1]*Department of Water Resources Engineering, Lund University, Lund, Sweden*

[2]*Department of Civil Engineering, Federal University of Pernambuco, Recife, Brazil*

[3]*Center for Technology, Federal University of Alagoas, Maceió, Brazil*

[4]*Centre for Environmental Policy, Imperial College London, London, UK*

ABSTRACT

Eastern Northeast Brazil (ENEB) generally experiences a high variability in precipitation in the dry season, with amplitudes that can overcome 500 mm. The understanding of this variability can help in mitigating the socio-economic issues related to the planning and management of water resources this region, which is highly vulnerable to drought. This work aims to assess spatio-temporal variability of precipitation during the dry season and investigate the relationships between climate phenomena and drought events in the ENEB, using univariate (Spearman correlation) and multivariate statistical techniques, such as Principal Component Analysis, Cluster Analysis, and Maximum Covariance Analysis. The results indicate that the variability of precipitation in the dry season can be explained mainly (62%) by local physical conditions and climate conditions have a secondary contribution. Further analysis of the larger anomalous events suggests that the state of Atlantic and Pacific oceans can govern the occurrence of those events, and the conditions of Atlantic Ocean can be considered a potential modulator of anomalous phenomena of precipitation in ENEB.

Keywords: drought, extreme events, climate indices, semiarid, Northeast Brazil

INTRODUCTION

Understanding climate variability is essential to the better quantification of processes related to hydrological cycle in order to improve long-term forecasting of extreme events such as droughts (Silverman and Dracup, 2000; Morid et al., 2007; Labat, 2010; Massei and Fournier, 2012; Räsänen and Kummu, 2013). Droughts have potential impacts on society, environment and economy (Krol et al., 2006; Santos et al., 2010; Hastenrath, 2012; Mazdiyasni and AghaKouchak, 2015). Thus, major efforts have been made in the last decade to understand the dynamics of intensity and frequency of droughts, either by using teleconnection concepts to find relations between climate and drought indices (Morid et al., 2007; Ganguli and Reddy, 2014), or by analyzing the variability of drought indices (Santos et al., 2010; AghaKouchak, 2015; Mazdiyasni and AghaKouchak, 2015; Xu et al., 2015).

Northeastern Brazil (NEB) is known for being exposed to major drought events, notably in its semiarid portion, with serious social and economic implications for of over 20 million people (IBGE, 2010). The worst performing economic and social indicators of Brazil can be found in this part of the country (UNPD, 2013; Sena et al., 2014; Cunha et al., 2015) due to the absence of resilient drought planning and management institutions. In general, NEB experiences high intensity of hydrological extreme events and a large interannual variability of precipitation (Giannini et al., 2004). Spatially, precipitation in NEB is heterogeneous, ranging from an annual total of 300 mm in the semiarid region to 2000 mm in the coastal zone (Amorim et al., 2014; Campos, 2015; Cunha et al., 2015).

Different precipitation regimes can be found across the NEB region. For instance, the Northern Northeast Brazil (NNEB) is characterized by a rainy season (between February and May) with a large interannual variability (Uvo and Berndtsson, 1996; Uvo et al., 1998) and no precipitation during the major part of the dry season (October–December). In contrast, the Eastern Northeast Brazil (ENEB) can receive up to 500 mm of precipitation during the dry season. The precipitation in the dry season of the ENEB is essential to socio-economic stability in the region as it provides the required water resources to maintain the local agro-pastoral activities, mainly rain fed agriculture, as well as in the combat of starvation and malnutrition. Irrespective of the level of precipitation in the wet season, the occurrence of below-average precipitation in the dry season, has major impacts on vital drinking and irrigation water. Nevertheless, there remains a limited understanding of the precipitation variability during the dry season in the ENEB.

It is well known that large-scale global climate phenomena such as the El Nino- Southern Oscillation (ENSO) and anomalous patterns of Sea Surface Temperature (SST) in the tropical Atlantic play an important role in the droughts at NNEB during the rainy season (Uvo et al., 1998; Andreoli and Kayano, 2004, 2006; Giannini et al., 2004). Giannini et al. (2004) showed that the influence of ENSO over the NNEB precipitation can be intensified or hindered by the condition of the SST anomalies in the tropical Atlantic, suggesting that there is no univocal relationship between large-scale anomalous SST patterns and droughts in NNEB (Kane, 1997; Andreoli and Kayano, 2006).

Although many studies have shown advances in the understanding of the NEB drought dynamics, these studies mainly focus on the wet season of the NNEB. The precipitation dynamics of the ENEB is poorly understood (Da Silva et al., 2010; Amorim et al., 2014), especially during the dry season. Therefore, this work aims to fill this knowledge gap by investigating the spatio-temporal relationships between climate phenomena and drought events in the ENEB using multivariate statistical techniques. This knowledge is a first step toward long-term drought forecasting in the ENEB to improve water resources planning and management in the region.

MATERIALS AND METHODS

Study Area

The Paraíba do Meio and Mundaú river basins are located in the ENEB, between 8°30′00″S to 10°00′00″S latitude and 35°30′00″W and 37°00′00″W longitude (Figure 1). The basins cover an area of 7283 km^2 approximately (Paraíba do Meio and Mundaú river basins drain areas of 3157 and 4126 km^2, respectively).

The basins have similar climate-physiographic properties that vary greatly from the outlets (coastal zone) to the headwaters. The lower part of the basins is characterized by a humid climate with tropical forest, different from the upper part of the basins, which are semiarid. The central part of the basins is characterized by transition climate and vegetation between semiarid and tropical climate. The average annual precipitation in the upper part of the basins is about 800 mm while in the lower part of basins it is about 2000 mm (COTEC Consultoria Técnica Ltd., 1999). The annual precipitation regime is characterized by a wet season (May to July) when 50% of the annual precipitation is concentrated, and a dry season between October and December (Figure 2).

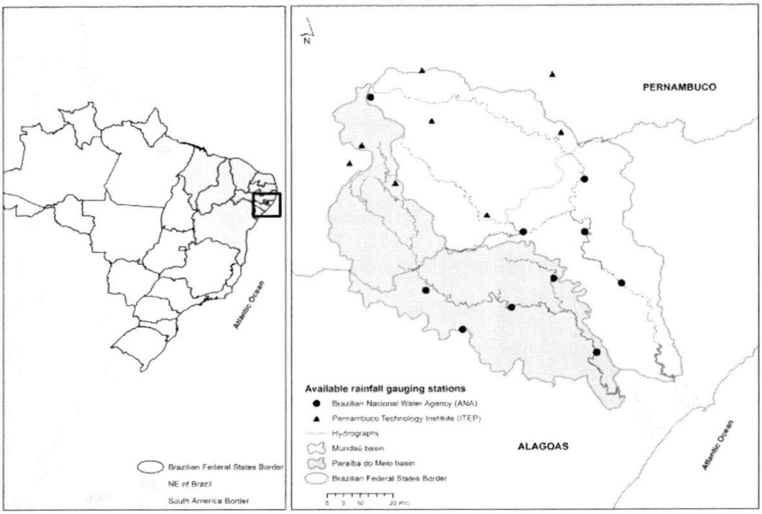

Figure 1. Paraíba do Meio and Mundaúriver basins located in ENEB.

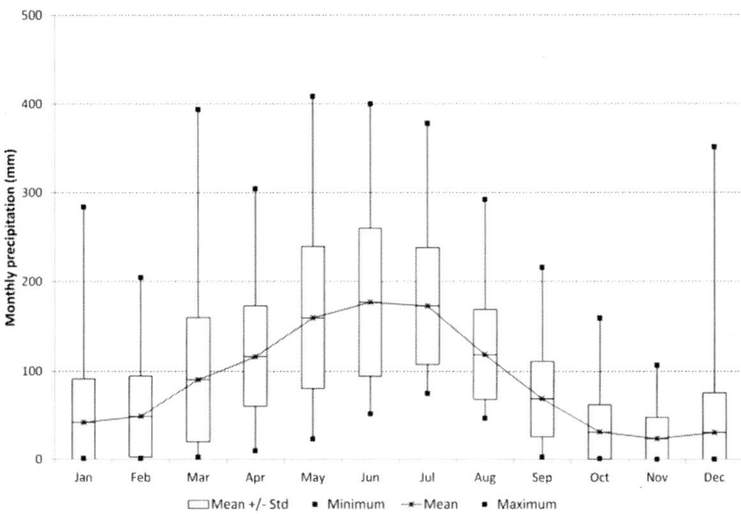

Figure 2. Monthly variability of precipitation in the study area. The three driest months occur from October to December, which presents an amplitude of accumulated precipitation up to 500 mm.

The total population in the Paraíba do Meio and Mundaú river basins is around 800 thousand inhabitants (IBGE, 2010) distributed among 35 municipalities. The livelihoods and economic activities are mainly livestock and rain fed agriculture (sugar cane and subsistence farming), which is highly correlated with variations in the available water resources. Therefore, climate variability and their impacts in water availability are significant for the livelihoods in the basins.

Data

The precipitation data in this study were collected from 18 well-distributed rainfall gauges, 10 belonging to the Brazilian National Water Agency (ANA) and 8 belonging to the Pernambuco Technology Institute (ITEP), on a daily time-scale from 1962 to 2008 (Figure 1). All daily precipitation data were qualitatively checked through consistency tests (Collischonn and Tucci, 2001). In addition, missing data in the individual rainfall gauges with gaps over 5% of the total sample were identified, and these gaps were filled using nearby stations up to 30 km through inverse distance square method (Xia et al., 1999).

From treated daily precipitation data, monthly time series of precipitation were derived, as well as the accumulated precipitation for the three driest months, October to December (OND). The OND precipitation time series were used to estimate 3-months Standardized Precipitation Index ($SPI\text{-}3_{OND}$, according to McKee et al., 1993) in order to determine the intensity of drought at this timescale (Table 1). It is worth pointing out that this index can be also used to determine periods of anomalously wet events, as SPI measures rainfall anomalies.

Table 1. Drought categorization according to the SPI values (adapted from Agnew, 2000).

Non exceedance probability	SPI	Drought category
0.05	>1.65	Extremely Wet (EW)
0.10	>1.28	Severely Wet (SW)
0.20	>0.84	Moderately Wet (MW)
0.60	> −0.84 and <0.84	Normal (N)
0.20	< −0.84	Moderate Drought (MD)
0.10	< −1.28	Severe Drought (SD)
0.05	< −1.65	Extreme Drought (ED)

Seven climate indices were used to investigate their possible relationships with drought events. These indices are related to Atlantic and Pacific conditions, such as the Atlantic Meridional Mode (AMM), which is divided into AMM_SST and AMM_Wind, Atlantic Multidecadal Oscillation (AMO), North Atlantic Oscillation (NAO), Pacific Decadal Oscillation (PDO), and El Niño Southern Oscillation (ENSO), which is measured by Southern Oscillation Index (SOI) and Niño 3.4 SST index. The climate indices were obtained at a monthly frequency from the database of the National Center for Atmospheric Research (NCAR) and National Oceanic and Atmospheric Administration (NOAA), using the same period for precipitation data. In addition, El Niño and La Niña events for OND were identified using a classification of cold and warm episodes considering the 3-month running mean by the Climate Prediction Center (CPC-NOOA).

AMM is a dominant source of variability in the tropical Atlantic (Chiang and Vimont, 2004), leading mode of basin-wide coupled ocean-atmosphere interaction between SST and low-level wind (Chiang and Vimont, 2004; Kossin and Vimont, 2007), which is historically known as interhemispheric or Atlantic dipole mode. AMO is a near-global scale mode of natural variability of the SST in the North Atlantic Ocean (Schlesinger and Ramankutty, 1994; Knight et al., 2006). NAO is a measure of the strength of the westerly winds blowing across the North Atlantic Ocean in the 40–60°N latitude belt (Greatbatch, 2000), producing large changes in the heat and moisture transport between Atlantic and the neighboring continents (Hurrell, 1995). PDO is an oceanic-atmospheric phenomenon and could be measured either by the intensity of the SST anomalies in Central North Pacific or a mean zonal Sea Level Pressure (SLP) anomalies over the North Pacific (Mantua et al., 1997). ENSO is considered as the dominant mode of the Pacific climate by an ocean-atmosphere coupled phenomenon associated with near-global climate variations (Rasmusson and Arkin, 1985; Gouirand and Moron, 2003; Räsänen and Kummu, 2013). SOI is the atmospheric component of the ENSO, calculated by the normalized difference between the intensity of the SLP at Darwin, Australia and at Tahiti (Trenberth, 1997; Singh, 2001; Labat, 2010). Niño 3.4 is a region in the central tropical Pacific where ENSO intensity can be measured by the intensity of the SST anomalies (Trenberth, 1997; Paz et al., 2011).

Spatial Patterns of OND Precipitation and Homogeneous Groups

Principal Component Analysis (PCA, e.g., Jackson, 1991) was applied using OND total precipitation (i.e., array with 47 years as rows and 18 rain gauges in the columns). This technique is useful in reducing the information in a large number of variables into a smaller set, and identifying the major part of the overall variance (Muñoz-Díaz and Rodrigo, 2004; Neto et al., 2014). The first and second components of the array of eigenvectors were evaluated to identify spatial patterns of rainfall co-variability. Physical factors that may explain the observed patterns in the first and second components are discussed in the conclusion section.

A cluster analysis was performed using the enough number of modes of the principal components to explain at least 75% of the total variance (Uvo, 2003). For this analysis the method of Ward (Uvo and Berndtsson, 1996; Muñoz-Díaz and Rodrigo, 2004) was used to identify Homogeneous Groups (HG) of rainfall stations with similar precipitation variability in the Paraíba do Meio and Mundaú river basins.

Spatio-Temporal Patterns of Droughts

The Homogeneous Groups (HG) determined by the Cluster Analysis were used to evaluate the temporal evolution and the percentage of the area in each HG affected by drought events considering different drought categories according to the SPI-3_{OND} values. Drought categories are based on Agnew (2000) and shown in Table 1.

Spatio-Temporal Relationships between Climate and Drought Events

A Spearman rank correlation was applied to investigate the relationship between the time series (scores) of the two first modes of the PCA (defined as the projection of the original data onto the new PCA axes) and the average of climate indices from October to December. The rank correlation was considered due to its capacity to measure both linear and non-linear correlations between variables (Paz et al., 2011). The statistical significance of the correlation coefficient was checked by means of *t*-test and 95% significance was used.

In addition, aiming to assess climate interactions linked to drought phenomena, we carried out a comparison analysis of the state of the climate indices during drought episodes classified as moderate, severe or extreme categories (Table 1). Moreover, a Maximum Covariance Analysis (MCA) was carried out between global sea surface temperature and the spatial average precipitation for the two HG determined by the cluster analysis.

The MCA is a multivariate analysis procedure based on the singular value decomposition of the cross-covariance matrix between fields of two datasets. It allows the isolation of sets of mutually orthogonal pairs of spatial patterns that maximize the squared temporal covariance between the two physical variables. The theory of MCA has been discussed in Bretherton et al. (1992), where it was then named as Singular Value Decomposition, and has been largely applied since (e.g., Uvo et al., 1998; Czaja and Frankignoul, 2002; Rana et al., 2012, among many others). Results from the MCA may be presented as correlation coefficients between the values of each grid point of one field and the expansion coefficients of the other field. For this application, the patterns showed on each mode indicates how well the pattern of the precipitation (SST) anomalies relates to the expansion coefficient of SST (precipitation).

All data were standardized before being used in the MCA with the objective of eliminating possible scale effects (Uvo, 2003; Muñoz-Díaz and Rodrigo, 2004), by means of the equation below:

$$Z_{i,j} = \frac{x_{i,j} - \overline{X_j}}{\sigma_j}, \tag{1}$$

where x_i is the observed value in the year i and rainfall station j, and Xj‾‾‾‾‾Xj‾ and σ_j are the average and the standard deviation, respectively, in each rainfall station j.

All calculations were performed using the software MATLAB®. The outputs maps were produced in ArcGIS® 10.2.

RESULTS

Spatial Patterns of OND Precipitation and Homogeneous Groups

The three first principal components together explain above 75% of the total variance of the OND precipitation. The first and the second principal components were found to be the most significant ones, both statistically and physically, explaining 62.59 and 8.63% of the total variance, respectively. The eigenvectors of the first principal component consist of positive values for the entire domain (Figure 3A), with higher values from outlet to center of the basins. A small reduction of the eigenvectors values in the upper part of the basins is noticed, mainly after a significant topographic barrier located at central eastern portion of Paraíba do Meio and Mundaú river basin. The eigenvector of the second principal component indicated a very clear separation between the upper and lower parts of the basins, consisting of both positive and negative values, respectively (Figure 3B), with a transition region in the central part of the basins. The cluster analysis (Figure 4) indicated two well-defined

Homogenous Groups (HG1 and HG2) of OND precipitation, following mostly the spatial pattern found in the second principal component. The classification with three HGs was rejected because the third HG covers a non-distinct area in HG1 composed by only four rainfall gauges.

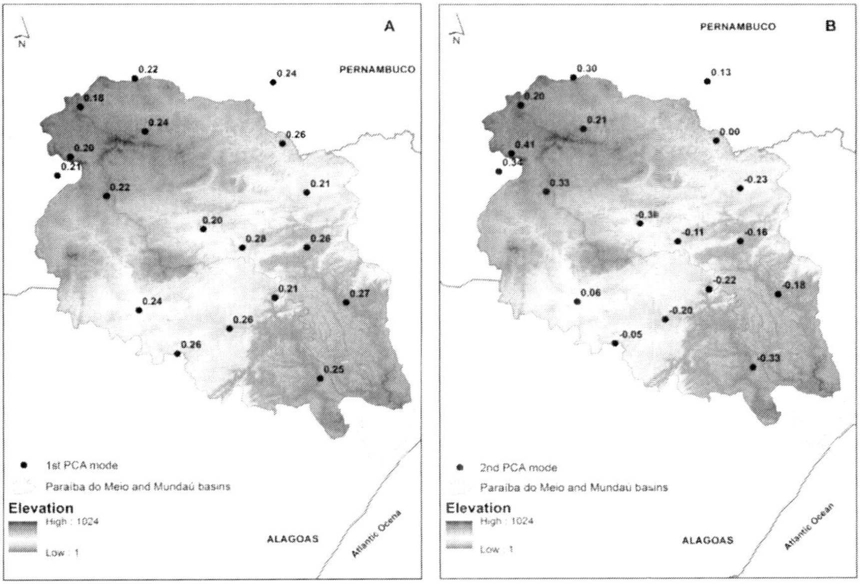

Figure 3. Eigenvectors of the co-variance matrix of the (A) first and (B) second PCA components of OND precipitation.

Spatio-Temporal Patterns of Droughts

Figure 5 shows a temporal evolution of the percentage of affected areas by different drought categories in the two defined Homogeneous Groups (HG1 and HG2). In general, the frequency of the events is similar in both Homogeneous Groups for drought events of larger spatial extension. For instance, the number of events affecting more than 50% of the area was 6 in HG1 and 8 in HG2. Considering the drought events that affected more than 70% of the area, only four events were observed for both Homogeneous Groups.

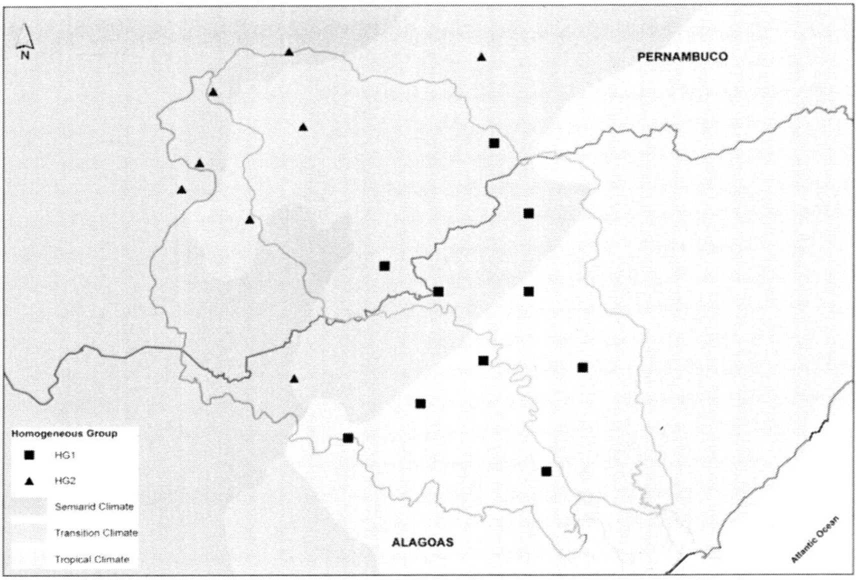

Figure 4. Cluster analysis of the spatial similarity of the OND precipitation (marks) and climatic classification (shades).

One isolated dry event of large spatial extension occurred in the 1970's, with affected area exceeding 65% of the total area. In the 80's, drought events were more frequently observed, with 3-years episodes affecting more than 50% of the HG1 and HG2. The major drought event of this decade took place in 1987, affecting the entire HG1 with 52% of this area classified in the extreme category of drought. The frequency of drought events had a remarkable increase in the 90's. However, in only 2-years isolated drought episodes (1992 and 1998) the affected area exceeded 50%. In 1998, specifically, the drought affected almost the whole study area, covering notably more than 80% of the area in HG2 as extreme category of drought.

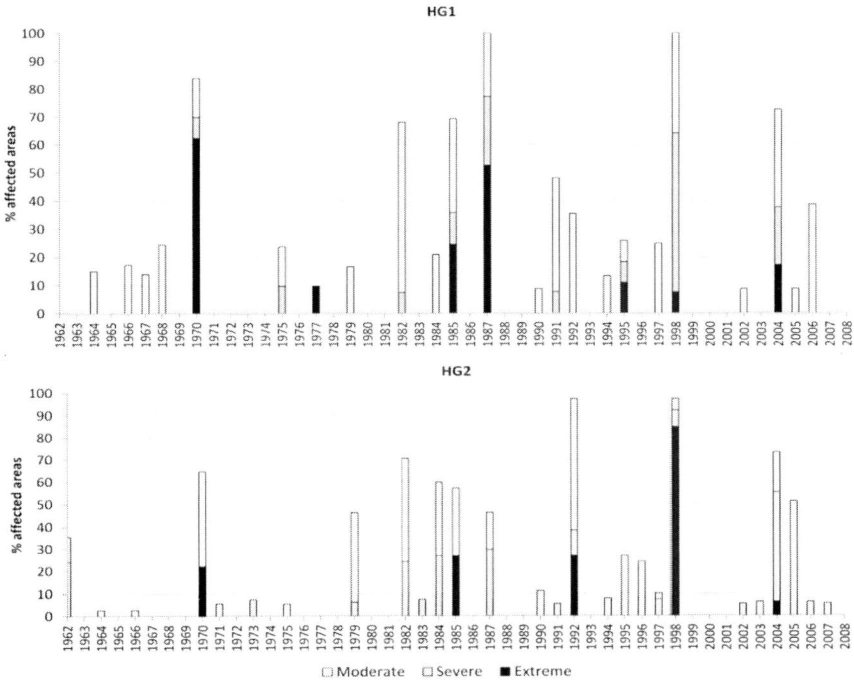

Figure 5. Temporal evolution of the percentage of affected areas by different drought categories of SPI3Oct-Dec, considering the two-defined Homogeneous Groups (HG1 and HG2) of the Cluster Analysis.

Spatio-Temporal Relationships between Climate Indices and Drought Events

The Spearman correlations between the two first PCA modes and climate indices are presented in Table 2. Note that only NAO showed a negative and statistically significant correlation with the second PCA mode. The first PCA mode was better correlated with AMM_Wind, NAO, and AMO, although such negative correlations were not statistically significant. We also found correlation values to be predominately negative between all climate indices and both PCA modes.

Table 2. Spearman correlation between the two first PCA modes and climate indices.

PCA Modes	Climate Index						
	NAO	PDO	AMO	SOI	NINO 3.4	AMM_wind	AMM_sst
1st	−0.1864	−0.0378	−0.1858	0.0944	−0.0165	−0.2143	−0.1995
2nd	−0.4393*	−0.1217	−0.1190	−0.0616	−0.0070	0.0113	−0.1651

*Indicates correlation values with 95% statistical significance.

A general perspective of the state of the most statistically significant climate indices found in the correlation analysis in years with anomalous OND precipitation is shown in Table 3. We have selected only events with large spatial extension in the Homogeneous Groups, since they are more related to large-scale global climate phenomena.

Table 3. State of the climate indices, in terms of anomaly values, for some anomalous episodes of OND precipitation that occurred over the Homogeneous Groups (HG1 and HG2).

	Climate index							Drought category	
	NAO	PDO	AMO	SOI	NINO 3.4	AMM_Wind	AMM_SST	HG1	HG2
DROUGHT YEARS									
1970**	−0.467	−1.063	−0.204	2.533	−1.203	0.727	0.597	ED	MD
1982*	1.260	0.127	−0.365	−4.433	3.058	−4.460	−4.397	MD	MD
1984**	0.520	0.703	−0.335	−0.433	−1.189	−5.580	−2.980	N	MD
1985	−0.447	−0.027	−0.270	−0.500	−0.601	−1.883	−1.070	MD	MD
1987*	−0.310	1.367	0.029	−0.833	1.854	−1.177	0.333	ED	SD
1992	0.690	0.797	−0.287	−2.000	0.116	0.087	−0.663	N	SD
1998**	0.480	−0.783	0.351	1.867	−1.599	−2.800	4.157	SD	ED
2004*	0.290	−0.303	0.247	−1.233	1.105	5.083	4.253	MD	MD
WET YEARS									
1963*	−0.350	−1.227	−0.052	−2.333	1.351	−1.673	−2.503	EW	EW
1965*	−0.547	−0.297	−0.205	−1.633	2.168	−1.680	−2.103	N	SW
1967	0.337	−0.363	−0.151	−0.900	−0.440	0.977	−1.473	N	SW
1974**	0.280	0.070	−0.419	0.267	−0.744	−3.607	−3.830	N	MW
1976*	−0.473	1.193	−0.366	0.167	1.227	−5.233	−0.663	SW	MW
1981	−0.863	0.550	−0.091	−0.133	0.074	−0.990	0.280	MW	MW
1986*	1.723	1.513	−0.342	−1.500	1.503	−6.570	−3.900	SW	N
1988**	0.287	−0.183	−0.146	2.433	−2.497	−2.727	−0.137	N	MW
1989	−0.113	−0.277	−0.123	−0.267	−0.136	0.080	0.643	MW	MW
1993	0.373	1.240	−0.262	−0.900	0.484	−7.197	−4.807	EW	MW
1999**	0.643	−1.970	0.026	1.933	−1.699	0.273	2.717	MW	N
2001	0.107	−1.187	0.236	−0.433	−0.185	6.223	4.627	MW	N

*Drought events were classified as in **Table 1**, according to Agnew (2000). Years with El Niño (*) and La Niña (**) phenomena were highlighted in the table, as well as the drought (white background) and wet (shading background) years.*

Although positive and negative phases of Niño 3.4 are highly related to El Niño and La Niña years, respectively, the most drought and wet events predominantly occur in anomalous state of ENSO phenomena, but these events are not necessarily linked to El Niño or La Niña years. In general, the anomalous events of OND precipitation were predominantly observed during the negative phase of the AMM_Wind, which also showed higher magnitudes in El Niño or La Niña years. In addition, the phases of the Niño 3.4 and AMM_Wind seem to regulate drought category during non-El Niño and non-La Niña conditions. For instance, in these years, drought and wet events occurred when Niño 3.4 and AMM_Wind exhibited the same and different phases, respectively. The driest (e.g., 1970, 1987, and 1998) and wettest (e.g., 1963, 1976, and 1993) episodes in terms of spatial extension and intensity occurred in years with the positive and negative phase of AMM_SST, respectively. However, the states of ENSO and AMM_Wind also can influence those driest and wettest events, mainly for less intense anomalies of AMM_SST (i.e., below to 1.00 in term of absolute values). For instance, in 1987 occurred an El Niño episode that likely favored a driest condition. In 1976, the intense and negative phase of AMM_Wind may have contributed for a wettest condition in the ENEB, overcoming the El Niño effect in this year. There is no well-defined relationship between drought categories and the state of PDO, NAO, and AMO.

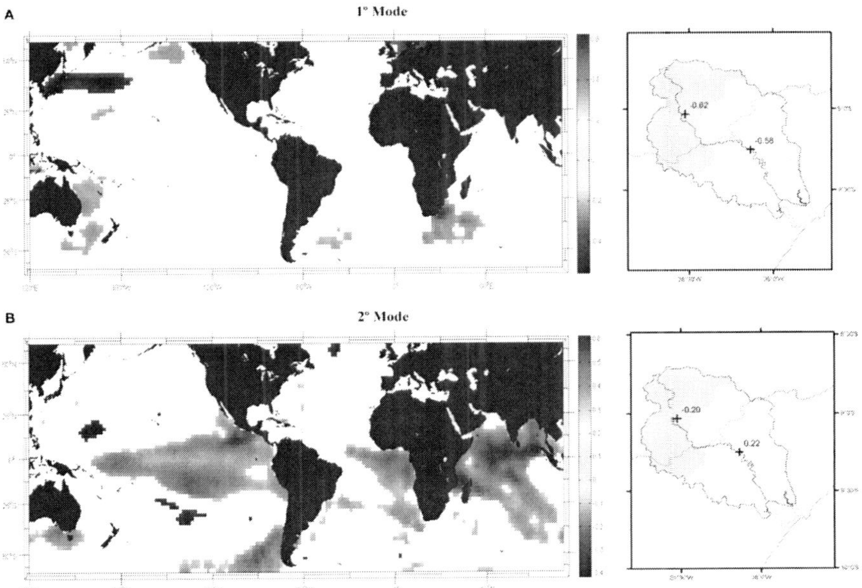

Figure 6. Heterogeneous correlation maps for the first (A) and second (B) modes of the MCA for OND global SST (left panel) and OND precipitation (right panel) for each HG. Only significant correlation at 95% are showed in the SST maps. Positive and negative correlation values are in red and blue colors, respectively.

Maximum Covariance Analysis

Results from the two first modes of the MCA are presented in Figure 6. The first mode, that has a squared covariance fraction (SCF) of 55.1%, shows that precipitation for the two HG are not correlated with any SST pattern connected to known climate variability phenomena. On the other hand, the SST pattern of the second mode (SCF of 28.1%) is clearly recognizable as ENSO and other climate variability patterns with higher statistically significant correlations along the Tropical Pacific Ocean, east Tropical Atlantic and Indian Ocean.

DISCUSSION

The two Homogeneous Groups defined by Cluster Analysis were closely related to the first and second PCA modes. The spatial structure presented by the eigenvectors of the first PCA mode corresponds closely to the spatial distribution of the physiographic characteristics in the region (topography barriers), suggesting that it represents the OND precipitation co-variability as determined by the influence of the topography on the small and synoptic scales meteorology. In addition, its PCA scores represent the temporal variability of the interaction of local and synoptic scales meteorology with the topography. This interpretation is supported by the lack of significant correlation between the scores and the time series of climatic indices presented in Table 1. The second PCA mode is related to the large-scale climatic conditions over the area. Its eigenvector spatial structure is in agreement with the climatic differences in the region, i.e., semiarid and tropical climate in the upper and lower part of the basins, respectively. Moreover, the temporal variability in this mode is correlated to global climatic phenomena by teleconnections as seen in Table 1. This connection is also supported by the MCA results that clearly shows SST patterns representative of known climate phenomena such as ENSO, correlated to precipitation of the two HG in the second mode, while the first mode does not present any SST recognized pattern.

Although the study area has been sub-divided in two distinct Homogeneous Groups, notably the drought events with larger spatial extension seems to affect both Groups similarly in terms of affected areas and intensity of drought. This suggests that global climate phenomena, which cover large geographic areas, can govern driest events in the study area.

It was also observed that a small area of the basins, located in the upper part of HG2, is frequently more affected by drought events of low spatial extension. This small area is entirely inserted in Brazilian semiarid, which is well known as a very vulnerable region to drought events (Krol et al., 2006; UNICEF, 2006; Mesquita and Bursztyn, 2014; Sena et al., 2014). The vulnerability in that region is mainly due its natural characteristics (i.e., lower rainfall amount in season wet, higher rates of annual evapotranspiration and crystalline substrate) to be prone of droughts (Krol et al., 2006; Campos, 2015). In addition, the underdevelopment conditions of the Brazilian semiarid in social and infrastructure aspects also can intensify the drought effects, especially for the rural population, where the economic activity is basically subsistence agriculture in an extreme poverty environment (Cunha et al., 2015). As an example, the most recent drought (from 2012 to 2013), reaching in economic losses costs of greater than US $4.6 billion due to cattle mortality, crop's damage and insurance claims (Campos, 2015; Cunha et al., 2015). This extreme drought event demanded a significant assistance from the Brazilian government to drought-affected population with drought relief programs and using tank trucks to draw water from reliable sources for public consumption (Campos, 2015). Thus, it is important to highlight that even though this small area of HG2 experiences less intense drought events than in HG1, the socio-economic impacts in this area are certainly more significant.

Larger anomalous episodes of OND precipitation seem to be more related to stronger anomalies of some analyzed climate indexes, suggesting a cause-and-effect relationship between climate aspects and driest/wettest years. Therefore, this understanding can mitigate regional vulnerability in the NEB and avoiding social-economics losses (Andreoli and Kayano, 2006; Amorim et al., 2014; Sena et al., 2014; Campos, 2015).

We showed that NAO might have a secondary role on the interannual variability of OND precipitation, since this index was better correlated to the second PCA mode and this mode only explains a small part of the total variance of OND precipitation in the study area. Although some studies have indicated that state of NAO can influence the conditions in the Tropical Atlantic (Greatbatch, 2000; Mo and Hakkinen, 2001; Czaja et al., 2002; Chiang and Vimont, 2004), for instance through an interhemispheric relationship between North and South Atlantic (Robertson et al., 2000; Mo and Hakkinen, 2001), our results find only a slight effect of North Pacific on OND precipitation in ENEB.

On the other hand, the state of climate indices, such as Niño 3.4, AMM_Wind and AMM_SST, seem to modulate the development of larger anomalous events of OND precipitation in terms of spatial extension and intensity, suggesting that conditions in the Tropical Pacific and Atlantic can play an important role in the occurrence of more extreme events in ENEB, in agreement with results presented by Andreoli and Kayano (2006) for the entire NEB and by Amorim et al. (2014) for ENEB.

Our results showed that the state of the Tropical Atlantic determines the occurrence of the larger anomalous events in ENEB and ENSO can intensify or inhibit the total of the OND precipitation. Indeed, the positive (or negative) phase of AMM_SST is closely related to the reduction (or enhancement) of precipitation over the NEB as expressed by Chiang and Vimont (2004). Thus, the state of AMM seems to be a potential indicator for larger anomalies of the OND precipitation. In addition, it is well known that ENSO strongly influence the variability of precipitation in NEB for the rainy season (Nobre and Shukla, 1996; Uvo et al., 1998; Giannini et al., 2004; Andreoli and Kayano, 2005), since in El Niño years the equatorial convection is displacement eastward, creating a subsidence area over NEB, which can inhibit precipitation in NEB. Our findings suggest an unclear interannual relationship between ENSO and anomalous precipitation events in ENEB during the dry season, as wet episodes may occur during El Niño years and dry ones during La Niña years. Such unexpected association between ENSO and OND precipitation in ENEB may be explained either by the influence of other factors (e.g., Tropical Atlantic SST condition) on OND precipitation anomalies, which may disrupt this interannual relationship (Kane, 2001), or due to the most prominent long-term relationship (Kane, 1999). Indeed, we only could find few years with strong ENSO anomalies (e.g., 1982 and 1999) in which Pacific conditions seem to have overcome the effect of Tropical Atlantic conditions on anomalies of precipitation in ENEB.

Notably, the larger anomalous episodes of OND precipitation were not well related to the state of long-term climate indices, such as PDO and AMO. This suggests deeper investigations are needed to get a better understanding of the relationship between OND precipitation and those indices, such as spectral analysis to track non-linear patterns in different time-frequencies.

AUTHOR CONTRIBUTIONS

DC, Study conception and design, Acquisition of data, Analysis and interpretation of data, Drafting of manuscript and Critical revision. TS, Study conception and design, Acquisition of data, Analysis and interpretation of data and Drafting of manuscript. CF, Study conception and design, Analysis and interpretation of data, Drafting of manuscript, Critical revision. KM, Analysis and interpretation of data and Critical revision. CU, Study conception and design, Analysis and interpretation of data and Critical revision.

CONFLICT OF INTEREST STATEMENT

The authors declare that the research was conducted in the absence of any commercial or financial relationships that could be construed as a potential conflict of interest.

ACKNOWLEDGMENTS

The first author would like to thank CAPES Foundation, Ministry of Education of Brazil by the scholarship granted in the process n° BEX 13470/13-2. We are also grateful to institutional grants provided by The Swedish Foundation for International Cooperation in Research and Higher Education (STINT) through Research Links Project, which funded the international cooperation between Lund University and Federal University of Alagoas and to the Swedish Research Council (Vetenskapsrådet) that partially finances KM and CU. The authors also would like to thank the anonymous referees, who helped to significantly improve the quality and the clarity of this manuscript.

REFERENCES

1. AghaKouchak, A. (2015). A multivariate approach for persistence-based drought prediction: application to the 2010–2011 East Africa drought. *J. Hydrol.* 526, 127–135.
2. Agnew, C. T. (2000). Using the SPI to identify drought. *Drought Netw. News* 12, 6–12.
3. Amorim, A. C. B., Chaves, R. R., and Silva, C. M. S. (2014). Influence of the tropical Atlantic Ocean's sea surface temperature in the eastern northeast Brazil precipitation. *Atmos. Clim. Sci.* 4, 874–883.
4. Andreoli, R. V., and Kayano, M. T. (2004). Multi-scale variability of the sea surface temperature in the Tropical Atlantic. *J. Geophys. Res.* 109, 1–12.
5. Andreoli, R. V., and Kayano, M. T. (2005). ENSO-related rainfall anomalies in South America and associated circulation features during warm and cold Pacific decadal oscillation regimes. *Int. J. Climatol.* 25, 2017–2030.
6. Andreoli, R. V., and Kayano, M. T. (2006). Tropical Pacific and South Atlantic effects on rainfall variability over Northeast Brazil. *Int. J. Climatol.* 26, 1895–1912.
7. Bretherton, C. S., Smith, C., and Wallace, J. M. (1992). An intercomparison of methods for finding coupled patterns on climate data. *J. Clim.* 5, 541–560.
8. Campos, J. N. B. (2015). Paradigms and public policies on drought in northeast Brazil: a historical perspective. *Environ. Manage.* 55, 1052–1063.
9. Chiang, J. C. H., and Vimont, D. J. (2004). Analogous Pacific and Atlantic meridional modes of tropical atmosphere-ocean variability. *J. Clim.* 17, 4143–4158.
10. Collischonn, W., and Tucci, C. E. M. (2001). Simulação Hidrológica de Grandes Bacias. *RBRH - Rev. Bras. Recur. Hídricos* 6, 95–118.

11. COTEC Consultoria Técnica Ltd. (1999). *Plano Diretor de Recursos Hídricos da Bacia do Rio Mundaú*. Recife: Governo do Estado de Pernambuco - Secretaria de Recursos Hídricos.
12. Cunha, A. P. M., Alvalá, R. C., Nobre, C. A., and Carvalho, M. A. (2015). Monitoring vegetative drought dynamics in the Brazilian semiarid region. *Agric. For. Meteorol.* 214-215, 494–505.
13. Czaja, A., and Frankignoul, C. (2002). Observed impact of Atlantic SST anomalies on the North Atlantic oscillation. *J. Clim.* 15, 606–623.
14. Czaja, A., Van der Vaart, P., and Marshall, J. (2002). A diagnostic study of the role of remote forcing in tropical Atlantic variability. *J. Clim.* 15, 3280–3290.
15. Da Silva, D. F., de Sousa, F. A. S., and Kayano M. T. (2010). Escalas temporais da variabilidade pluviométrica na bacia hidrográfica do rio mundaú. *Rev. Br. Meteorol.* 25, 324–332.
16. Ganguli, P., and Reddy, M. J. (2014). Ensemble prediction of regional droughts using climate inputs and the SVM-copula approach. *Hydrol. Process.* 5009, 4989–5009.
17. Giannini, A., Saravanan, R., and Chang, P. (2004). The preconditioning role of tropical Atlantic variability in the development of the ENSO teleconnection: implications for the prediction of Nordeste rainfall. *Clim. Dyn.* 22, 839–855.
18. Gouirand, I., and Moron, V. (2003). Variability of the impact of El Nino -southern oscillation on sea-level pressure anomalies over the North Atlantic in January to March (1874-1996). *Int. J. Climatol.* 23, 1549–1566.
19. Greatbatch, R. J. (2000). The north Atlantic oscillation. *Stoch. Environ. Res. Risk Assess.* 14, 213–241.
20. Hastenrath, S. (2012). Exploring the climate problems of Brazil's Nordeste: a review. *Clim. Change* 112, 243–251.
21. Hurrell, J. W. (1995). Decadal trends in the North Atlantic Oscillation: regional temperatures and precipitation. *Science* 269, 676–679.
22. IBGE (2010). *Instituto Brasileiro de Geografia e Estatística. Características Gerais da População, Religião e Pessoas Com Deficiência. CENSO DEMOGRÁFICO 2010*. Rio de Janeiro: IBGE.
23. Jackson, J. E. (1991). *A User's Guide to Principal Components.* New York, NY: Wiley & Sons, Inc.
24. Kane, R. P. (1997). Prediction of droughts in north-east Brazil: role of ENSO and use of periodicities. *Int. J. Climatol.* 17, 655–665.
25. Kane, R. P. (1999). Rainfall extremes in some selected parts of central and South America: ENSO and other relationships reexamined. *Int. J. Climatol.* 19, 423–455.
26. Kane, R. P. (2001). Interrelationship and ENSO relationship of Fortaleza rainfall in different seasons. *Rev. Br. Geofísica* 19, 199–214.
27. Knight, J. R., Folland, C. K., and Scaife, A. A. (2006). Climate impacts of the Atlantic Multidecadal Oscillation. *Geophys. Res. Lett.* 33, L17706.
28. Kossin, J. P., and Vimont, D. J. (2007). A more general framework for understanding atlantic hurricane variability and trends. *Bull. Am. Meteorol. Soc.* 88, 1767–1781.
29. Krol, M., Jaeger, A., Bronstert, A., and Güntner, A. (2006). Integrated modelling of climate, water, soil, agricultural and socio-economic processes: a general introduction of the methodology and some exemplary results from the semi-arid north-east of Brazil. *J. Hydrol.* 328, 417–431.
30. Labat, D. (2010). Cross wavelet analyses of annual continental freshwater discharge and selected climate indices. *J. Hydrol.* 385, 269–278.
31. Mantua, N. J., Hare, S. R., Zhang, Y., Wallace, J. M., and Francis, R. C. (1997). A pacific interdecadal climate oscillation with impacts on salmon production. *Bull. Am. Meteorol. Soc.* 78, 1069–1079.
32. Massei, N., and Fournier, M. (2012). Assessing the expression of large-scale climatic fluctuations in the hydrological variability of daily Seine river flow (France) between 1950 and 2008 using Hilbert-Huang Transform. *J. Hydrol.* 448–449, 119–128.

33. Mazdiyasni, O., and AghaKouchak, A. (2015). Substantial increase in concurrent droughts and heatwaves in the United States. *Proc. Natl. Acad. Sci. U.S.A.* 112, 11484–11489.
34. McKee, T. B., Doesken, N. J., and Kleist, J. (1993). "The relationship of drought frequency and duration to time scales," in *AMS 8th Conference of Applied Climatology* (Anaheim, CA), 179–184.
35. Mesquita, P. S., and Bursztyn, M. (2014). "Climate variability in semi-arid brazil?: food insecurity, agricultural production and adjustment to perceived changes," in *International Policy Center for Inclusive Growth* (Bras lia: United Nations Development Programme).
36. Mo, K. C., and Hakkinen, S. (2001). Interannual variability in the tropical Atlantic and linkages to the Pacific. *J. Clim.* 14, 2740–2762.
37. Morid, S., Smakhtin, V., and Bagherzadeh, K. (2007). Drought forecasting using artificial neural networks and time series of drought indices. *Int. J. Climatol.* 27, 2103–2111.
38. Muñoz-Díaz, D., and Rodrigo, F. S. (2004). Spatio-temporal patterns of seasonal rainfall in Spain (1912-2000) using cluster and principal component analysis: comparison. *Ann. Geophys.* 22, 1435–1448.
39. Neto, J. L. R., Fragoso, C. R. Jr., Malhado, A. C. M., and Ladle, R. J. (2014). Spatio-temporal variability of chlorophyll-a in the coastal zone of northeastern Brazil. *Estuaries Coasts* 38, 72–83.
40. Nobre, P., and Shukla, J. (1996). Variations of sea surface temperature, wind stress, and rainfall over the tropical Atlantic and South America. *J. Clim.* 9, 2464–2479.
41. Paz, A. R., Uvo, B. C., Bravo, J. M., Collischonn, W., and Da Rocha, H. R. (2011). Seasonal "Precipitation forecast based on Artificial neural networks," *Computation Methods for Agricultural Research: Advances and Applications*, eds H. A. Prado, H. Chaib Filho, and A. J. B. Luiz (New York, NY: Hershey), Chapter. 16, 326–354.
42. Rana, A., Uvo, C. B., Bengtsson, L., and Sarthi, P. P. (2012). Trend analysis for rainfall in Delhi and Mumbai, India. *Clim. Dyn.* 32, 45–56.
43. Räsänen, T. A., and Kummu, M. (2013). Spatiotemporal influences of ENSO on precipitation and flood pulse in the Mekong River Basin. *J. Hydrol.* 476, 154–168.
44. Rasmusson, E. M., and Arkin, P. A. (1985). "Interannual climate variability associated with the El Nino/Southern oscillation," in *Coupled Ocean-Atmosphere Models*, ed J. C. J. Nihoul (Amsterdam: Elsevier Science Publishers), 289–302.
45. Robertson, A. W., Mechoso, C. R., and Kim, Y.-J. (2000). The influence of atlantic sea surface temperature anomalies on the North Atlantic Oscillation. *J. Clim.* 13, 122–138.
46. Santos, J. F., Pulido-Calvo, I., and Portela, M. M. (2010). Spatial and temporal variability of droughts in Portugal. *Water Resour. Res.* 46, 1–13.
47. Schlesinger, M. E., and Ramankutty, N. (1994). An oscillation in the global climate system of period 65–70 years. *Nature* 367, 723–726.
48. Sena, A., Barcellos, C., Freitas, C., and Corvalan, C. (2014). Managing the health impacts of drought in Brazil. *Int. J. Environ. Res. Public Health* 11, 10737–10751.
49. Silverman, D., and Dracup, J. A. (2000). Artificial neural networks and long-range precipitation prediction in California. *J. Appl. Meteorol.* 39, 57–66. doi:
50. Singh, O. P. (2001). Multivariate ENSO index and Indian monsoon rainfall: relationships on monthly and subdivisional scales. *Meteorol. Atmos. Phys.* 78, 1–9.
51. Trenberth, K. E. (1997). The definition of El Ni??o. *Bull. Am. Meteorol. Soc.* 78, 2771–2777.
52. UNICEF (2006). *The United Nations Children's Fund. UNICEF in the Brazilian semi-arid. National Pact a World for Children and Adolescents in the Semi-Arid.* Brasília: UNICEF.
53. UNPD (2013). *United Nations Development Programme. Índice de Desenvolvimento Humano Municipal Brasileiro. Atlas do Desenvolvimento Humano no Brasil.* Brasilia: UNPD.

54. Uvo, C. B. (2003). Analysis and regionalization of northern European winter precipitation based on its relationship with the North Atlantic oscillation. *Int. J. Climatol.* 23, 1185–1194.
55. Uvo, C. B., Repelli, C. A., Zebiak, S. E., and Kushnir, Y. (1998). The relationships between tropical Pacific and Atlantic SST and northeast Brazil monthly precipitation. *J. Clim.* 11, 551–562.
56. Uvo, C. R. B., and Berndtsson, R. (1996). Regionalization and spatial properties of Ceará state rainfall in Northeast Brazil. *J. Geophys. Res.* 101, 4221–4233.
57. Xia, Y., Fabian, P., Stohl, A., and Winterhalter, M. (1999). Forest climatology: estimation of missing values for Bavaria, Germany. *Agric. For. Meteorol.* 96, 131–144.
58. Xu, K., Yang, D., Yang, H., Li, Z., Qin, Y., and Shen, Y. (2015). Spatio-temporal variation of drought in China during 1961–2012: a climatic perspective. *J. Hydrol.* 526, 253–264.

CHAPTER 14

Analysis of Convective Thunderstorm Split Cells in South-Eastern Romania

Daniel Carbunaru,[1,2] Sabina Stefan,[1] Monica Sasu,[1,2] and Victor Stefanescu[1,2]

[1]*Department of Atmospheric Physics, Faculty of Physics, University of Bucharest, P.O. Box MG-11, 077125 Bucharest, Romania*
[2]*National Meteorological Administration, Bucuresti-Ploiesti Avenue, No. 97, 013686 Bucharest, Romania*

ABSTRACT

The mesoscale configurations are analysed associated withthesplitting process of convective cells responsible for severe weather phenomena in the south-eastern part of Romania. The analysis was performed using products from the S-band Doppler weather radar located in Medgidia. The cases studied were chosen to cover various synoptic configurations when the cell splitting process occurs. To detect the presence and intensity of the tropospheric jet, the Doppler velocity field and vertical wind profiles derived from radar algorithms were used. The relative Doppler velocity field was used to study relative flow associated with convective cells. Trajectories and rotational characteristics associated with convective cells were obtained from reflectivity and relative Doppler velocity fields at various elevations. This analysis highlights the main dynamic features associated with the splitting process of convective cells: the tropospheric jet and vertical moisture flow associated with the configuration of the flow relative to the convective cells for the lower and upper tropospheric layers. These dynamic characteristics seen in the Doppler based velocity field and in the relative Doppler velocity field to the storm can indicate further evolution of convective developments, with direct implications to very short range forecast (nowcasting).

1. INTRODUCTION

In south-eastern Romania, in the convective season (from May to September), severe weather phenomena develop frequently and evolve [1–5], sometimes leading to significant damages. Analysis of severe convective events and their structure, at least for certain classes of phenomena (isolated convective cells), taking into account the mesoscale configuration can improve the nowcasting procedures. Using an S-band Doppler radar (10 cm wavelength), we see that the mesoscale phenomena (supercellular thunderstorms) have certain features during their dynamic evolution. Of great importance in understanding the evolution of supercells is the splitting process (separation of the convective cells in two other ones rotating opposite to one another, namely, cyclonic and anticyclonic), a process that takes place during the evolution of convective developments and is closely linked to the state of mesoscale dynamic configuration, especially vertical wind shear [6, 7]. This process has been studied both qualitatively [8–18] and by numeric analysis [19–21], highlighting the interaction between horizontal vorticity from the vertical profile of horizontal wind and the updraft.

Depending on the curvature of the hodograph, the development of convective cells is favoured when the propagation is either to the right, or to the left oftroposphericshear; both cells may be favoured as well, when the hodograph is linear. Along with this particular interaction, there may also exist secondary ones [22] that lead to the diminishing of the cell that propagates to the left of shear vector, in the Northern Hemisphere. Observations of the split process when the Doppler radar is used [23–32] highlight the cyclonic and anticyclonic rotation, within the field of Doppler velocity, associated with convective cells compatible with vertical shear.

The above studies also showed the asymmetry between the cells that propagate to the right and those that propagate to the left of the tropospheric shear vector in the Northern Hemisphere, the latter generally has a shorter lifespan than the former, but is associated with severe hail.

In some cases the inflow at low levels and the outflow at upper levels [33] to and from the storm were seen with the aid of the Doppler relative velocity field; these two flows are in direct correlation with the intensity of updraft.

In this paper, several cases are analysed which featured the splitting process of strongly convective cells (supercells). The basic criteria in choosing these cases are the different mesoscale configuration (various directions of the tropospheric flow). The study focused on the analysis from fields and products of the meteorological Doppler radar that can provide relevant data on the severity of convective phenomena.

This paper is structured as follows. The first part contains the radar data that were used during the analysis of splitting processes, and the methodology thereof. At the second part, several case studies are presented that emphasize certain characteristics of the mentioned process. The paper ends with conclusions derived from these case studies, namely, mesoscale dynamic features and some particularities of convective cells trajectories. Since most of the convective cells appearing in the radar fields featured supercell traits, the common elements that were identified could be used in nowcasting.

2. DATA AND METHODS

For the dynamic mesoscale analysis and that of the splitting process, the WSR-98D Doppler radar in S band (10 cm) was used, located in Medgidia (south-eastern Romania) (Figure 1). The range of the radar is shown in Figure 1.

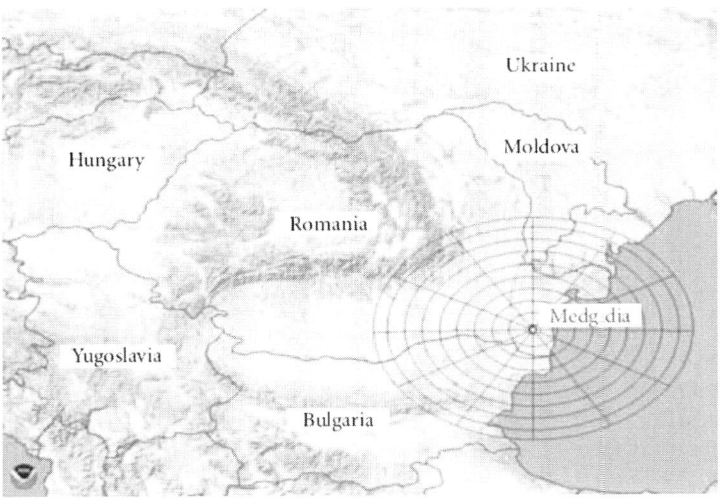

Figure 1: The spatial domain of radar analysis (geographic projection). The radar range is 230 km, the spatial resolution is 1 km, and the temporal resolution is 6 minutes, between two scans. NOAA Weather and Climate Toolkit (http://www.ncdc.noaa.gov/oa/wct/).

The splitting process was analyzed in the fields of radar reflectivity, Doppler base velocity, and that of Doppler relative velocity at different scan elevations, in order to see the formation of cyclonic and anticyclonic vortices. The Doppler relative velocity field is obtained from the Doppler base velocity field minus the average velocity of all detected storms and represents Doppler velocity field of the reference storm [34]. The Principal Users Processor (PUP) software (METSTAR WSR-98D PUP version 10.8.R) was used. The study of Doppler velocity field at various elevations, in parallel with the Vertical Azimuth Display (VAD) Wind Velocity product resulted in information about the vertical shear of horizontal velocities and tropospheric jets. The VAD Wind Velocity is a Doppler weather radar product which gives the vertical profile of horizontal wind vectors above radar station [34–36]. Strong shear in the wind field assumes the existence of tropospheric low-level jet (LLJ) or upper-level jet (ULJ). Also the Doppler relative velocity field gave information about the low and upper relative flow associated to the storms in radar field. The presence of relative low-level flow was analysed (arrow-shaped tube, in the lower part of Figures 2(a) and 2(b)) and the relative upper-level flow (arrow-shaped tube, at top of Figures 2(a) and 2(b)). For the first three analysed cases, vertical vorticity profiles were made, further to be compared with circular movement associated with supercells (horizontal circles with cyclonic and anticyclonic vorticity, in Figure 2).

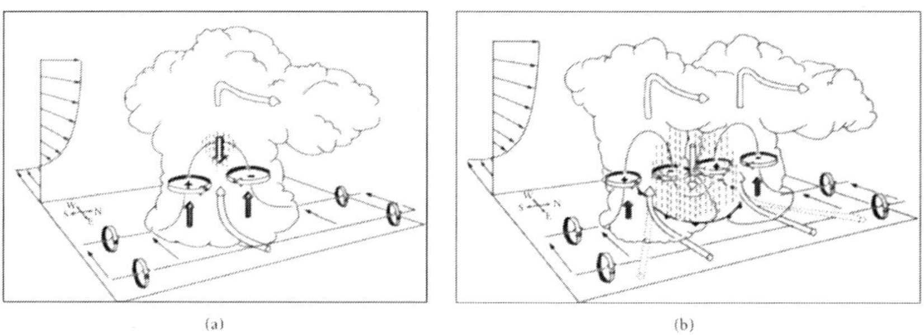

Figure 2: Tiling of the horizontal vorticity due to updraft (a). Negative vorticity maxima (on the left of shear) and positive vorticity maxima (on the right of shear) and (b) the convective cell split in two other cells with opposite vorticity (from Lin 2007, adapted by Klemp 1987, and from Rotuno 1981).

The evolution of convective cells was observed in the reflectivity field at elevations of 0.5 and 1.5 degrees, also in the Doppler radial velocities field and in the Doppler relative velocity field at elevations of 0.5, 1.5, 2.4, and 3.4 degrees.

The evolution in time and space of each convective cell that was characterised by the splitting process was also observed. Identification of splitting was done by tracking both reflectivity and cyclonic or anticyclone rotation, as seen in Figures 3(a) and 3(b). Figure 3(c) shows how the vertical vorticity of rotational movements was calculated.

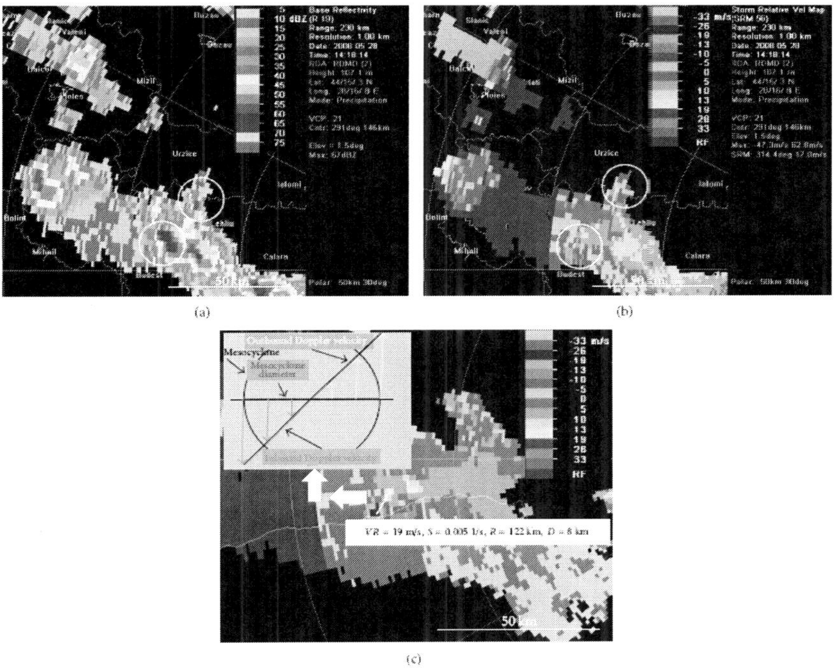

Figure 3: Radar reflectivity at 1.5-degree elevation (a), relative velocities at 1.5-degree elevation, (b) the white circles mark cyclonic and anticyclonic rotation associated with the two cells after the split, and shear due to the cyclonic and anticyclonic rotation (c).

Thus, for each rotational motion, maximum inbound and outbound velocities were identified in the Doppler relative velocity field at elevation where rotational movement had maximum intensity. The PUP application was used. For the actual calculation, radar algorithm is dividing the difference between the maximum inbound velocity and the maximum outbound velocity, by the distance between the two (mesocyclone diameter in Figure 3(c)).

Trajectories of convective cells in each case were obtained using polar coordinates, relative to the radar position, of the cells before and after the splitting. For the three cases studied in detail, the result was a map showing the individual characteristics of convective developments, taking into account the stage before splitting and the after-splitting stage (which is associated with the cyclonic and anticyclone rotation). Vertical variation of the horizontal velocity (vertical shear) was derived from the VAD Wind Profile algorithm and vertical profiles were then constructed from wind shear at particular moments, representative of the evolution of vertical shear during the life of the convective cell. The vertical domain associated with the -coordinate was separated into a domain with directional shear (in the lower tropospheric layer) and a domain with nondirectional shear (in the middle tropospheric level). This division was possible because the dynamic configuration near the radar site, at least in the three cases studied, has a nondirectional shear above 1800 m associated with an ULJ, and directional shear below 1800 m. In the first three cases, this dynamic configuration is qualitatively compatible with Doppler velocity field at 3.4-degree elevation.

To analyse the presence of the cell splitting process of convective cells in south-eastern Romania and its role in supercell evolution, cases of severe weather were analysed where this process was shown clearly in the radar fields. For the first three cases, in the reflectivity field the convective cells were identified that split, and those that resulted from the split. Using the Doppler base velocity field at 3.4-degree elevation one can tell if there is a quality equivalence between the vertical profile of horizontal velocity near the radar station and the mesoscale flow. Trajectories of split cells were rebuilt, to identify some of their peculiarities. The trajectories were compared with the precipitation field, as obtained from WSR-98D S-band radar precipitation algorithm (Precipitation Processing System, PPS [37]).

The vertical profiles associated with horizontal vorticity were compared with vertical vorticities specific to cyclonic or anticyclonic rotation. In the last two cases, the presence of an upper-level jet was analysed, and the intensity of relative flow was examined.

3. RESULTS AND DISCUSSIONS

3.1. The First Studied Case, May 28, 2008

A cyclone located in north-east Romania contributed to the instability conditions present in the Doppler radar area of the Medgidia site. Starting at 1130 UTC, on an NW-SE intense tropospheric circulation, convective cells developed in SE Romania that were split and became supercells. The most severe supercells were the ones with cyclonic rotation, such as those having the V shape in reflectivity field. Besides these, multicells systems were present in the radar field, with a lower degree of severity.

The temporal evolution of mesoscale flow is represented in Figure 4, between 1406 UTC and 1506 UTC, as seen in the VAD Wind Profile product and in the base Doppler velocities field. The vertical wind profile was characterized by intense mesoscale flow from the north west to south east, above 1800 m (6 KFT) (Figure 4(a), in the white box). Below this level, from the VAD Wind Profile and base Doppler velocities, one can notice the anticyclonic rotation of velocity vector (Figure 4(a), in the yellow box).

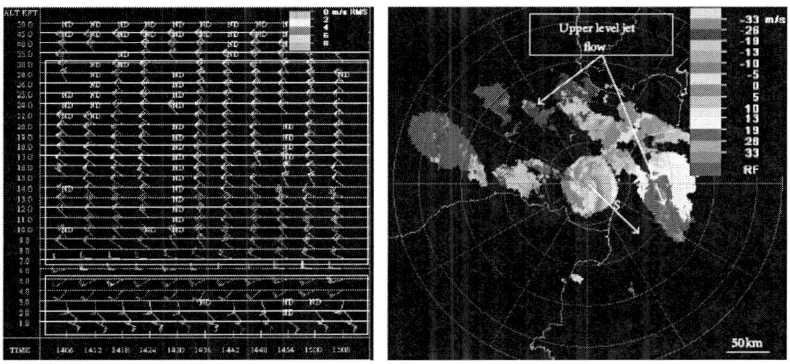

(a) (b)

Figure 4: May 28, 2008, vertical profile of horizontal wind (VAD Wind Profile) for the 1406 UTC–1506 UTC (a) and (b) Doppler velocities at 3.4-degree elevation, at 1341 UTC. S vector represents the tropospheric mean shear at 1800–6000 m. The flow is from north west and the jet is present at 5000–6000 m.

At the 5000 m level, Figure 4(a), the values of velocity reach 35 m s^{-1} growing to 40–45 m s^{-1} at 35 KFT (10500 m). The colour for the wind vectors represents the level of confidence in the VAD Wind Velocity data; the level of confidence is larger in the lower troposphere (green, dispersion of 0–2 m s^{-1}) and in the upper troposphere (yellow, dispersion of 2–4 m s^{-1}). A larger dispersion of velocity is recorded in the lower layers (red, dispersion of 4–6 m s^{-1}), which can be associated with different gust fronts caused by convective developments. In the same figure, both the anticyclonic rotation of the velocity vector in the lower two km and the upper tropospheric jet are consistent with the warm sector of the cyclone located north east of Romania. This vertical configuration of the horizontal velocity is also associated with increased thermodynamic instability in the troposphere below approximately 500 hPa; the main factor for the increase in conditional instability is moisture advection in the lower layers. However, the presence of a vertically expanded upper tropospheric jet is associated with ageostrophic velocities, which are the result of horizontal variation of the geostrophic velocities that lead to changes in the thermal gradient within the middle troposphere [18].

Vertical wind profile configurations of the mesoscale flow can be seen in the Doppler velocity field at 3.4-degree elevation (Figure 4(b)), and they are qualitatively equivalent with the vertical wind profile given by the VAD Wind Profile product. This equivalence is the result of the zero isodope analysis [34] in the Doppler velocity field. The zero isodope represents the area where the Doppler velocities are zero (gray colours, in Figure 4(b)). Near the radar site (in the boundary layer) the isodope is S shaped, compatible with the velocities profile (within the yellow box, in Figure 4(a)). Going away from the radar (increasing the height), this has a linear shape on the NE-SW direction and is compatible with the upper tropospheric jet direction (Figure 4(a)). The intensity of the jet stream is obtained by measuring the Doppler velocities at 90 degrees from the isodope, having the same height (same distance from the radar in the Doppler velocity field).

The dynamic field of the Doppler velocities was maintained over the interval with the mentioned convective phenomena (a few hours). In Figure 5(a) the reflectivity field is shown at 1500 UTC at 1.5-degree elevation, and in Figure 5(b) the Doppler relative velocity field is shown at 2.4-degree elevation at 1500 UTC. The reflectivity field at each elevation has a 1 km^2 resolution with a coverage area of 230 km radius around the radar site and the corresponding product is built every 6 minutes. The angle of the radar azimuth beam is about 0.97 degrees, so that 366 azimuths are needed to cover the entire radar area. At 1.5-degree elevation (Figure 5(a)), as the height of the radar echo increases along with the distance to the site at 230 km from the site the height is 9.6 km. In Figure 5(a), the Doppler velocities field relative to the storm is at 2.4-degree elevation, and at the 230 km the height is 13.2 km.

In the reflectivity field, six convective cells can be seen, resulting from three convective cells that had split, and in the relative Doppler velocity field the rotation can be seen, associated with resulted convective cells. A fundamental feature is the relative jet at 100–1000 m with velocities of up to 29 m s^{-1}. This relative jet is the storm input flow and can only be observed in the storm's reference frame; it may be associated with the moisture inflow. Above 6000 m there is a relative flow with velocities up to 11 m s^{-1}.

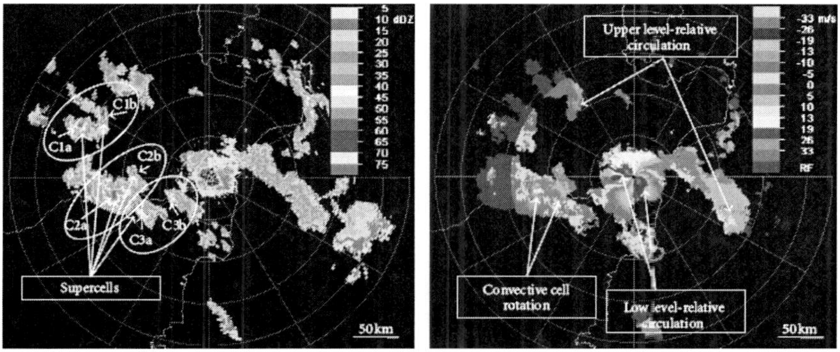

Figure 5: May 28, 2008. (a) radar reflectivity at 1.5-degree elevation, at 1500 UTC; (b) relative velocities at 2.4 degrees elevation, at 1500 UTC.

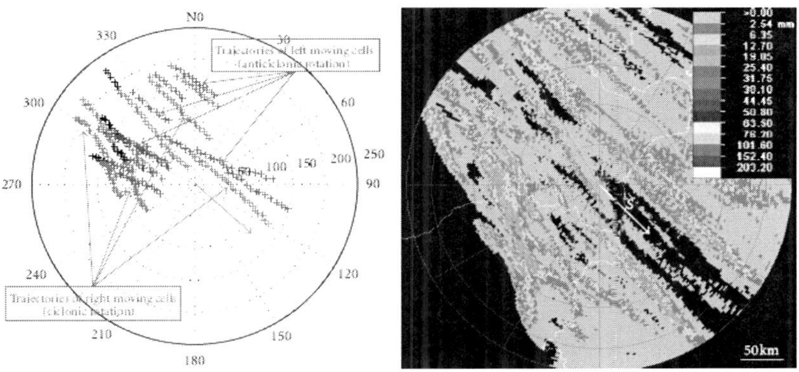

Figure 6: May 28, 2008, the convective cells evolution for 0900 UTC–2000 UTC using the Medgidia Doppler radar in (a). The center of images is the radar position. S vector represents the mean shear at 1800–6000 m altitude. Red stands for cyclonic rotation, blue for anticyclonic, and black signifies cells before split. The distance is measured in km. The accumulated precipitations (mm), for 0900 UTC–2000 UTC (b).

Trajectories of convective cells that had split during evolution present symmetry to the mean shear in the 1800–6000 m layer (blue arrow), with cells characterised by cyclonic rotation left of the vector and anticyclone rotation right of it (Figure 6(a)). In Figure 6(a), advection of convective cells by the tropospheric circulation is on the NW-SE direction. In this case, because the flow is parallel wih the shear vector on the mean troposhere, the convective cells with cyclonic rotation (red) deviate to the right and the ones with anticyclonic rotation (blue) to the left of this direction. Following the cell evolution, the intensity of the cyclonic rotation is larger than that of anticyclonic rotation. The cause for this difference can be associated with directional shear in the 300–1800 m layer and is favoured by advection of vorticity, whose vector is parallel to the wind relative to the storm [17, 18].

The splitting process may occur several times in the cells following from a previous split. For multiple splits, cells are favoured whose movement occurs on the left of the mean tropospheric shear. One possible explanation is that these cells draw more vorticity from their surroundings, vorticity which is perpendicular to the relative storm flow [18].

Cumulative rainfall in the evolution area, associated to the convective cells, is presented in Figure 6(b). The maximums are associated with convective cells, appearing as lines with symmetrical shear. Trajectories associated with more precipitation are to the right of shear in the 1800–6000 m layer. The amount of precipitation is amplified by the intersection of convective cells paths (cells with cyclonic rotation intersect those with anticyclonic rotation); indeed, in Figure 6(a), these intersections are more likely to the right of the mean shear vector.

Vertical wind variation associated with VAD Wind Profile, Figure 4(a), is represented for 6 times in Figure 7. Note the presence of the tropospheric jet in the 1800–9000 m layer, with values of about 40 m s^{-1} at 9000 m. The vertical profile of horizontal vorticity versus vertical vorticity associated with cell rotation is given in Figure 8; here, vorticity was calculated every 305 m on the -axis. One can see several maximum values of vorticity. The highest values are found in the 2000–4000 m layer. The horizontal vorticity of the environment (Figure 8, blue chart) was superimposed over cyclonic or anticyclonic vertical vorticity associated with mesocyclones (Figure 8, vertical lines) in the radar area. The parallel lines with vertical axes signify the vorticity of mesocyclones calculated with the radar algorithm in Figure 3(c). Vorticity was plotted both with a plus and minus sign for symmetry, to be more easily compared with the horizontal vorticity of the environment.

Above 1800 m, there is strong intensification of wind, associated with the presence of a ULJ from south east; this ULJ reaches 35 m s^{-1} at 5000 m. The dynamic field that can be viewed in the Doppler velocities, Figure 9(b), retained its structure between 1030 UTC and 1800 UTC.

Radar reflectivities show four secondary convective cells that are resulted from the splitting of two initial cells (Figure 10(a)). Three of them are supercells, and one is nonsevere. In the field of relative Doppler velocities a relative jet is present whose speed is up to 16 m s^{-1} at 1000 m, while above 5000 m intense circulation is present, with up to 16 m s^{-1}, (Figure 10(b)).

The refletivity field and the relative velocity characteristics are the same as in the pevious case.

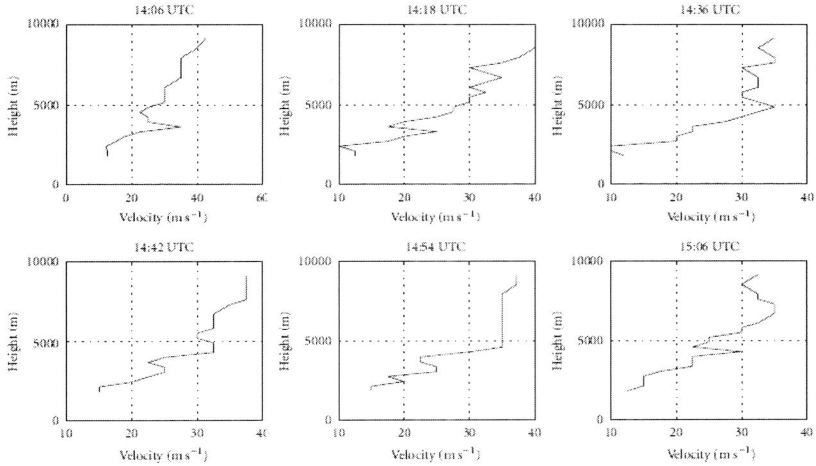

Figure 7: May 28, 2008, vertical velocity size distribution for 1800–9000 m altitude at 1406 UTC–1506 UTC.

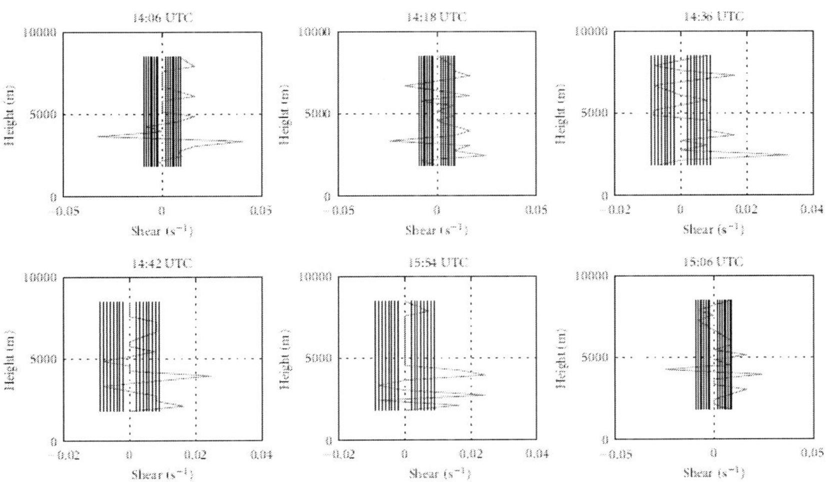

Figure 8: May 28, 2008, vertical shear distribution for 1800–9000 m layer, at 1406 UTC–1506 UTC (blue chart). Shear is calculated every 305 m.

Horizontal shear distribution associated with convective cells (vertical lines).

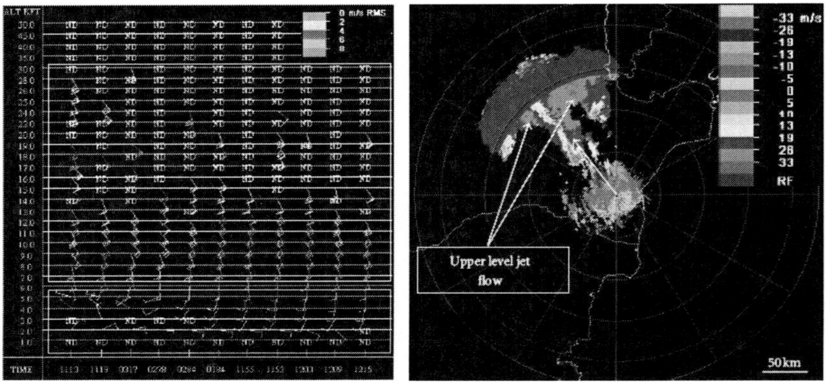

(a) (b)

Figure 9: June 25, 2009, vertical profile of horizontal wind (VAD Wind Profile) for 1113 UTC–1215 UTC, using the Medgidia Doppler radar in S band in (a) and (b) Doppler velocities at 3.4 degrees elevation at 1341 UTC. The vector represents the mean shear for 1800–5000 m; the flow is from southeast and the jet is present at 4000–6000 m.

Severe storms moving towards the left of mean shear (anticyclonic rotation) were characterized by high intensity and shorter lifespan and were associated with the presence of Three-Body Scatter Spike (TBSS) radar artifact (Figures 10(c) and 10(d)). This artifact is a radar signature in reflectivity field, generally a 10–30 km long, aligned radially downrange from a strong echo region in reflectivity field (>60 dBZ) associated with a severe storm. It is caused by non-Rayleigh radar microwave scattering or Mie scattering and is asociated with high probability of large hail [38–40].

In Figure 11 three vertical velocity profiles are presented, at certain times representative for this case. Here, the tropospheric ULJ has 35–40 m s^{-1} at 6000 m. The vorticity profile is shown in Figure 12 (blue chart).

The mesocyclone vorticities associated with convective cells that splited are represented by parallel lines with the vertical axes (Figure 12). Note that most values of vorticity lie between −0.005 s^{-1} to −0.015 s^{-1} and 0.005 s^{-1} to 0.015 s^{-1}. There is also rotation with values of shear around 0.04 s^{-1}. In a first approximation, the vorticities of the vertical wind profile are comparable with those of the mesocyclone.

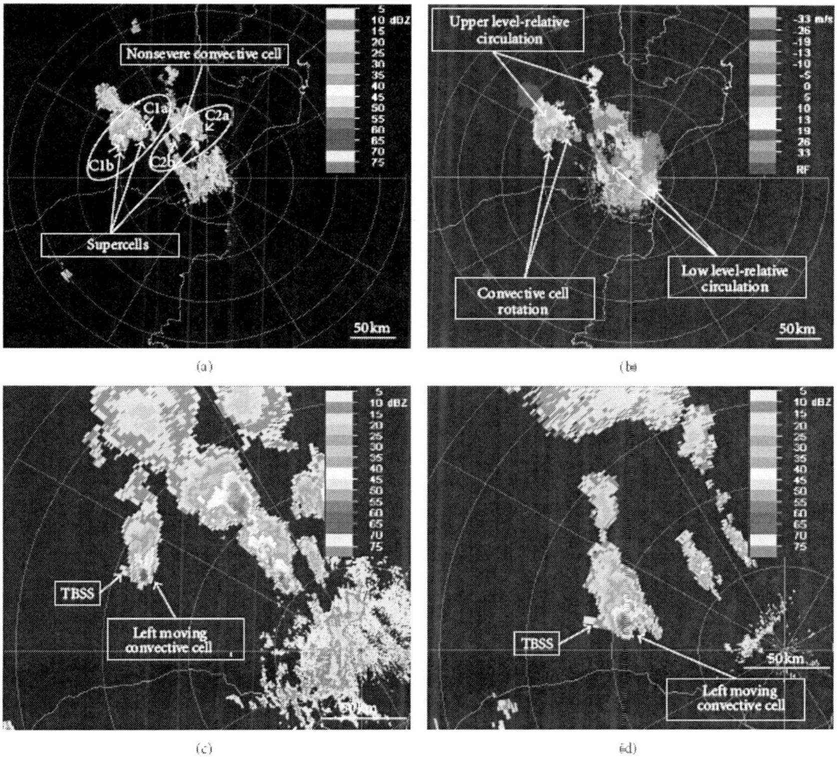

Figure 10: June 25, 2009, radar reflectvities at 2.4-degree elevation, at 1209 UTC (a). Doppler relative velocities at 2.4-degree elevation at 1209 UTC (b). Radar reflectvities at 1.5-degree elevation, at 1257 UTC, (c) and radar reflectvities at 2.4-degree elevation, at 1523 UTC (d).

Trajectories of convective cells resulted from the split, present symmetry to the mean shear on the 1800–6000 m layer, with cells characterized by cyclonic rotation to the left of the vector and cells with anticyclonic rotation to the right (Figure 13(a)).

The evolution of these cells shows that the intensity of cyclonic rotation is greater than that of anticyclonic rotation. The precipitation distribution function, by shear in the 1800–5000 m layer is represented in Figures 13(b), 13(c), and 13(d). Grouping in three-hour intervals was made in order to highlight the symmetry of precipitation distributions from shear vector. Large amounts were recorded at the intersection of several trajectories caused by several splits and also to the right of shear.

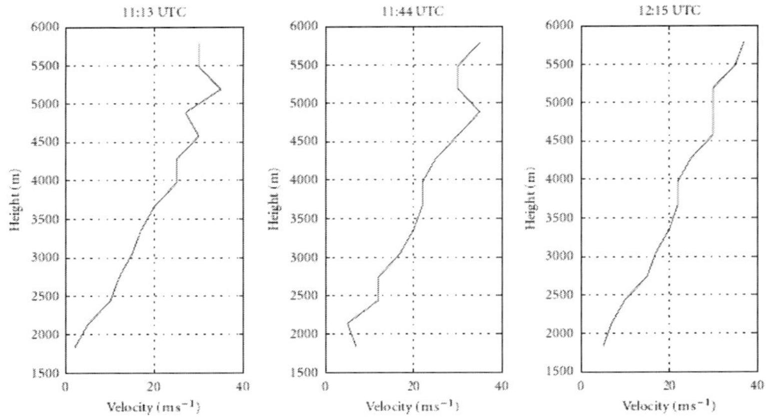

Figure 11: June 25, 2009, vertical velocity size distribution for 1800–6000 m, at 1113 UTC–1215 UTC.

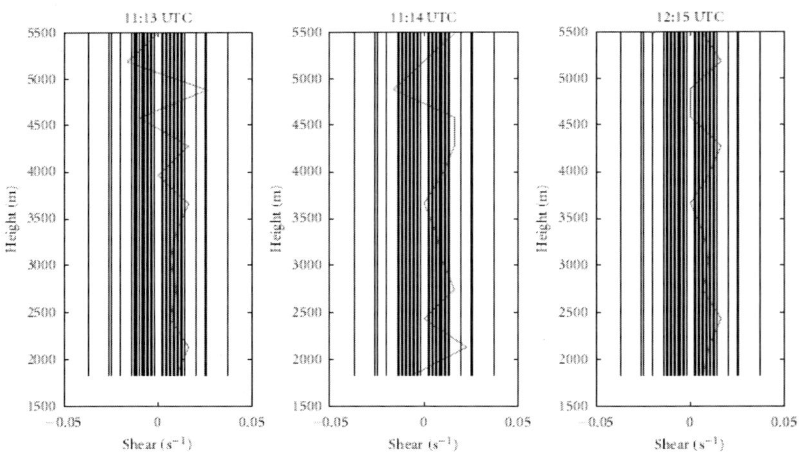

Figure 12: June 25, 2009, vertical shear distribution for 1800–6000 m, at 1113 UTC–1215 UTC (blue chart). Shear is calculated every 305 m. Horizontal shear distribution associated with convective cells (vertical lines).

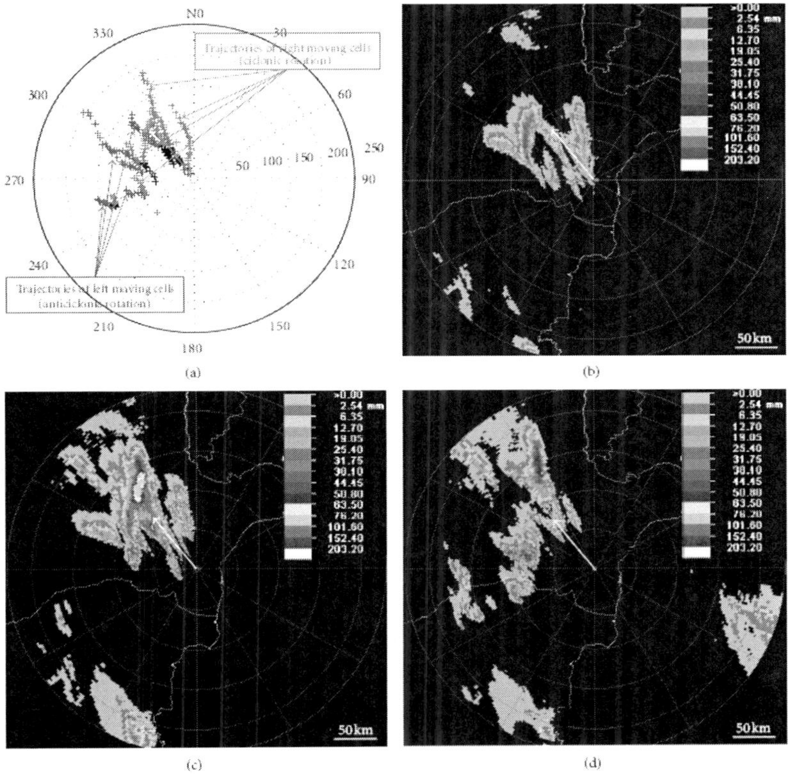

Figure 13: June 25, 2009, the convective cells evolution at 1030 UTC–1800 UTC using the Medgidia Doppler radar in (a). The white vector represents the mean shear for 1800–5000 m altitude. Red is the cyclonic rotations and blue is the anticyclonic rotations. The distance is in km. Accumulated precipitations are for every three hours: (b) 1040 UTC–1340 UTC, (c) 1205 UTC–1505 UTC, and (d) 1405 UTC–1705 UTC.

3.3. The Third Studied Case, August 11, 2003

In this case, starting at 1300 UTC, in SE Romania and NE Bulgaria, during the fast evolution of a trough which moved from W to E, convective cells have developed with supercell characteristics, visible in the radar-swept area.

About this event it can be readily said that, depending on the wind distribution, vertical shear, and time of day, two dynamic behaviors are distinguishable. As such, in the interval 0400 UTC–1115 UTC, the vertical velocity profile is characterized by a moderate mesoscale flow above 1500 m (Figure 14(a)). The radar reflectivity at 1.5-degree elevation is in echoes with a strip form, perhaps associated with conditional symmetric instability [41, 42] or with an internal-gravitational wave [18] (Figure 14(b)). Here, no splitting process occurred.

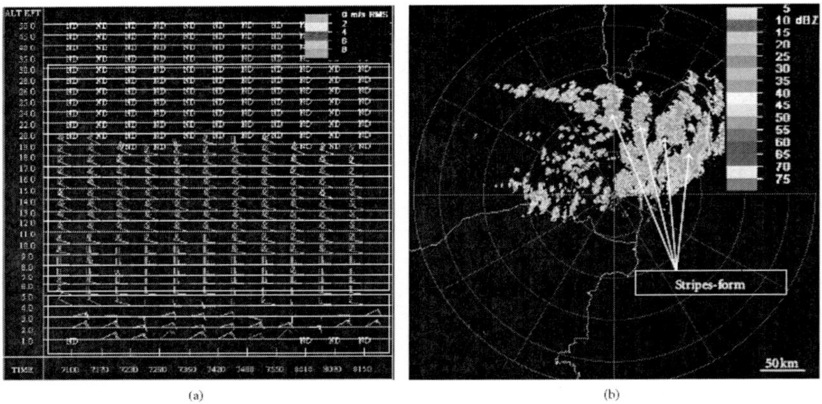

Figure 14: August 11, 2003, (a) vertical profile of horizontal wind (VAD Wind Profile) for 300–5800 m, at 0710 UTC–0815 UTC; (b) reflectivity field at 1.5-degree elevation.

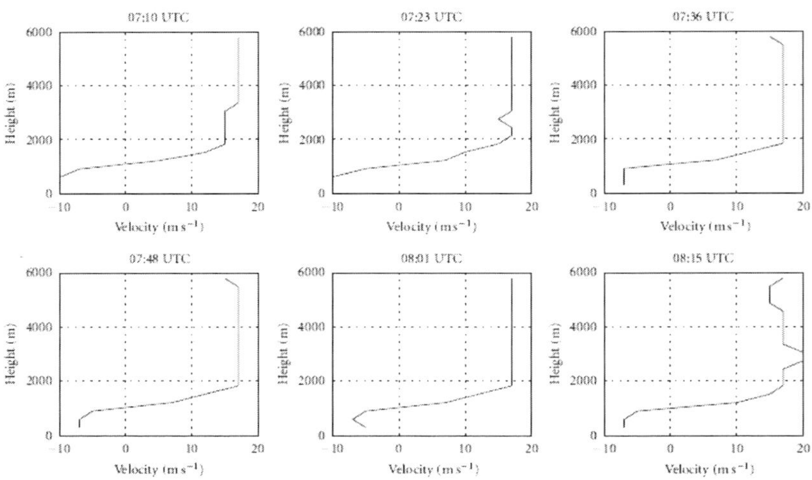

Figure 15: August 11, 2003, vertical velocity size distribution for 300–5800 m, at 0710 UTC–0815 UTC.

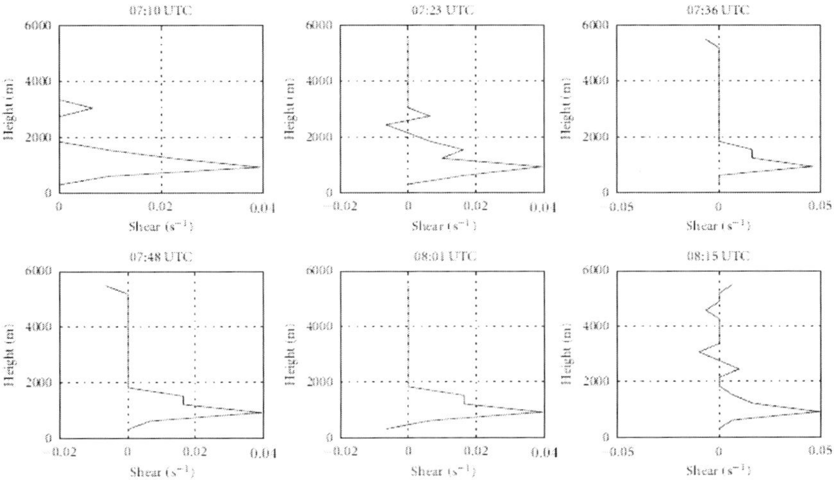

Figure 16: August 11, 2003, vertical shear distribution for 300–5800 m, at 0710 UTC–0815 UTC. Shear is calculated every 305 m.

As it can be seen in Figure 15, wind velocities have a maximum at 20 m s^{-1} at 2000 m. The vorticity distribution, shown in Figure 16, has a pronounced maximum of 0.04 s^{-1} at 1000 m.

Between 1328 UTC and 1432 UTC, the mesoscale dynamic state changes, a ULJ being present in the high troposphere, that can be seen in the vertical profile of horizontal wind (Figure 17(a)). Between 304 and 1500 m, neither the measurements of wind nor the VAD Wind Profile are reliable. Here, the influence of convective phenomena that can generate gust fronts is present. In Figure 18 the presence of the tropospheric jet is shown by velocities of 37 to 40 m s^{-1} at 6700 m. Due to the tropospheric jet, the reflectivity field has changed, emphasizing now the splitting process both in radar reflectivities and Doppler relative velocities (Figures 17(b) and 17(c)).

Mesocyclone vorticities associated with convective cells that split are represented by parallel lines with the vertical axes (Figure 19). Note that most of vorticity values lie between −0.001 s^{-1} and −0.009 s^{-1}, and between 0.001 s^{-1} and 0.009 s^{-1}. The maximum vorticity here had a value around 0.011 s^{-1}.

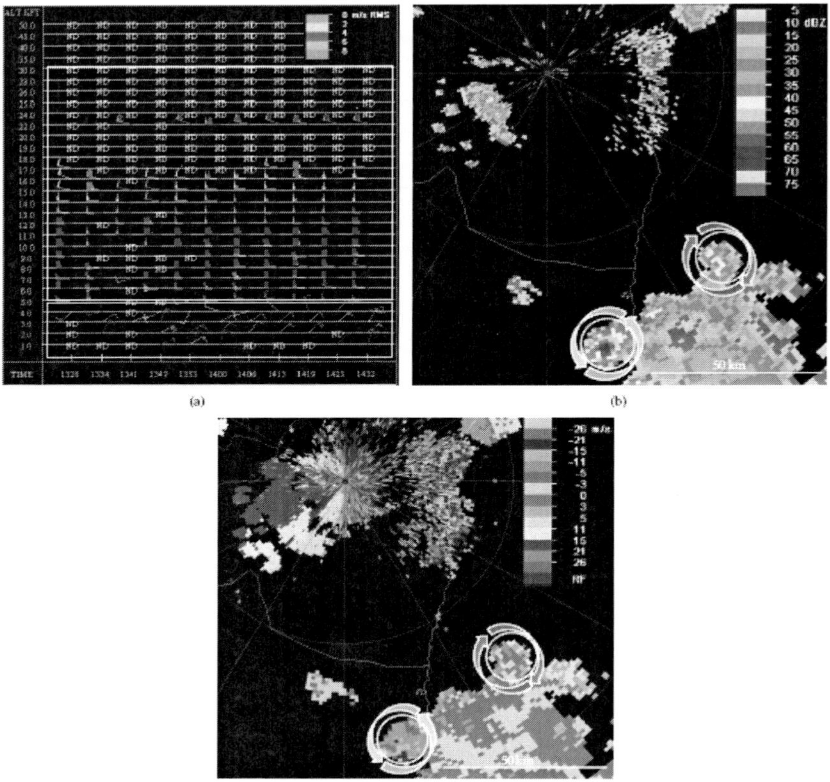

Figure 17: August 11, 2003, (a) vertical profile of horizontal wind (VAD Wind Profile) for 1500–6700 m, at 1328 UTC–1432 UTC; (b) radar reflectivity at 1.5-degree elevation, at 1400 UTC and (c) Doppler relative velocities at 1.5-degree elevation, at 1400 UTC.

The Doppler relative velocities to the elevation of 1.5 degrees, Figures 20(a) and 20(b), show a relative jet in the low levels, whose speed is up to 23 ms^{-1} at 500 m, while the relative flow at higher levels (above 6000 m) gets to 18 $m\,s^{-1}$. Spatial distribution of convective cell trajectories before and after splitting is shown in Figure 21(a). There is overlapping of the mean tropospheric shear vector on the 1500–6700 m layer. Symmetric trajectories to the mean shear are seen. In the precipitation field (Figure 21(b)), cells having higher water amounts are to the right side of the mean shear.

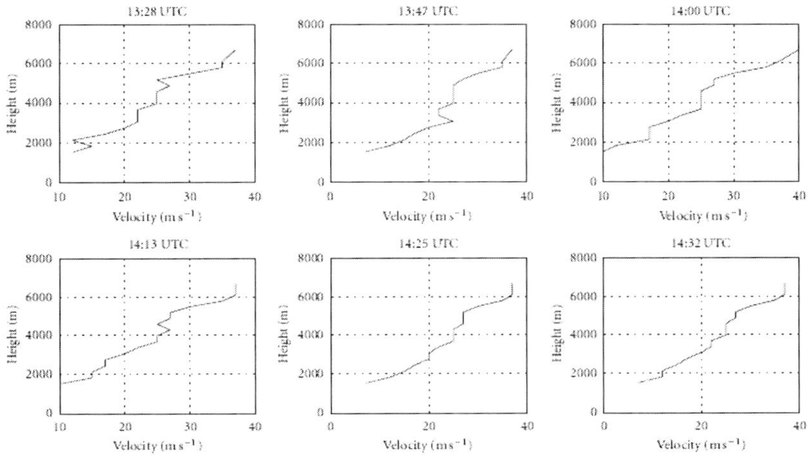

Figure 18: August 11, 2003, vertical velocity size distribution for 1500–6700 m, at 1328 UTC–1432 UTC.

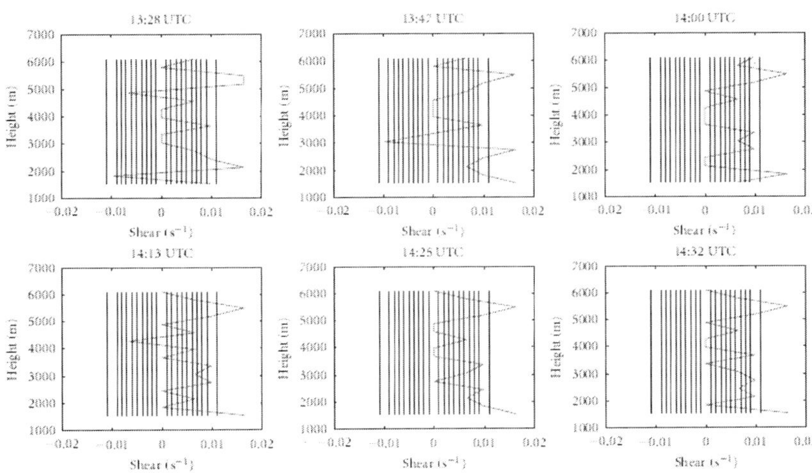

Figure 19: August 11, 2003, vertical shear distribution for 1500–6700 m, at 1328 UTC–1432 UTC (blue chart). Shear is calculated every 305 m. Horizontal shear distribution associated with convective cells (vertical lines).

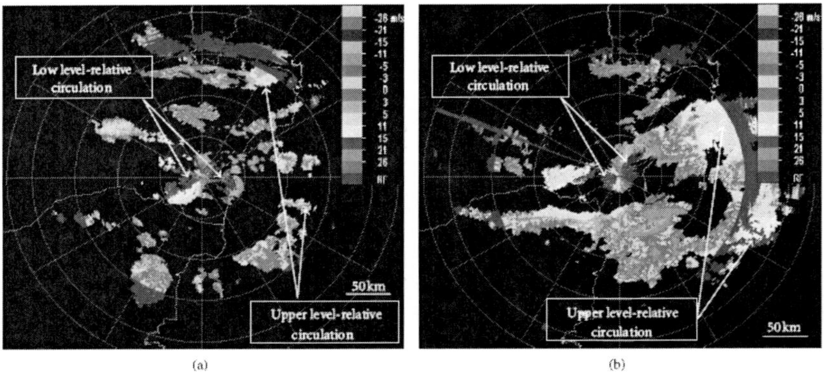

Figure 20: August 11, 2003, Doppler relative velocities, at 1432 UTC (a) and 1712 UTC (b) at 1.5-degree elevation.

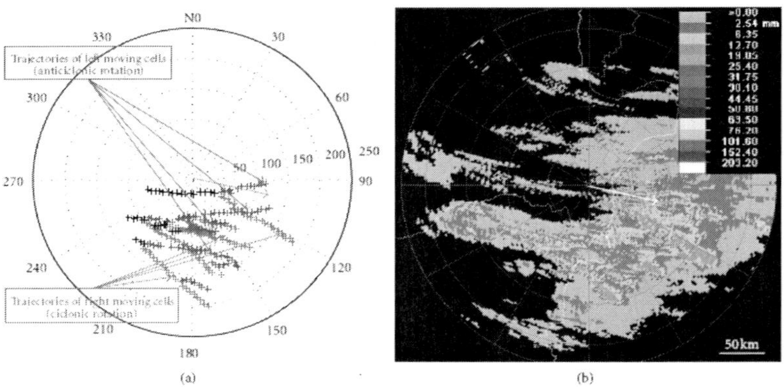

Figure 21: August 11, 2003, (a) the convective cells distribution, between 1126 UTC and 1925 UTC, using the Medgidia Doppler radar. The center of the image marks both the radar site and the origin of coordinates. The vector represents the mean shear for 1500–6700 m. Red stands for cyclonic rotation, blue for anticyclonic, and black signifies cells before split. The distance is measured in km. (b) Accumulated precipitation in mm, between 1547 UTC and 1847 UTC.

3.4. Other Cases (September 27, 2004; June 23, 2009)

In the first case, a quasi-stationary trough of a big amplitude was present to the west of Romania. The first severe convective cells were seen by the radar at 1020 UTC in Bulgaria and were advected in Romania by the tropospheric circulation. The convective cells which later split were

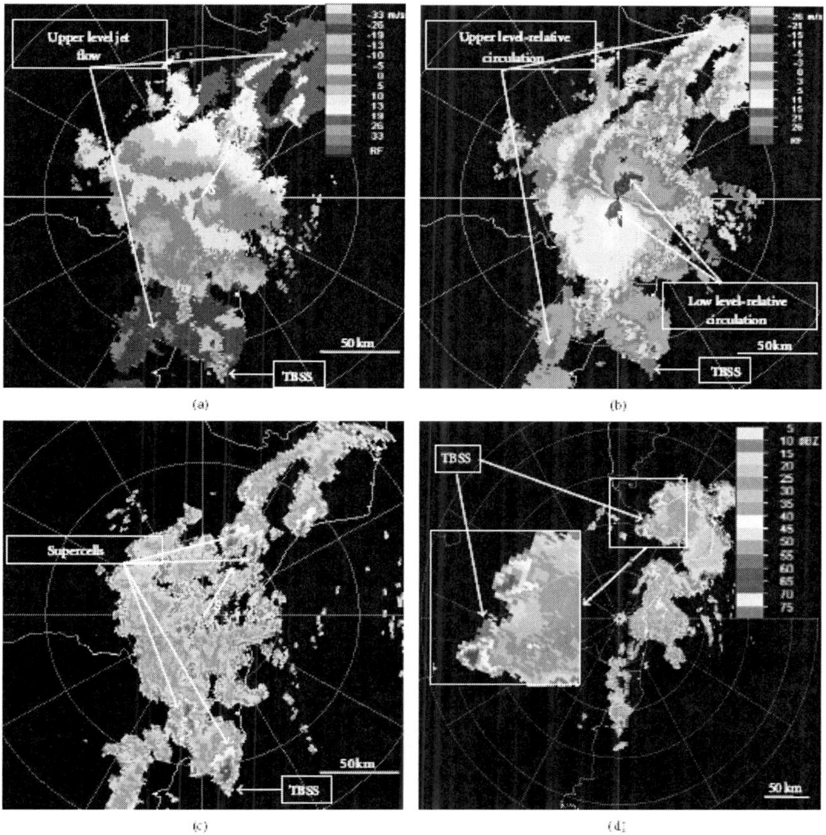

Figure 22: September 27, 2004: (a) Doppler velocities at 1347 UTC, at 3.4-degree elevation; (b) Doppler relative velocities at 1347 UTC, at 3.4-degree elevation; (c) reflectivity at 1347 UTC, at 1.5-degree elevation.

initiated nearby the radar site at 1220 UTC. In this case, the convective cells with anticyclonic rotation had a bigger life time and a greater severity as well.

In the second case, the mesosynoptic configuration was modulated by the cyclone west of Romania. The circulation due to this cyclone was southern over the south east of Romania in the boundary layer as well as at higher altitude. In the first part of the day, the Doppler radar caught a mesoconvective system with supercells. After the convective system dissipated, other new convective cells have been initiated and split; the most severe ones of them have cyclonic rotation.

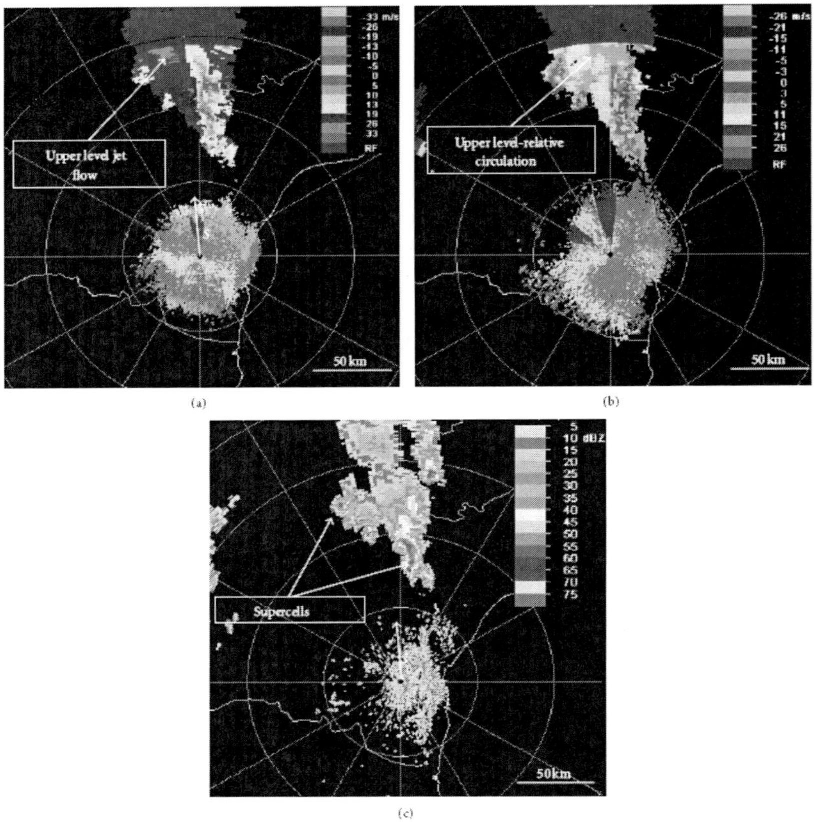

Figure 23: June 23, 2009: (a) Doppler velocities, at 1004 UTC at 3.4-degree elevation; (b) Doppler relative velocities, at 1004 UTC at 2.4-degree elevation; (c) reflectivity, at 1004 UTC at 1.5-degree elevation.

As in the cases above, for these two other cases the main feature associated with the mesoscale configuration was the ULJ, whose speed exceeded 33 m s^{-1}. In the first case (September 27, 2004), intense flow is visible at lower levels in the field of Doppler relative velocity, and in upper layers at approximately 11–15 m s^{-1}, as visible in Figure 22(b).

The convective cells in Romania that evolved to the left of tropospheric shear were associated with nonnegligible probability of hail as indicated by the presence of Three-Body Scatter Spike (Figure 22(d)). Cyclonic shear at low levels (Figure 22(a)) can be a direct indication of the severity of convective cells that propagate to the left of the shear vector. The cells in Bulgaria show strong propagation to the right of the tropospheric shear and also showing the Three-Body Scatter Spike (Figure 22(c)). For these, the splitting process was not seen. In the second case (June 23, 2009), intense flow is not present at lower levels in the field of Doppler velocity relatively, but a relatively high speed of flow of about 15 m s^{-1} is seen at upper levels (Figure 23(b)). The radar reflectivity field at 1.5-degree elevation (Figures 22(c) and 23(c)) highlights the splitting process; the secondary cells evolve to the right (cyclonic rotation) and to the left (anticyclone rotation) of the shear vector S in the 1800–6000 m layer.

4. CONCLUSIONS

This study shows the splitting process in SE Romania, linked to convective developments which can be seen using the S-band Doppler radar. This process is important as a dynamic one, because it can modulate the energy transfer between the synoptic scale and the mesoscale. The main mesosynoptic conditions associated with the generation of this dynamic process can be distinguished, and also its evolution. Information about all of these can be used as an additional tool in nowcasting.

From the analysis of convective cells resulted from the split it can be said that the splitting process of convective cells observed in radar fields indicates a corresponding severe weather event. As a fundamental condition, splitting of convective cells occurred when an upper level jet (30 m s^{-1}–40 m s^{-1}) was detected in the high troposphere (above 6000 m). In the first stages of the convective cells, the presence of the jet stream in the upper troposphere and especially its vertical extent lead them to be strongly advected by the flow. Also, the presence at lower levels of a directional variation in the horizontal wind vector leads to a stronger relative flow. In the case of a tropospheric configuration that favours the splitting process, the lower relative flow has a high intensity and is opposite to the upper relative flow; this configuration is seen in the relative Doppler velocities field at different elevations. In all the three detailed cases, below 1800 m the horizontal wind vector had a rotation associated to the asymmetry of the splitting process, and also to the strong relative flow between 100 and 1000 m, providing a powerful moisture flow through updraft and strong vorticity advection.

All the analysed cases have shown the independence of the splitting process to the atmosphere circulation. This is characteristic to baroclinic waves with a west-to-east evolution, and also to cyclones near Romania: north-west circulation for May 1–28, 2008, south-east circulation for June 2–25, 2009, west circulation for August 3–11, 2003, south-south-east circulation for September 27, 2004, and south circulation for June 23, 2009.

The largest amounts of precipitation are related to cell cross-trajectories associated with the splitting process and lie to the right of shear in 1800–6000 m layer (Figures 6, 13, and 21). The splitting process may lead to a mesoscale convective system. If this happens the secondary cells interact constructively afterwards; this will happen to the right side of the mean tropospheric shear.

The local maxima of vertical vorticity (Figures 9, 13, and 21) are most positive (to the right side of the 0,0 coordinate). Keeping the same sign of horizontal vorticity is essential in formation of cyclonic and anticyclonic vortices, confirmed by Figure 1. The magnitude of the horizontal vorticity associated with vertical wind profiles and that of vertical vorticity for cyclonic and anticyclonic rotation are comparable; the majority of values is lying in the $0.001\,s^{-1}$–$0.009\,s^{-1}$ interval (in a first approximation, these magnitudes verify the dynamic process associated to the formation of cyclonic and anticyclonic vortices).

Both the June 25, 2009 and in the September 27, 2004 cases highlight the TBSS associated with convective cells (anticyclonic rotation) with a motion to the left of the shear vector in the mean troposphere, which means that these cells have attributes that indicate the presence of severe hail. An explanation could be the fact that the convective cells with motion left to the mean shear are more isolated than those with motion to the right of it. The former cells are moving to the zone where the values of geopotential and temperature are falling. Here, the thermal gradient is perpendicular to the shear vector (thermal wind relation), something that can accelerate the increase in thermodynamic instability.

To have a good image of the splitting process, in the next studies observational data from meteorological stations and soundings would have to be used.

REFERENCES

1. A. Stan-Sion and C. Soci, "Supercell environment in Southeastern Romania," in Proceedings of the 5th Annual Meeting of the European Meteorological Society (EMS '05), 7th European Conference on Applications of Meteorology (ECAM '05), pp. 12–16, Utrecht, The Netherland, September 2005.
2. A. Stan-Sion and B. Antonescu, "Mesocyclones in Romania—characteristics and environments," in Proceedings of the 23rd AMS Conference on Severe Local Storms (SLS '06), Louis, Mo, USA, November 2006.
3. A. Bell, D. Carbunaru, B. Antonescu, and S. Burcea, "The Black Sea influence in convective storms initiation in South Eastern Romania," in Proceedings of the 34th AMS Conference on Radar Meteorology, Williamsburg, Va, USA, October 2009.
4. A. Bell, "Mesoscale environment for tornadic supercells in SE Romania," in Proceedings of the 6th European Conference on Radar in Meteorology and Hydrology: Advances in Radar Technology, Sibiu, Romania, September 2010.

5. B. Antonescu, D. Carbunaru, M. Sasu, S. Burcea, and A. Bell, "A climatology of supercells in Romania," in Proceedings of the 6th European Conference on Radar in Meteorology and Hydrology: Advances in Radar Technology, Sibiu, Romania, September 2010.
6. C. A. Doswell III, On the Use of Hodographs—Vertical Wind Profile Information in Severe Storms Forecasting, U. S. Dept. Of Commerce/NOAA, National Severe Storms Laboratory and National Weather Service, Southern Region, Fort Worth, Tex, USA, 1988.
7. M. J. Bunkers, "Vertical wind shear associated with left-moving supercells," Weather and Forecasting, vol. 17, no. 4, pp. 845–855, 2002.
8. R. Rotunno, "On the evolution of thunderstorm rotation," Monthly Weather Review, vol. 109, pp. 577–586, 1981.
9. R. Rotunno and J. B. Klemp, "The influence of shear-induced pressure gradient on thunderstorm motion," Monthly Weather Review, vol. 110, no. 2, pp. 136–151, 1982.
10. E. A. Brandes, "Vertical vorticity generation and mesocyclone sustenance in tornadic thunderstorms: the observational evidence," Monthly Weather Review, vol. 112, no. 11, pp. 2253–2269, 1984.
11. R. Davies-Jones, "Streamwise vorticity: the origin of updraft rotation in supercell storms," Journal of the Atmospheric Sciences, vol. 41, no. 20, pp. 2991–3006, 1984.
12. J. B. Klemp, "Dynamic of tornadic thunderstorms," Annual Review of Fluid Mechanics, vol. 19, pp. 369–402, 1987.
13. R. Davies-Jones, "Linear and nonlinear propagation of supercell storms," Journal of the Atmospheric Sciences, vol. 59, no. 22, pp. 3178–3205, 2002.
14. R. P. Davies-Jones, "Reply to comment on Linear and nonlinear propagation of supercell storms," Journal of the Atmospheric Sciences, vol. 60, pp. 2420–2426, 2003.
15. R. Rotunno and M. L. Weisman, "Comment on Linear and nonlinear propagation of supercell storms," Journal of the Atmospheric Sciences, vol. 60, pp. 2413–2419, 2003.
16. J. M. L. Dahl, Supercells—Their Dynamics and Prediction, Free University of Berlin, Institute of Meteorology, Department of Theoretical Meteorology, 2006.
17. Y.-L. Lin, Mesoscale Dynamics, Cambridge University Press, 2007.
18. P. Markowski and Y. Richardson, Mesoscale Meteorology in Midlatitudes, Royal Meteorological Society, 2010.
19. J. B. Klemp and R. B. Wilhelmson, "Simulation of right- and left-moving storms produced through storm splitting," Journal of the Atmospheric Sciences, vol. 35, pp. 1097–1110, 1978.
20. M. L. Weisman and J. B. Klemp, "The dependence of numerically simulated convective storms on vertical wind shear and buoyancy," Monthly Weather Review, vol. 110, no. 6, pp. 504–520, 1982.
21. M. L. Weisman and J. B. Klemp, "The structure and classification of numerically simulated convective storms in directionally varying wind shears," Monthly Weather Review, vol. 112, no. 12, pp. 2479–2498, 1984.
22. L. D. Grasso, "The dissipation of a left-moving cell in a severe storm environment," Monthly Weather Review, vol. 128, no. 8, pp. 2797–2815, 2000.
23. D. L. Andra Jr., "Observation of an anticyclonically rotating severe storm," in Proceedings of the 17th Conference onn Severe Local Storms, pp. 186–190, American Meteorological Society, St. Louis, Mo, USA, 1993.
24. R. P. Kleya, "A radar and synoptic scale analysis of a splitting thunderstorm over north-central Texas on November 10, 1992," in Proceedings of the 17th Conference On Severe Local Storms, pp. 211–213, American Meteorological Society, St. Louis, Mo, USA, 1993.
25. F. H. Glass and S. C. Truett, "Observation of a splitting severe thunderstorm exhibiting both supercellular and multicellular traits," in Proceedings of the 17th

Conference On Severe Local Storms, pp. 224–228, American Meteorological Society, St. Louis, Mo, USA, 1993.
26. R. A. Brown and R. J. Meitín, "Evolution and morphology of two splitting thunderstorms with dominant left-moving members," Monthly Weather Review, vol. 122, pp. 2052–2067, 1994.
27. G.W. Carbin, "Analysis of a splitting severe thunderstorm using the WSR-88D," NWS Eastern Region WSR-88D Operational Note 08, National Weather Service, Eastern Region Headquarters, Bohemia, NY, USA, 1997.
28. L. D. Grasso and E. R. Hilgendorf, "Observations of a severe left moving thunderstorm," Weather and Forecasting, vol. 16, no. 4, pp. 500–511, 2001.
29. J. P. Monteverdi, W. Blier, G. Stumpf, W. Pi, and K. Anderson, "First WSR-88D documentation of an anticyclonic supercell with anticyclonic tornadoes: the Sunnyvale-Los Altos, California, tornadoes of 4 May 1998," Monthly Weather Review, vol. 129, no. 11, pp. 2805–2814, 2001.
30. J. F. Dostalek, J. F. Weaver, and G. L. Phillips, "Aspects of a tornadic left-moving thunderstorm of 25 may 1999," Weather and Forecasting, vol. 19, no. 3, pp. 614–626, 2004.
31. D. T. Lindsey and M. J. Bunkers, "Observations of a severe, left-moving supercell on 4 May 2003," Weather and Forecasting, vol. 20, no. 1, pp. 15–22, 2005.
32. R. Edwards and S. J. Hodanish, "Photographic documentation and environmental analysis of an intense, anticyclonic supercell on the Colorado plains," Monthly Weather Review, vol. 134, no. 12, pp. 3753–3763, 2006.
33. E. N. Rasmussen and J. M. Straka, "Variations in supercell morphology. Part I: observations of the role of upper-level storm-relative flow," Monthly Weather Review, vol. 126, no. 9, pp. 2406–2421, 1998.
34. National Oceanic Atmospheric Administration (NOAA), http://www.wdtb.noaa.gov/courses/dloc/index.html.
35. R. M. Lhermitte and D. Atlas, "Precipitation motion by pulse Doppler radar," in Proceedings of the 9th Conference on Radar Meteorology, pp. 218–223, American Meteorological Society, 1961.
36. K. A. Browning and R. Wexler, "The determination of kinematic properties of a wind field using Doppler radar," Journal of Applied Meteorology, vol. 7, pp. 105–113, 1968.
37. R. A. Fulton, P. Breidenbach, D. J. Sco, D. A. Miler, and T. O'Bannon, "The WSR-98D rainfall algorithm," Weather Forecast, vol. 13, pp. 337–395, 1988.
38. D. S. Zrnic, "Three-body scattering produces precipitation signature of special diagnostic value," Radio Science, vol. 22, no. 1, pp. 76–86, 1987.
39. J. B. Wilson and D. Reum, "The flare echo: reflectivity and velocity signature," Journal of Atmospheric and Oceanic Technology, vol. 5, pp. 197–205, 1988.
40. L. R. Lemon, "The Radar "Three-Body Scatter Spike": an operational large-hail signature," Weather and Forecasting, vol. 13, pp. 327–340, 1988.
41. D. A. Bennetts and J. C. Sharp, "The relevance of conditional symmetric instability to the prediction of mesoscale frontal rainbands," Quarterly Journal Royal Meteorological Society, vol. 108, no. 457, pp. 595–602, 1982.
42. D. M. Schultz and J. A. Knox, "Banded convection caused by frontogenesis in a conditionaly, symmetrically, and inertialy unstable environment," Monthly Weather Review, vol. 135, pp. 2095–2110, 2007.

CHAPTER 15

A New Approach Method of CH_4Emission Estimation from Landfills Municipal Solid Waste (MSW)

Danila Vieru

Romanian Expert on Environmental Issues, Bucharest, Romania

ABSTRACT

The CH_4 is one of the six Greenhouse Effect Gases (GEG) that is mentioned in the Kyoto Protocol. The GEG is generated by the anthropic activities which are conducive to climate changes if their management is not conducted in a proper way. The main purpose of the environment policy is the reduction of the GEG emission. It is well-known that the CH_4 gas emission from municipal solid waste MSW landfills is responsible for 4 ÷ 5% of the total Greenhouse Effect. It is necessary to have a practical method to calculate the quantitative CH_4 gas emission, in order to apply an efficient management of the CH_4 gas emission from MSW landfills, conforming or non-conforming. This method has to be transparent, credible, coherent, and applicable both for conforming and non-conforming MSW deposits. This paper proposes a new estimation calculation method of the CH_4 gas emission from all MSW deposits in Romania. The IPCC group of experts has made recommendations related to the estimation of CH_4 but the European Union (EU) admits that the environmental conditions are not the same in every Member State. The annual evolution of CO_2 for the CH_4 gas emission at every MSW location is valuable information for the Environment

Authority with a view to realistic environmental planning and for an efficient policy to be applied in order to reduce the greenhouse effect of MSW landfills.

Keywords: Ecological Condition, GEG, Landfill, MSW, Urban Area

1. INTRODUCTION

In May 2013, the United Nations (UN) adopted the KYOTO Protocol [1] relating to the pollution emission agents and the transfer registers (based on the so called PRTR Protocol or Kiev Protocol) [2] together with the UN Convention on climate changes.

This Convention is referring, among others, to the landfills having a daily activity of more than 10,000 tons/day MSW which amounts to more than 450,000 tons/year. For these *MSW* landfills, starting with 2007, the individual CH_4 emission rate [3] has to be calculated and the results have to be communicated to the public. EU has adopted the European Emission Register in order to be in conformity with the PRTR Protocol. This Register provides some criteria to be fulfilled: transparency, coherence, the possibility to compare results. These criteria are a condition for the calculated results to be accepted into a national data base. Romania has adopted the UN PRTR Protocol and for the MSW landfills with more than 10,000 tons/day, the CH_4 emission will be included in a register. The Member State governments have to report all aspects related to the Climate Changes [3] to an inter-governmental group.

It is very clear that a method to estimate the CH_4 methane gas emission from MSW landfills is absolutely necessary [3] .

This method has to cover the calculation of the CH_4 emission from both conforming and non-conforming MSW Romanian landfills [4] [5] . This method was applied for the CH_4 emission calculation of 13 MSW landfills—conforming and non-conforming. In this paper the calculated values for CH_4 emission [4] [5] [6] and the equivalent CO_2 for 1 non-conforming and for 2 conforming landfills are presented.

Analyzed landfills are located in Satu Mare, Ilfov and Bucharest municipality, Romania. The proposed method has a high degree of efficiency.

The CH_4 emission calculus for those 13 Municipal landfills (msw) and the drawing up adjacent graphics related to the equivalent of CO_2 demonstrate that the GEG is present. The Romanian Environmental Authorities have to act on this matter and to acknowledge about the GEG intensity and its duration [7] , in the same time.

The Proposed method allows us the quantitative evaluation of CH_4 emission to be used as a natural energy source. Within the actual management of wastes only the sort of wastes having economical energy value is applied, according to the Europe Council provisions. It is to be mentioned also that only 20% of the generated wastes is sorted. In the deposit body, they are not included: metallic wastes, plastics, tires, recyclable wood or with energetic value, paper wastes and recyclable cartoon. It is to be mentioned also that, from information delivered by the local source, within the landfill body they are not included: inert wastes (construction and demolition), plastics, soils and stones, asbestos; the total contents of these wastes are not considered to be more than 10%.

I have to make a remark: the drawing up graphics were obtained by manual calculation rather than using specific software.

2. PRESENT SITUATION

All types of wastes were deposited together [4] , in specially designated MSW deposit areas, those coming from the anthropic activities as well as those generated by the agriculture and live-stock farm activities, e.g. animal and bird dejections. The bio-degradable wastes (rubbish) generated by intensive agriculture have to be taken into consideration as well.

The problem of the global warming and the obligation to apply the Kyoto Convention requirements involve the fulfillment of the rules regarding the limitation of the MSW gas emission [7] and the prohibition to have MSW landfills which do not comply with the rules of environmental protection [2] .

Since 1999 Romania has started to have MSW landfills, in ecological condition, in accordance with the European regulation in the field, and, from 2007, when Romania adhered to the European Union (EU), all the MSW landfills have to respect, strictly, the EU legislation, as provided within the 75/442/CE Directives [5] [8] provisions.

This Directive [5] [8] was adapted [4] to the Romanian legislation by Government Decision [4] order no. 349/2005.

3. ESTIMATIVE METHODS FOR CH_4 GAS EMISSION CALCULATION

The quantity of the CH_4 gas emission from MSW landfills can be estimated, by calculus applying two methods, as follows:

METHOD No. 1

IPCC 2006 Method-Default Method (DM).

This method supposes that a non-dangerous MSW deposit will generate [9] [10] , within a year, a certain quantity of CH_4 and, in the next year, it will be a new amount of CH_4 This method will not take into consideration the hypothesis that an MSW deposit is a conglomerate mixed wastes one (see Table 1). Another factor to be taken into consideration is the time-the basic factor for GES emission [10] . Different MSW components are gradually, deteriorated in time, so CH_4 and CO_2 as well as the non-methane gases, and are generated.

In order to illustrate results due to the method 1 use, the conform MSW calculus equations regarding CH_4 emission [10] [11] will be indicated, as follows.

These calculus equations are:

Table 1. The percentage (%) composition of the MSW landfills.

Description of the composition of *MSW* landfills	Percentage [%]
Bio from kitchen, cantina's + animal manures, bio-wastes + market wastes + street wastes	51.2 ÷ 60
Rubbish from gardens and parks	16 ÷13
Paper + cartoon non-recyclable	14.2 ÷12.2
Non-recyclables woods and straws	3 ÷4,1
Non recyclable-textiles	2.6 ÷1.3
Sludge	1 ÷3
Industrial wastes (similar to home wastes) + sterilized medical wastes	12 ÷6.4

$$CH_{4(Gg/year)} = \left[\left(MSW_f * MSW_{F*}L_0\right) - R\right] * \left[\left(1 - 0X\right)\right]$$

where:

- L_0-CH_4 generated potential $\left(G_gC/G_gMSW\right)$ which depends by the MSW morphological composition it will be calculated by using the following relation, [7] [11] ;
- R-CH_4 recovered at the inventory year of $\left(G_gC/G_gMSW\right)$, the recommended value, supposing that CH_4 is burned and not collected; if not, the recovered quantity of CH_4 calculated by using this method will be reduced from the CH_4 generated quantity.
- 0X-oxide factor having a fractionary values-0 for non-conforming deposits and 0.1 for the well arrangements (conforming) deposits.

$$L_{0(GgC/GgMSW)} = \left[MCF * DOC_f * F * 6/12\right]$$

CH_4 generated potential, where:

- MCF-CH_4 correction factor, whose values are dependent by the location and the management of MSW;
- DOC_f-the DOC dissimilated fraction-0.55 having values within the interval 0.5 ÷ 0.6;
- F- CH_4 fraction part-from deposit gas (LFG) [5] , given value is 0.5;
- 16/12-the C conversion coefficient within CH_4;

The Dissolved Organic Carbon (DOC) is determined [11] [12] by using the relation:

$$DOC_{(GgC/GgMSW)} = \left(0.4 * A\right) + \left(0.17 * B\right) + \left(0.15\left(C\right) + \left(0.3 * D\right)\right)$$

where:

- A-the MSW fraction represented by paper and non-reciclable textiles [6] [10] [13] [14] .
- B-the MSW fraction represented by garden and parks wastes, and other bio-degradable organic wastes, excepted food wastes [6] [10] [13] [14].
- C-the MSW fraction represented by food wastes and other bio-degradable wastes, [6] [10] [13] [14] ;
- D-the MSW fraction represented by woods or straw wastes, [IPCC], [6] [10] [13] [14] ;

This method has the following difficulty:

- -Don't take into consideration that in the last 6 months deposited MSW are not degradable
- -The CH_4 emission quantity is very high (inadmissible)

It is supposed that a MSW landfill will generate, within a year, a certain amount of CH_4 gas emission which can be estimated [10] . This method doesn't take into consideration the hypothesis that a MSW landfill is a mixed conglomerate of wastes (rubbish).

Another factor to be considered is the time which is the basic factor for CH_4 gas emission [10] . Different components of the MSW landfill are, gradually, degraded in time, and CH_4, other gases are produced [6] .

METHOD No. 2

I developed a new calculation method for the methane gas emission estimation, from the Romanian waste landfills [7] [11] , method called: "DANILA VIERU METHOD FOR A CONFORMING AND NON-CONFORMING MSW LANDFILLS CH_4 GAS EMISSION ESTIMATION IN ROMANIA, BY CALCULUS".

According to the above- mentioned method, it is assumed that the waste (rubbish) from MSW landfills will be gradually degraded [11] based on the following factors [10] [12] :

Ÿ Structure of the wastes (rubbish) composition;

Ÿ Environmental factors existing in that area;

Ÿ The thickness of the waste (rubbish) layer;

Ÿ The compacting grade (level);

Ÿ The depth of the place where the MSW is located;

Ÿ Time passed from the first deposition of wastes (rubbish).

Due to the time factor, this method was called: "DANILA VIERU METHOD FOR CONFORMING AND NON-CONFORMING MSW LANDFILLS CH_4GAS EMISSION ESTIMATION IN ROMANIA, BY CALCULUS".

The IPCC-International Experts Group on Climate Change makes recommendation [9] related to the use of some coefficients concerning the estimation of CH_4 gas emission from MSW landfills but no to the use a specific calculus formula.

In the case of a MSW conglomerate landfill, having a broad range of types and amounts of wastes (rubbish), Romania did not possess an adequate (proper) formula for the MSW CH_4 gas emission estimation up to the year of 2012. The statistics of the wastes (rubbish), under the rule of the

Regulations no. 2150/2002 on waste statistics [17] do not solve the problem of the composition of the waste (rubbish) from MSW. The use of waste statistics assumes that the waste (rubbish) should be analyzed by means of a representative sample of economic operators and human agglomeration [12] .

Taking into consideration that every district of Romania has approx. 200 economic operators and urban agglomeration we shall have approximately 8400 economic operators, in total [9] .

Approximately 500,000 economic operators are assumed to be in the country which means that statistics representation will cover only 1.6% of the total country economic operators. This fact is quite unacceptable.

DESCRIPTION OF"DANILA VIERU METHOD FOR CONFORMING AND NON-CONFORMING MSW LANDFILLS CH_4 GAS EMISSION ESTIMATION IN ROMANIA, BY CALCULUS"

The method: "Danila Vieru method for conforming and non-conforming MSW landfills CH_4 gas emission estimation, in Romania, by calculus", makes use of the following formula:

$$CH_4\left(Gg/\text{year}\right)_T = Q_{\text{mswdegrad}.T} * \%TDOC_{\text{dissolved}.T} * DOC_f * 16/12 * F * F_r, \quad (1)$$

This formula (equation) has some advantages, e.g.:

1) The hierarchy [6] of degraded MSW, IN TIME, under the environmental factors [atmospheric precipitations, annual average temperature, alternating periods of rain and drought, freezing and non-freezing periods, the degree of MSW compression, the thickness of waste (rubbish) layers, etc. [13] ;

2) The use of time periods for the degradation of MSW;

3) The use of IPCC recommendation related to the application of the methodology calculation formula of CH_4 gas emission from MSW landfills;

4) Taking into consideration the specific environmental conditions of every district of Romania;

5) The specific economic conditions of every district, such as: industrial development, hand-made production, various branches of agriculture, etc. are taken into consideration;

It is well-known that CH_4 methane is a specific gas, and its contribution (percentage) to global warming is about 4 ÷ 5% so that the need for the quantification of CH_4 gas emission is imperative. In the meantime, measures to reduce the contribution of the CH_4 gas emission from MSW landfills have to be taken into account.

In July 16, 2009, due to the presence of non-conforming MSW landfills in Romania, some of them are closed while others will be in transition periods, in the case of MSW landfills, the emission of CH_4methane gas will continue even after the closing period of non-conforming MSW landfills until approximately the year 2017. Before wastes (rubbish) are deposited within the body of MSW and a rational sorting have to be are done.

After the closure of MSW landfills, the quantity of the CH_4 gas emission will decrease but still will continues to exist [14] . Following the legal conditions for opening a new MSW landfill it is absolutely necessary to know the evolution of CO_2 (in equivalent), the location of the new MSW landfill and the potential impact over the environment. As it is known, in approximately years, the warming effect will be intensified due to the collection of the gas MSW landfill.

In my opinion, the above mentioned remarks should be taken into consideration when a CH_4methane gas emission calculus formula is applied, for the entirely territory of Romania.

4. EXAMPLE OF CALCULUS, METHODOLOGY—THE ASSESSMENT

Basic consideration:

a) The percentage composition of MSW landfill body is in accordance with the data provisions given in Table 1.

b) The wastes (rubbish) from the MSW landfill body are gradually degraded in accordance with the environment conditions;

c) To calculate the quantity of CH_4 gas emission from degraded MSW, at the year of calculation, the

IPCC recommended values [9] have been taken into consideration.

d) The MSW degraded quantity has the same percentage composition as the MSW landfill body;

e) The MSW degraded quantity generates DOC-Dissolved Organic Carbon, and, as a consequence, the CH_4 gas emission is produced.

f) The MSW degraded quantity calculated, in the year T, is given by the expression: $Q_{mswdegrad.T}$

Within Table 1 the waste composition, as% from total, was established following information delivered by:

- Local Environmental Authorities, in accordance with the Regalement of the Council of Europe no. 2150/2002 and the European Parliament information with referring to the waste statistics (November 25/2002) [17] . For example, for the Region 8 Bucharest Ilfov-landfill Chiajna, the information delivered (see Figure 1, also) are: "Methane Vol.?54.4%, Carbon Dioxide Vol.?38.1%, Oxygen Vol.? 1.3%, Nitrogen Vol.?6.1%, etc. As an important remark, within the year 2011 about 7.5 million cubic meters of Methane gas has been extracted."
- Direct observation done at the MSW landfills location with referring to the wastes composition;
- Direct information delivered by local authorities regarding annually collected wastes quantities and the way of the management;
- Information delivered by the MSW landfills administrators related to the collection area, quantities and type of wastes included in MSW.

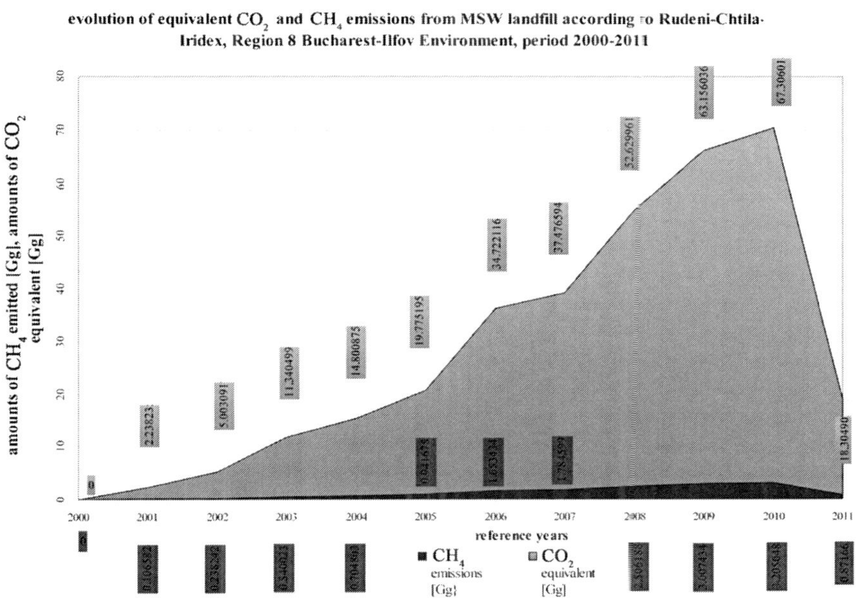

Figure 1. The Evolution of CO_2 (equivalent) and CH_4 emission from the landfill Rudeni-Chitila-Iridex, Environmental Reg. 8, Bucharest, Ilfov District, in the period: 2000 ÷ 2011.

Table 2 presents the composition of the MSW landfills wastes, located within 3 environmental regions areas-region 8 Bucharest Ilfov, Satu Mare County and Bihor county. It is to be mentioned that the Waste composition, as a conglo- merate landfill, is subjected to the environment factors, and as a consequence, the LFG gas (mainly, CH_4) is generated, covering the total lifetime of the deposit.

5. THE EVALUATION OF $Q_{MSWDEGRAD.T}$ IN THE T YEAR OF CALCULATION

To determine the MSW degraded quantity, in the first year of emission, the following formula has been used:

$$Q_{mswdegrad.T} = \left[\left(Q_{msw.T} + Q_{msw.T-1}\right)\right] * \left[\left(1 - \exp(-Kt)\right)\right][Gg], \tag{2}$$

After the first year, the calculation formula became:

$$Q_{mswdegrad.T} = \left[\left(Q_{msw.T} + Q_{mswundegra.T-1}\right)\right] * \left(1 - \exp(-Kt)\right][Gg], \tag{3}$$

where:

- $Q_{msw.T}$-MSW, the amount deposited in the account, [Gg];
- $Q_{mswT-1)}$-MSW deposited one year ago; [Gg];
- $Q_{mswundergradT-1}$-the remaining amount of MSW degraded after year calculation [Gg];
- K: is the degradation rate of MSW. This factor depends on waste composition and site conditions, and describes the degradation process rate. The IPCC Guidelines [9] give, for K, a very wide range of values between 0.005 and 0.4.
- t: time of degradation
- t: time of wastes degradation within deposit body; during calculation process, t is replaced with relation (13 – m)/12 or (25 – m)/12, where m re- present the no. of months when msw wastes were degraded within deposit body, at the calculation year. m?within the interval 7 ≤ m ≤ 12, m- within the interval 7 ≤ m ≤ 18, represents no. of months when 45% of the wastes is degraded in the proportion of 45%. The m values are established in accordance with the deposit nomograme, based on the deposit equation $-3x + 7 = 13 - x$, [15] . The deposit

Table 2. The MSW percentage (%) composition within the deposit body in some environmental Romanian regions.

Environmental region	Types of wastes					
	Bio from kitchen, cantina's + animal manures, bio-wastes + market wastes + street wastes	Rubbish from gardens and parks	Paper + cartoon non-recyclable + Non recyclable-textiles	Non-recyclables woods and straws	Sludge	Industrial wastes (similar to home wastes) + sterilized medical wastes
	All composition% according to the information provided by the Local Environmental Authorities and direct observations from storage place					
Region 8 Environmental Bucharest-Ilfov	51.20	16.00	16.80	3.00	1.00	12.00
Satu Mare county	58.00	13.00	10.30	6.00	1.50	11.20
Bihor county	60.00	11.12	10.88	6.50	2.00	9.50

equation has an unique solution, but in every year has another expression i.e. in the year $2-11x+7=25-x$ [15] , in the year $3-19x+7=37-x$, [14] etc. How to drawing up the Nomograme [15] of the MSW deposit will be explained in another paper.

- T-represent the year of calculation not the current calendar year.

A certain MSW deposited quantity remains undegraded every year [8] [12] . This quantity will be taken into consideration in the next year as the $Q_{mswundegrad.T}$.

This quantity can be estimated by using the formula:

$$Q_{mswundegrad.T} = \left(Q_{msw.T} + Q_{mswT-1}\right) - Q_{mswdegrad.T}\left[Gg\right], \quad (4)$$

The calculation of the total Dissolved Organic Carbon?($TDOC_{dissolved.T}$)-quantity from MSW degraded, at the year T, $Q_{mswdegrad.T}$ has been done by means of the following formula

$$TDOC_{dissolved.T} = \sum\left[A+B+C+D+E+G\right]\left[Gg\right], \quad (5)$$

where:

A = DOC generated by $Q_{mswdegrad.T}$ which contains% $MSW_{biodegrad}$ stated;

$$A = Q_{mswdegrad.T} * \%Q_{mswbiodehrad.T} * k_0, \left[Gg\right], \quad (6)$$

k_0: in accordance with [9] DOC generation ratio by% $MSW_{biodegrad.degrad.T}$, deposited;

B = DOC generated by $Q_{msw(G+P)degrad.T}$ which contains $\%MSW_{(G+P)}$, stated;

$$B = Q_{mswdegrad.T} * \%Q_{msw(G+P)degrad.T} * k_1 [Gg], \tag{7}$$

k_1: in accordance with [9] , DOC generated ratio by% $MSW_{(G+P)degrad.T}$, deposited;

C = DOC generated by $Q_{mswdegrad.T}$ which contains $\%MSW_{H+C+text.}$, stated;

$$C = Q_{mswdegrad.T} * \%Q_{msw(H+C+text.)degrad.T} * k_2, [Gg], \tag{8}$$

k_2: in accordance with [9] , DOC generated ratio by $\%MSW_{(H+C+ text.)degrad.T}$, deposited;

D = DOC generated by $Q_{mswdegrad.T}$ which contains $\%MSW_{(wood+straw)}$, stated

$$D = Q_{mswdegrad.T} * \%MSW_{(wood+straw)degrad.T} * k_3, [Gg], \tag{9}$$

k_3: in accordance with [9] , DOC generated ratio by% $MSW_{(wood+strawdegrad.)T}$, deposited;

E = DOC generated by $Q_{msw\ degrade.T}$ which contains $\%MSW_{sludge}$, stated;

$$E = Q_{mswdegrad.T} * \%MSW_{sludg.degrad.T} * k_n, [Gg], \tag{10}$$

k_n: in accordance with [9] , DOC generated ratio by% $MSW_{sludg.degrad.T}$, deposited;

G = DOC generated by $Q_{mswdegrad.T}$ which contains $\%MSW_{industry}$, stated;

$$G = Q_{mswdegrad.T} * \%Q_{mswind.degrad.T} * k_4, [Gg], \tag{11}$$

k_4: in accordance with [9] , DOC generated ratio by% $MSW_{ind.degrad.T}$, deposited.

The total composition of MSW wastes within the body can be changed annually, at two years, at three years or five years depending on the best environment information detained.

% $TDOC_{dissolved.T}$ is the ratio $(TDOC_{dissolved.T})/(Q_{msw\ taken\ into\ consid.T})$ because DOC is distributed within total wastes deposited but it is considered to be generated only by $Q_{mswdegrad.T}$ and it is determined by using the following formula:

$$\%TDOC_{dissolved.T} = (TDOC_{dissolved.T})/(Q_{msw\ taken\ into\ consid.T})[\%] \tag{12}$$

where $Q_{msw\ taken\ into\ consid.T}$ is calculated by using the relation:

$$Q_{msw\ taken\ into\ consid.T} = Q_{msw.T} + Q_{msw\ undergrad.T-1}, [Gg] \quad (13)$$

- DOC_f = fraction [%] of DOC dissolved under anaerobic conditions (taking into consideration the environmental condition from landfill) which generated CH_4.

The calculus can be done in this way:

- Empirical [16] by using the formula: 0.014 T + 0.28, where T?is the annual average temperature, in C^0, in the district where MSW is located.

By using IPCC recommended values for the temperate-continental zones, in Eastern and Central Europe, [5] [9] we found the following percentage values: 50%, 55%, 60% and 77%.

If we take into consideration the Romanian districts climate zone conditions the recommended values (as percentage) are to be: 43%, 45%, 50%, 55% and 60%.

- 1.3333(16/12) is the conversion factor of the carbon from CH_4 emission.

- F-MSW landfill CH_4 gas emission correction factor and depends on the management of landfill; this factor assumes the compacting level of the solid municipal waste (rubbish) MSW landfill body and its values are:

a) 0.4 ÷ 0.5-if *MSW* landfill is not compacted;

b) 0.6 ÷ 0.7-if the *MSW* landfill is compacted by means of a compactor and a bulldozer;

c) 0.8 ÷ 0.9-if *MSW* landfill is compacted with two bulldozers and two compactors. It is to be observed that there is not value 1 because there are no perfect ways of *MSW* management.

- F_r-is a correction factor of CH_4 gas emission fraction from gas deposit [Landfill Gas- *LFG*], according to the IPPC recommended values; these values of F_r are within interval 40 ÷ 60%,

Taking into consideration the above formula and using adequate input data, the graphical representations for the evolution of the equivalent CO_2 of MSW landfills [4] [9] [13] —Landfill Rudeni-Chitila-Iricex, Landfill Vidra-Ecosud are presented in Figure 1 and Figure 2.

The evolution of the equivalent CO_2 for a non-conforming MSW landfill is presented in Figure 3. It is to be observed that the CH_4 gas emission continues, after the closing date?the year 2010, as shown.

Wastes deposited quantities (msw) within deposit body are shown in Table 3. These quantities, due to "m" values, according to the Nomograme [15] , generated CH_4 quantities as presented within Figure 1, with the following signi ficance [9] [13] [15] : in the year 2011 there were collected 7.5 million cubic meters of CH_4 which have been used for green energy production.

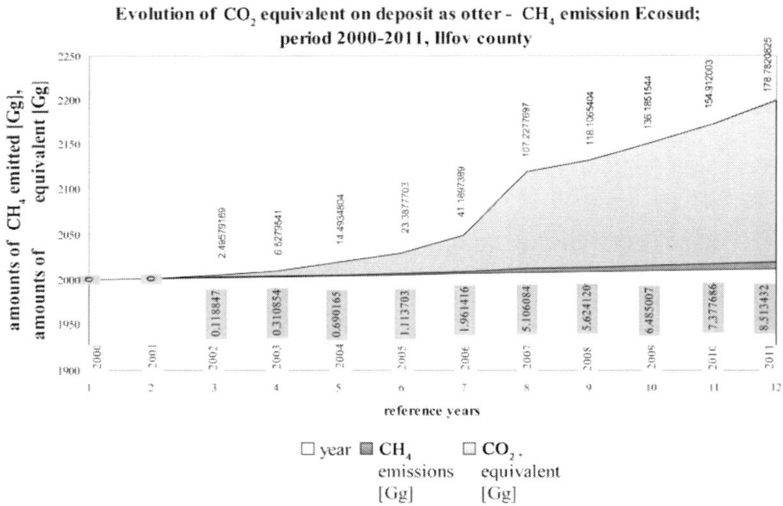

Figure 2. Evolution of CO_2 (equivalent) and CH_4 Methane gas emission from the landfills Vidra-Ecosud, Ilfov District, in the period: 2000 ÷ 2011.

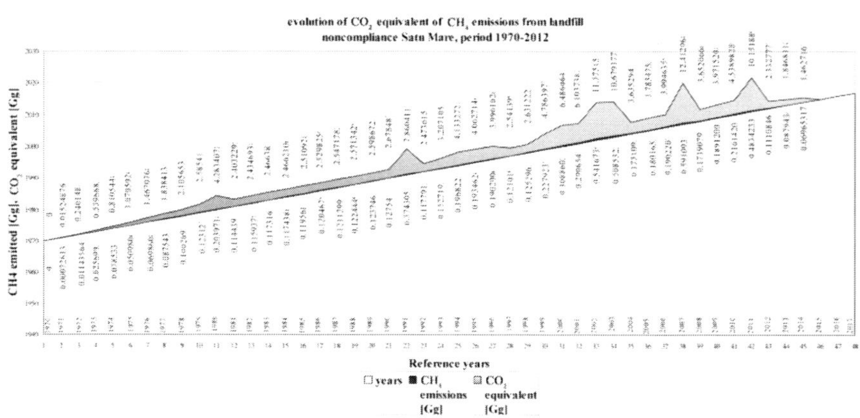

Figure 3. The MSW landfill disposal time period: 1970 ÷ 2015 lasting for CH_4 gas emission, after disposal was completed. The percentage composition of MSW may be changed, year by year. The sludge from MSW can be taken into consideration, separately or may be incorporated within bio-degraded waste (rubbish).

Table 3. Present the MSW wastes deposited within the body, for the period 2000 ÷ 2011.

Landfill (*MSW*, Chitila-Iridex, environmental Region 8 Bucharest-Ilfov											
Year of storage											
2000	2001	2002	2003	2004	2005	2006	2007	2008	2009	2010	2011
Quantities of wastes (*MSW*) stored [*Gg*]											
43.536	361.15	361.65	309.42	349.46	384.45	367.98	245.49	448.69	434.85	425.52	361.00

For the period 2000 to 2011, the percentage (%) of MSW composition has been considered, as shown in Table 1. Plastic wastes, inert waste, construction and demolition have not to be taken into consideration because they will not affect the CH_4 gas emission [8] [14] .

The data were confirmed by collection data.

6. A CASE STUDY

Within 2000 ÷ 2011 period (see Figure 1) quantities belonging to the interval 250 ÷ 400 Gg, there were deposited, annually. The GEG Effect has been intensified has been intensified, so that in the year of 2011 and a quantity of 7.5 million cubic meters of CH_4 has been used for electric energy production. As a direct consequence the GEG Effect decreased considerably, see Figure 1.

For the period 2000 to 2011, the CH_4 calculated values of gas emissions are presented in Figure 1, by using Formula (1):

$$CH_{4(Gg/year)T} = Q_{mswdegrad.T} * \%TDOC_{dissolved.T} * DOC_f * 16/12 * F * F_r, \quad (1)$$

Using some indicators related to the MSW landfills CH_4 gas emission, a calculation model is presented below. These indicators are those recommended by IPCC group of experts, group for the Central and Eastern Europe, [9] as follows:

$Q_{msw2000} = 43,536\ [Gg]$ MSW landfill deposited at the Year 2000;

$Q_{msw2001} = 361,157[Gg]$ MSW landfill deposited at the Year 2001;

$$Q_{mswdegrad.T} = \left[\left(Q_{msw.T} + Q_{msw.T-1}\right)\right] * \left[\left(1 - \exp(-Kt)\right)\right][Gg], \quad (2)$$

At the starting year of CH_4 emission within the Equation (2) can be used the expression:

$$\left[\left(1-\exp\left(-K\left(13-m/12\right)\right)\right)\right] \quad [3]$$

where m represents the number of months in which maximum 45% of deposited MSW are degradeted, $7 \le m \le 12$ [3] .

After the emission starting the expression $\left[\left(1-\exp\left(-K\left(25-m\right)/12\right)\right)\right]$, [15] , $7 \le m \le 18$ [15] can be used. m -the number of months is allocated to the MSW Nomograme [15] .

$$Q_{mswdegrad.2001} = \left[\left(Q_{msw2000} + Q_{msw2001}\right)\right] * \left[\left(1-\exp\left(-K\left(13-m\right)/12\right)\right)\right], [Gg]$$

$Q_{msw\ degrad.T}$ is degraded quantity, Equation (2) Which generated **DOC** (Organic Carbon Dissolved), and, finally, CH_4, at the year 2001.

$K = 0,4,\ m = 9,\ 7 \le m \le 12$ [3] , no. of months for the period 2000-2001, when MSW are degradeted, according to the MSW landfill Nomograme [15] .

$$Q_{mswdegrad.2001} = \left(43.536 + 361.157\right) * \left(1-0.8781\right); [Gg]$$

$$Q_{mswdegrad.2001} = 49,332 [Gg], \text{calculated by using the } Eq. \quad (3)$$

By using Equation (4) the $Q_{msw\ undegrad.2001}$ is calculated.

$$Q_{msw\ undegrad.T} = \left(Q_{msw\ T} + Q_{msw\ T-1}\right) - Q_{msw\ degrad.T}, [Gg] \quad (4)$$

$Q_{msw\ undegrad.2001} = \left(361.157 + 43.536\right) - 49.332$, [Gg], calculated by using the Equation (4)

$Q_{msw\ undegrad.2001} = 355.361$ [Gg], MSW quantity remained un-degraded in the end of 2001.

By using formula shown below the percentage of %TDOC has been determined:

$$\%TDOC_{dissolved.T} = \left(TDOC_{dissolved.T}\right) / \left(Q_{msw\ taken\ into\ consid.T}\right) [\%] \quad (12)$$

$\text{TDOC}_{dissolced T}$ – Total Organic Dissolved Carbon (DOC) $[Gg]$ was determined such as:

$$\text{TDOC}_{\text{dissolved.2001}} = \sum[A + B + C + D + E + G],[Gg], \tag{5}$$

The terms A, B, C, D, E, G are calculated at the year 2001, by using adequate equations

$$A = Q_{mswdegrad.T} * \%Q_{mswbiodehrad.T} * k_0,[Gg], \tag{6}$$

$$A_{2001} = Q_{msw\ \text{degrad.2001}} * \%Q_{msw\ \text{biodegrad.2001}} * k_0,[Gg]$$

$k_0 = 0.185$, the bio-degradable wastes DOC generation ratio, is in accordance with [IPCC, 2006], Chapter V, wastes;

$$Q_{msw\ \text{degrad.2001}} = 49.332[Gg]$$

$$\%MSW_{\text{biodegrad}} = 51.2$$

$$A_{2001} = 49.332 \times 0.512 \times 0.185 = 4.673,[Gg]$$

$$B = Q_{mswdegrad.T} * \%Q_{msw(G+P)degrad.T} * k1,[Gg],$$

$$B_{2001} = Q_{msw\ \text{degrad.2001}} * \%Q_{msw(G+P)\text{degrad.2001}} * k_1,[Gg] \tag{7}$$

$k_1 = 0.1$, the park and garden wastes DOC generation ratio, in accordance with [IPCC, 2006], Chapter V wastes;

$\%Q_{msw\ G+P} = 16$

$$B_{2001} = 49.332 * 0.16 * 0.1 = 0.789,[Gg]$$

$$C = Q_{msw\ \text{degrad},T} * \%Q_{msw\ H+C+\text{text.degrad.}T} * k_2,[Gg]$$

$$C_{2001} = Q_{msw\ \text{degrad.2001}} * \%Q_{msw(H+C+\text{text.})\text{degrad.2001}} * k_2,[Gg] \tag{8}$$

$k_2 = 0.06$, the papers + cartoon + textiles wastes DOC generation ratio, in accordance with [IPCC, 2006], Chapter V, wastes;

$$\%Q_{H+C+\text{text.degrad.2001}} = 16.8$$

$$C_{2001} = 49.332 \times 0.168 \times 0.06 = 0.497(Gg)$$
$$D = Q_{msw\ degrad.T} * \%MSW_{(Wood+straw)degrad.T} * k_3, [Gg]$$
$$D_{2001} = Q_{msw\ degrad.2001} * \%MSW_{(wood+straw)degrad.2001} * k_3, [Gg] \quad (9)$$

$k_3 = 0.03$, the wood + straw wastes DOC generation ratio in accordance with [IPCC, 2006], Chapter V, wastes;

$$\%MSW_{wood+straw.2001} = 3$$

$$D_{2001} = 49.332 * 0.03 * 0.03 = 0.044, [Gg]$$
$$E = Q_{msw\ degrad.T} * \%MSW_{sludg.degrad.T} * k_n, [Gg]$$
$$E_{2001} = Q_{mswdegrad.2001} * \%MSW_{sludg.2001} * k_n, [Gg] \quad (10)$$

$k_n = 0,185$, the containing sludge wastes DOC generation ratio in accordance with [IPCC, 2006], Chapter V, wastes;

$$\%MSW_{sludg.degrad.2001} = 1$$

$$E_{2001} = 49.332 * 0.01 * 0.185 = 0.091, [Gg]$$
$$G = Q_{msw\ degrad.T} * \%Q_{msw\ ind.degrad.T} * k_4, [Gg]$$
$$G_{2001} = Q_{msw\ degrad.2001} * \%Q_{msw\ ind.degrad.2001} * k_4, [Gg] \quad (11)$$

$k_4 = 0.09$, the industrial wastes (similar to home wastes) DOC generation ratio, in accordance with [IPCC, 2006], Chapter V, wastes;

$$\%Q_{msw\,ind.degrad.2001} = 12$$

$$G_{2001} = 49.332 * 0.12 * 0.09 = 0.533, [Gg]$$

$$\mathrm{TDOC}_{dissolved.T} = \sum [A + B + C + D + E + G], [Gg] \quad (5)$$

$$\mathrm{TDOC}_{dissolved.T} = 4.673 + 0.789 + 0.497 + 0.044 + 0.094 + 0.533 = 6.133 [Gg]$$

$$\%\mathrm{TDOC}_{dissolved.T} = (\mathrm{TDOC}_{dissolved.T}) / (Q_{msw\ taken\ into\ consid.T}) [\%] \quad (12)$$

$$\%TDOC_{dissolved.2001} = (TDOC_{dissolved.2001})/(Q_{msw\ taken\ into\ consid.2001})[\%]$$

$$Q_{msw\ taken\ in\ to\ consid.T} = Q_{msw\ T} + Q_{msw\ undegrad\ T-1}[Gg] \quad (13)$$

$$Q_{msw\ taken\ in\ to\ consid.2001} = Q_{msw2001} + Q_{msw\ undegrad.2001}[Gg]$$

$$Q_{msw\ taken\ in\ to\ consid.2001} = 361.157 + 43.536 = 404.693,[Gg]$$

$$\%TDOC_{2001} = 6.133/404.693 = 0.01515;\ 1.52\%\ \text{respectively;}$$

The CH_4 gas emission quantity at the year 2001 is calculated by applying the $Eq.(1)$, as follows:

$$CH_{4emission/2001} = 49.332 * 0.0152 * 1.3333 * 0.5 * 0.6 * 0.5 = 0.14997[Gg]$$

where:

49.332 [Gg] is MSW degraded quantity at the year 2001 which generated DOC and, later on, CH_4methane gas [5] [7] [10] ;

- 1.52% is the percentage% TDOC within landfill body;
- 0.5 represent DOC_f taking into consideration the existing condition from the analyzed emission;
- 1.3333 (16/12) represent C from CH_4;
- 0.6 represents the management level of the analyzed MSW landfill, at the year 2001;
- 0.5 represents the% content of CH_4 Methane gas within Landfill Gas (LFG).

It is to be observed that the CH_4 gas emission increased gradually, but not suddenly, in accordance with the environmental condition of the landfill location [6] . A certain wastes (rubbish) quantity of MSW landfill will remain un- degraded and will be taken into consideration in the next year, so the process of MSW degraded will generate again DOC, and, as a consequence, CH_4 Methane gas:

$$CO_{2equivalent2001} = CH_{4emitted2001} * 21 = 0.14997 = 3.14937[Gg]$$

At the year 2011, for the same MSW landfill-Chitila-Rudeni-Iridex, the quantity of CH_4 emission will be [8] [12] [15] ;

$Q_{msw2011} = 361.000[Gg]$ MSW, deposited

$Q_{mswundegrad.2010} = 496.989[Gg]$, the quantity of MSW landfill undegraded, remained from the year 2010;

$$Q_{msw\ \text{taken in to consid}.T} = Q_{msw\ T} + Q_{msw\ \text{undegrad}.T-1}[Gg] \quad (13)$$

$$Q_{msw\ \text{taken into consid}.2011} = 361.000 + 496.989 = 857.989,[Gg]$$

MSW landfill deposited taken into consideration for the calculus of $Q_{msw\ \text{degrad}.2011}$:

By using the Formula (2)

$$Q_{mswdegrad.T} = \left[\left(Q_{msw.T} + Q_{mswundegrad.T-1}\right)\right] * \left[\left(1 - \exp(-Kt)\right)\right][Gg], \quad (3)$$

$K = 0,4;\ m = 7$ in accordance with MSW deposit nomograme [3] [13] [17] .

$$Q_{msw\ \text{degrad}.2011} = 387.125[Gg]$$

the non-degraded quantity of MSW remained in the end of the year 2011; the $Eq.(4)$ is used:

$$Q_{mswundegrad.T} = \left(Q_{mswT} + Q_{mswundegrad.T-1}\right) - Q_{msw\ \text{degrad}.T}, [Gg] \quad (4)$$

$$Q_{mswundegra.2011} = 857.989 - 387.125 = 470.864,[Gg]$$

By using the Equation (12), the percentage $\%\text{TDOC}_{\text{dissolved}.T}$ has been calculated, as follows:

$$\%\text{TDOC}_{\text{dissolved}.T} = \left(\text{TDOC}_{\text{dissolved}.T}\right) / \left(Q_{msw\ \text{taken into consid}.T}\right)[\%] \quad (12)$$

$\text{TDOC}_{\text{dissolved}2011}, [Gg]$ was calculated by using the Equation (5)

$$\text{TDOC}_{\text{dissolved}.T} = \sum[A + B + C + D + E + G], [Gg] \quad (5)$$

The parameters-A, B, C, D, E, F, G, are determined at the year 2011, by using corresponding equations.

$$A = Q_{msw\ degrad.T} * \%Q_{msw\ biodegrad.T} * k_0, [Gg] \tag{6}$$

$$A_{2011} = Q_{msw\ degrad.2011} * \%Q_{msw\ biodegrad.2011} * k_0, [Gg]$$

$k_0 = 0.185$ the biodegradable DOC generation ratio, in accordance with [IPCC, 2006], Chapter V, wastes.

$$Q_{msw\ degrad.2011} = 387.879[Gg]$$

$$\%MSW_{biodegrad} = 51.2$$

$$A_{2011} = 387.125 * 0.512 * 0.185 = 36.685, [Gg]$$

$$B = Q_{msw\ degrad.T} * \%Q_{msw(G+P)degrad.T} * k_1, [Gg] \tag{7}$$

$$B_{2011} = Q_{mswdegrad.2011} * \%Q_{msw(G+P)degrad.2011} * k_1, [Gg]$$

$k_1 = 0.1$, parks and garden wastes DOC generation ratio in accordance with [IPCC, 2006], Chapter V, wastes [7] [9] [10] .

$$\%Q_{msw(G+P)} = 16$$

$$B_{2011} = 387.125 * 0.16 * 0.1 = 6.194[Gg]$$

$$C = Q_{msw\ degrad,T} * \%Q_{msw(H+C+text).degrad.T} * k_2, [Gg] \tag{8}$$

$$C_{2011} = Q_{msw\ degrad.2011} * \%Q_{msw(H+C+text.)degrad.2011} * k_2, [Gg]$$

$k_2 = 0.06$, the papers + cartoon + textiles wastes DOC generation ratio in accordance with [IPCC, 2006], Chapter V, wastes.

$$\%Q_{(H+C+text.)degrad.2001} = 16.8$$

$$C_{2011} = 387.125 * 0.168 * 0.06 = 3.902[Gg]$$

$$D = Q_{msw\ degrad.T} * \%MSW_{(Wood+straw)degrad.T} * k_3, [Gg] \tag{9}$$

$$D_{2011} = Q_{msw\ degrad.2011} * \%MSW_{(wood+straw)degrad.2011} * k_3, [Gg]$$

$k_3 = 0.03$, the wood + straw wastes DOC generation ratio in accordance with [IPCC, 2006], Chapter V, wastes.

$$\%MSW_{wood+straw.2001} = 3$$

$$D_{2011} = 387.125 * 0.03 * 0.03 = 0.384[Gg]$$

$$E = Q_{msw\ degrad.T} \times \%MSW_{sludg.degrad.T} \times k_n, [Gg] \tag{10}$$

$$E_{2011} = Q_{mswdegrad.2011} * \%MSW_{sludg.2001} * k_n, [Gg]$$

$k_n = 0.185$, wastes (containing sludge) DOC generation ratio in accordance with [IPCC, 2006], Chapter V, wastes [9] [10] [17] .

$$\%MSW_{sludg.degrad.2001} = 1$$

$$E_{2011} = 387.125 * 0.01 * 0.185 = 0,716, [Gg]$$

$$G = Q_{msw\ degrad.T} * \%Q_{msw\ ind.degrad.T} * k_4, [Gg] \tag{11}$$

$$G_{2011} = Q_{msw\ degrad.2011} * \%Q_{msw\ ind.degrad.2011} * k_4, [Gg]$$

$k_4 = 0.09$, MSW landfill containing industrial wastes (similar to home wastes) DOC generation ratio, in accordance with [IPCC, 2006], Chapter V wastes,

$$\%Q_{msw\ ind.degrad.2011} = 12$$

$$G_{2011} = 387.125 * 0.12 * 0.09 = 4.181, [Gg]$$

$$\text{TDOC}_{dissolved.T} = \sum [A + B + C + D + E + G], [Gg] \tag{5}$$

$$\text{TDOC}_{dissolved.2011} = 36.685 + 6.194 + 3.902 + 0.384 + 0.716 + 4.181 = 52.062, [Gg]$$

$$\%TDOC_{dissolved.T} = (TDOC_{dissolved.T})/(Q_{msw\ taken\ into\ consid.T})[\%] \quad (12)$$

$$\%TDOC_{dissolved.2011} = (TDOC_{dissolved.2011})/(Q_{msw\ taken\ into\ consid.2011})[\%]$$

$$Q_{msw\ taken\ in\ to\ consid.T} = Q_{msw\ T} + Q_{msw\ undegrad.T-1}, [Gg] \quad (13)$$

$$Q_{msw\ taken\ in\ to\ consid.2011} = Q_{msw2011} + Q_{msw\ undegrad.2010} [Gg]$$

$$Q_{msw\ taken\ in\ to\ consid.2011} = 361.000 + 496.989 = 857.989[Gg]$$

$\%TDOC_{2001} = 52.062/857.989$ 0.0607; 6.07%, respectively.

The quantity of CH_4 in the year 2011 gas emission is calculated by applying Formula (1) as follows:

$$CH_{4emission/2011} = 387.125 * 0.0607 * 1.3333 * 0.5 * 0.9 * 0.5 = 7.0494, [Gg]$$

where:

Ÿ 387.125 [Gg] is MSW degraded quantity of in 2011 which generated DOC and, later on, CH_4

Ÿ Methane gas; 6.07%, is the percentage % TDOC within landfill body;

Ÿ 0.5 represent DOC_f taking into consideration existing condition from the analyzed emission;

Ÿ 1.3333 (16/12) represent C from CH_4;

Ÿ 0.9 represents the management level of the analyzed MSW landfill, in the year 2001;

Ÿ 0.5 The content $[\%]$ of CH_4 methane gas within Landfill Gas (LFG).

It is to be observed that the CH_4 gas emission increased gradually, but not suddenly, in accordance with the environmental condition of the landfill location [6] . A certain waste (rubbish) quantity of MSW landfill will remain un-de- graded and will be taken into consideration in the next year, so the process of MSW degraded will generate again DOC, and, as a consequence, CH_4 Methane gas:

$$CO_{2equivalent2011} = CH_{4emitted2011} * 21 = 148.037[Gg]$$

It is to be observed that the CH_4 gas emission increased gradually, but not suddenly, in accordance with the environmental condition of the landfill location [4] [6] . A certain waste (rubbish) quantity of MSW landfill will remain un- degraded and will be taken into consideration in the next year, so the process of MSW degraded will generate again DOC.

The sludge from MSW can be taken into consideration, separately or may be incorporated within bio-degraded waste (rubbish).

7. CONCLUSIONS

This article doesn't comment on the present calculation model but rather draws the attention to a more adapted to the real conditions estimation, by calculus, of the CH_4 gas emission from the actual MSW landfills in Romania, which have to be estimated by the end of 2017. Even if deposited MSW quantities were up to 30 (Gg), in the beginning of 1979 and reached 90 (Gg) in 2010, the evolution of CO_2 exists and has to be known by the Romanian authorities.

It is considered that this estimation has to be determined up to the life-end of the considered landfill. As an example, at the existing MSW landfill, in the Satu Mare County, the evolution of the equivalent CO_2 for a period of 42 years up to 2010 when it was closed is presented. The authorities have to inform the public about the evolution of the equivalent CO_2 for existing MSW landfill and also for the location of the new MSW landfills.

On the other hand, for the non-hazardous MSW landfills having a capacity between $350 \div 450$ [Gg] it was observed that the top management of this MSW landfills issued estimated quantities of CH_4 gas at unrealistic values, sometimes more than two times lower with respect to the real one, estimated by usual calculation models.

To reduce the greenhouse effect, the evolution of the equivalent CO_2 for the existing MSW landfills in Romania has to be estimated in such a way as to be useful for an applicable environmental planning in accordance with the government's and the European policy in the field of environmental protection. Other gas emissions such as: NON-METHANE ORGANIC COMPOUNDS, NO_2, NO_x, Polycyclic aromatic hydrocarbons, HFC, PFC have not been taken into consideration.

The real estimation of the CH_4 emission quantity from MSW landfills, in Romania, will contribute to a better environmental planning and a better understanding of the contribution that different gases have on the general warming effect and climate changes greenhouse effect.

Finally, it is to be noted that the calculation of the CH_4 emission quantity, by using the Danila Vieru's Method, (Formula1), will help Romanian environmental authorities to implement the legal and right decisions regarding the adequate moment when the collected CH_4 emission can be burned, and thus be used in an economical manner.

The proposed method could be applied for the CH_4 emission calculation at MSW landfills quantities between 100 ÷ 200 (Gg/y) e.g. the Satu Mare non-conforming MSW landfill (see Figure 3).

This method which was verified for Romanian landfills could be easily adapted for other countries too, paving the way for a real estimation of the methane gas emission, as real as possible.

The proposed method can be applied either to the MSW landfills which respect legal providing and those (MSW) which not respect such provisions. The quantitative CH_4 estimation is beneficial for the Environmental Authorities but also for the potential investors interested in the CH_4management. It is to be noted that potential investors have to know the emission quantity and its duration. After MSW depositing is over, it is absolutely necessary to the time-dura- tion when the emission is stopped. In the same time, after the CH_4 emission is over, the resulted compost should be of interest for the farmers.

ACKNOWLEDGEMENTS

The author would like to express thanks for their support and understandings to the staff members of the Environmental Protection Agency (EPA) and to the local and regional subsidiaries of ETA's for their help and suggestions.

The author would like to express deep thanks and gratitude, including for the moral support, to Prof. Dr. Eng. Vladimir Rojanski for his advices in order to complete my work in order to estimate the CH_4 gas emission, by a calculus formula, both for conforming and non-conforming MSW solid landfills.

CITE THIS PAPER

Vieru, D. (2017) A New Approach Method of CH_4 Emission Estimation from Landfills Municipal Solid Waste (MSW). Atmospheric and Climate Sciences, 7, 191-209.

REFERENCES

1. KYOTO Protocol, Kyoto, Japan, December 1997, Convention on Climate Change, entered into Force on February (2005).
2. Kiev Protocol on Pollutant Release and Transfer Registers-UNECE, October 2009.
3. Richard Pelt, C.W., et al. (1998) User's Manual Landfill Gas Emission Model. EPA (Environmental Protection Agency), Washington DC, 20460.
4. Romania Government Decision (order) no.349/2005 for the landfills.
5. Council Directive 99/31/EC of 26 April 1999 on the landfill of waste entered into force on 16.07.1999, Bruxelles, Belgium.
6. Riitta Pipatti, C.S., et al. (2006) Waste Generation, Composition and Management Data. Chapter 2, IPCC, Geneva, Switzerland.
7. Jensen, J.E.F. and Pipatti, R. (2006) CH4 Emissions from Solid WASTE Disposal. IPCC/OECD/IEA, Geneva, Switzerland.
8. Directive 75/442/EEC of 15 July 1975 on waste, replaced by Directive 2008/98/ EC of the European Parliament and of the Council of 19 November 2008 on waste and repealing certain Directives.
9. IPCC Guidelines for National Greenhouse Gas Inventories (2006) User's Manual Landfill Gas Emissions Model. Volume 5 Waste, Version 2.0, Geneva, Switzerland. [Citation Time(s):17]
10. Pipatti, R., Silva Alves, J.W., et al. (2006) IPCC Guidelines for National Greenhouse Gas Inventories, Chapter 3: Solid Waste Disposal. IPCC, Geneva, Switzerland. [Citation Time(s):13]
11. Bergman, H., et al. (2004) Methane Emissions from Solid Waste Disposal Sites. Technical Bulletin No. 872, IPCC Guidelines, Reference Manual, part 6.2.2.4. Geneva, Switzerland.
12. Heijo Scharff, J.J. (2006) Applying Guidance for Methane Emission Estimation for Landfills. Emission Estimation Technique Manual for Municipal Solid Waste (MSW) Landfills, Version 2.0. Assendelft, the Netherlands.
13. Cai, B.-F., Liu, J.-G., Gao, Q.-X., et al. Estimation of Methane Emissions from Municipal Solid Waste Landfills in China Based on, Point Emission Sources Chinese Academy for Environmental Planning.
14. Ranjan, M.R., Ramanathan, A.L., Tripathi, A. and Jha, P.K. (2014) Landfill Mining: A Case Study from Ghazipur Landfill Area of Delhi. International Journal of Environmental Sciences, 4, 919-925.
15. Small Mathematics Encyclopedia (with a Forward by Academician Gheorghe Mihoc). Technical Publishing House, p. 93, Bucharest, 1980.
16. Tabasaran, O. (1981) Gas Production from Landfill. In: Bridgewater, A.V. and Lidgren, K., Eds., Household Waste Management in Europe, Economics and Techniques, Van Nostrand, Reinhold Co., New York, 159-175.
17. Regulation (EC) No. 2150/2002 of the European Parliament and of the Council of 25 November 2002 on waste statistics.

CHAPTER 16

Future Changes in Global Precipitation Projected by the Atmospheric Model MRI-AGCM3.2H with a 60-km Size

Shoji Kusunoki

Climate Research Department, Meteorological Research Institute, 1-1, Nagamine, Tsukuba, Ibaraki 305-0052, Japan

ABSTRACT

We conducted global warming projections using the Meteorological Research Institute-Atmospheric General Circulation Model Version 3.2 with a 60-km grid size (MRI-AGCM3.2H). For the present-day climate of 21 years from 1983 through 2003, the model was forced with observed historical sea surface temperature (SST). For the future climate of 21 years from 2079–2099, the model was forced with future SST projected by conventional couple models. Twelve-member ensemble simulations for three different cumulus convection schemes and four different SST distributions were conducted to evaluate the uncertainty of projection. Annual average precipitation will increase over the equatorial regions and decrease over the subtropical regions. The future precipitation changes are generally sensitive to the cumulus convection scheme, but changes are influenced by the SST over the some regions of the Pacific Ocean. The precipitation efficiency defined as precipitation change per 1° surface air temperature warming is evaluated. The global average of precipitation efficiency for annual average precipitation was less than the maximum value expected by thermodynamical theory, indicating that dynamical atmospheric circulation is acting to reduce the conversion efficiency from water vapor to precipitation. The precipitation efficiency by heavy precipitation is larger than that by moderate and weak precipitation.

Keywords: precipitation; global warming projection; uncertainty of projection; precipitation efficiency

1. INTRODUCTION

Uncertainty in future climate change projected by climate models originates from four main factors: emission scenario, model structure, internal natural variability and initial condition and external forcing and boundary condition [1]. For a given specific emission scenario, uncertainty of future projection is often evaluated as a spread of responses by multiple models. A single model can be used to estimate the spread of internal natural variability if it is integrated from multiple initial conditions. The spread of responses by a single model can be also estimated by implementing multiple versions of physical processes into the model [2].

It is natural to assume that future projection by models with higher reproducibility of observed climate is more reliable than projections by models with lower reproducibility. In principle, the reproducibility of precipitation climatology improves when we use a model with higher horizontal resolution. Studies by [3,4,5] (Table S1) have revealed that the realistic reproduction of summertime precipitation over East Asia requires an atmospheric model with higher horizontal resolution. Therefore, we have been conducting a series of global warming projection experiments using the Meteorological Research Institute-Atmospheric General Circulation Model Version 3 (MRI-AGCM3) with 20-km and 60-km grid sizes [3,4,6,7,8,9,10,11,12,13,14] (Table S1). These models have relatively higher horizontal resolution compared with global atmospheric models generally used in climate variability studies.

Since the MRI-AGCM3 model has only the atmosphere and lacks the oceans, we have to specify sea surface temperature (SST) to conduct the global warming projection. We adopted a 'time-slice experiment' [15] where the model was forced by prescribed external boundary conditions and forcing. For the present-day climate, observed SSTs were given to the model. This kind of model experiment is conventionally called the Atmospheric Model Intercomparison Project (AMIP)-type simulation, which is a standard method to evaluate the performance of atmospheric models. For the future climate, future SSTs projected by the conventional atmosphere-ocean general circulation model (AOGCMs) are given to the model.

One of the striking advantages of time-slice experiment is that we can increase the horizontal resolution of atmospheric part, because we can save computer resources allocated for the ocean part. Another advantage is that the present-day climatology simulated by an atmospheric model tends to be better than that by an AOGCM, because SST prescribed in the present-day climatology is observation. Errors in SST observation are generally smaller than those simulated by AOGCMs. SST simulated by AOGCM has inevitable errors leading to the distortion of model climatology.

On the other hand, the disadvantage of the time-slice experiment is the lack of air-sea interaction. Future change in the Indian monsoon is influenced by the SST-convection relationship, which would be erroneously distorted if SST feedback from the atmosphere is not taken into account [16]. Tropical cyclone simulated by the atmospheric model tends to show overestimation of intensity [17] and erroneous northward shift of the existence area in mid-latitudes [18], owing to the lack of cooling by mixing of the ocean surface layer.

Despite this weakness in the time-slice experiment, we selected the option to use the higher horizontal resolution version of the atmospheric global model, which enables us to project the small spatial scale structure of future climate change without any dynamical downscaling using regional climate models. Moreover, we are collaborating with scientists investigating the impact of global warming in the fields of natural disaster prevention, water resource management, agriculture and irrigation. Some of their tools, such as the river discharge model and the flood analysis model, require very high horizontal resolution forcing as input data. This is another reason to introduce a higher horizontal resolution model by the time-slice experiment to meet the requirements of impact assessment scientists.

Simulated precipitation change by the atmospheric model is sensitive to prescribed SST [19]. The dependence of precipitation change on SST was investigated using predicted future SST by some individual AOGCMs participating in the third phase of the Coupled Model Intercomparison Project (CMIP3), as well as the SST of the multi-model ensemble (MME) mean of the CMIP3 AOGCMs [6,7,8,9] (Table S1).

Simulated precipitation change by the atmospheric model is also susceptible to physical processes, such as the cumulus convection scheme implemented in the models. Using the MRI-AGCM3.2S (20 km) and 3.2H (60 km), [9] have conducted a set of ensemble simulations for three different cumulus convection schemes and four different SSTs obtained from the cluster analysis of 18 CMIP3 AOGCMs. They found that the uncertainty of precipitation over the southern part of Asia originates in the difference among cumulus convection schemes, while that over the Maritime Continent (MC) is caused by the difference among SSTs. The dependence of precipitation change on SST was further investigated using future SSTs obtained from the fifth phase of the Coupled Model Intercomparison Project (CMIP5) AOGCMs [11,12,13,14] (Table S1). However, the dependence of precipitation change on the cumulus convection scheme has not yet been fully examined using future SSTs of CMIP5 AOGCMs.

The purpose of this study is to answer the following questions. Firstly, do our models globally perform better than the atmospheric models of other institutions with respect to annual average precipitation, as well as extreme precipitation events? Secondly, how will precipitation change globally with respect to annual average precipitation, as well as extreme precipitation events? Thirdly, how large is the uncertainty of future change? Simulated precipitation by the atmospheric model is apparently sensitive to SST and the cumulus convection scheme. The works in [9,14] have quantitatively evaluated the relative contributions of SST and cumulus convection to the uncertainty of future precipitation change, but their target regions were limited to Asia. Therefore, we need to extend the target region to the whole globe in order to get the big picture of precipitation change from the global perspective. Fourthly, how will precipitation change globally in response to the increase of surface air temperature? In global warming projections, models consistently project a global precipitation increase that is slower than the increase in moisture expected from thermodynamic theoretical estimate [20,21]. We will investigate how the conversion rate of precipitation from water vapor depends on the metrics of precipitation, the cumulus convection scheme and SST.

2. MODELS AND EXPERIMENTAL DESIGN

2.1. The Global Atmospheric Model

The global atmospheric model used in this study is the MRI-AGCM3.2H. This model was cooperatively developed by the Japan Meteorological Agency (JMA) and the MRI [22]. The horizontal grid size of the model is 60 km. Hereafter, we call this model as the 60-km model in this paper. The model has 60 levels reaching to 0.01 hPa top equivalent to an altitude of approximately 80 km. We also used the same version with different horizontal resolution models of the MRI-AGCM3.2S (20 km) and the MRI-AGCM3.2L (180 km) in order to compare the reproducibility of present-day climatology (Table S2).

As for the cumulus convection scheme, we implemented the "Yoshimura scheme" (YS; [23]) which is modified from the Tiedtke scheme [24]. Simulated precipitation by models largely depends on the kind of cumulus convection scheme implemented in the model. Therefore, in order to evaluate the sensitivity of simulated precipitation on the cumulus convection scheme, we further utilized the Arakawa–Schubert (AS) scheme [25] and the Kain–Fritsch (KF) scheme [26], as well as the Yoshimura scheme. These versions of the models were already used for ensemble simulations in previous studies that analyzed future changes in precipitation over Asia [9,10,14].

The Tiedtke, YS and AS schemes are all categorized as classical mass-flux type cumulus schemes. In the Tiedtke scheme, only a single convective updraft is calculated in a single grid point, but is represented as a more detailed entraining and detraining plume. In the AS scheme, on the other hand, multiple convective updrafts with different heights are explicitly calculated within a single grid cell, although each updraft is a more simplified entraining plume. YS can be regarded as a kind of hybrid of the Tiedtke and AS schemes. YS represents all top-level cumulus plumes by interpolating two convective updrafts with maximum and minimum rates of turbulent entrainment and detrainment [27]. YS performs better than the AS scheme in simulating precipitation in the tropics [22] and summer precipitation over East Asia [5]. Because the AS scheme was the default convection scheme for the MRI-AGCM3.0 and 3.1 models, an option for choosing the AS scheme was kept in the MRI-AGCM3.2. The KF scheme was originally developed to simulate the mesoscale convective system (MCS) in mid-latitudes. The KF scheme is based on a one-dimensional cloud model in which the exchange of mass between cloud and environment is modulated through entrainment and detrainment. Since the KF scheme well simulates MCSs, it is implemented in the operational regional model of the JMA. Therefore, we also included the KF scheme in the options of the convection scheme in the MRI-AGCM3.2.

2.2. Sea Surface Temperature and Sea Ice

We adopted a time-slice experiment [15] where the 60-km model was forced by prescribed external boundary conditions and forcing. For the present-day climate from the year 1983–2003 (21 years), we integrated the model giving the observed historical sea surface temperature (SST) and observed sea ice concentration provided by the Hadley Centre sea ice and sea surface temperature data set version 1 (HadISST1) [28].

For the future climate from year 2079–2099 (21 years), the boundary SST data were constructed by combining three components: (i) future change in the MME mean of SST projected by the CMIP5 multi-model model (28 models) dataset; (ii) the linear trend in the MME mean of SST projected by the CMIP5 multi-model dataset; and (iii) the detrended interannual observed SST anomalies for the period from year 1979–2003. Future change in the MME mean of SST was taken from the difference between the historical experiments and the future simulation. We integrated the model under the Representative Concentration Pathway (RCP) 8.5 emission scenario [29] for the future climate. Future sea ice concentration was obtained in a similar manner. For further details, refer to [30].

Considering that the response of atmospheric models is intrinsically susceptible to the SST prescribed, we assessed the dependence of future precipitation change on the geographical distribution of SST. For this purpose, we have separated 28 CMIP5 models into three groups with the clustering analysis based on the future change in annual mean SST projected by CMIP5 models over the tropics (Figure 1a–d). Figure 1e depicts the deviation of Cluster 1 from the average of all models (C0; Figure 1a). Cluster 1 (C1; Figure 1e) is distinguished by larger warming in the Southern Hemisphere as compared to the Northern Hemisphere; Cluster 2 (C2; Figure 1f) is characterized by large warming in the central Pacific in the tropics; Cluster 3 (C3; Figure 1g) is distinguished by large warming around Japan. The future sea ice concentration was calculated according to the clustering of SST. For more details, see [31]. With these four different SST distributions, as well as three different cumulus convection schemes, we can evaluate the spread of the model response, which can be interpreted as the kind of reliability information on future precipitation change.

2.3. Other External Forcings

Observed historical concentrations of greenhouse gases (GHG) such as carbon dioxide and methane were given for the present-day climate simulations. The RCP8.5 emission scenario was applied to the future climate simulations. We used three-dimensional natural and anthropogenic aerosol distributions simulated by the MRI-Earth System Model (MRI-ESM; [27]) based on historical and the Special Report on Emission Scenario (SRES) A1B scenario [32] aerosol emission data. We included aerosols erupted only from Mt. Pinatubo in the year 1991. The three-dimensional distributions of stratospheric ozone calculated by the MRI-Chemical Transport Model (CTM) (MRI-CTM; [33]) based on historical and A1B scenario aerosol emission data were prescribed.

Aerosol and stratospheric ozone distributions simulated by the MRI-ESM and MRI-CTM assuming the RCP8.5 scenario were not available at the time when we started a set of global warming projections in this study. Table 1 summarizes the experimental designs and definitions of the simulation names.

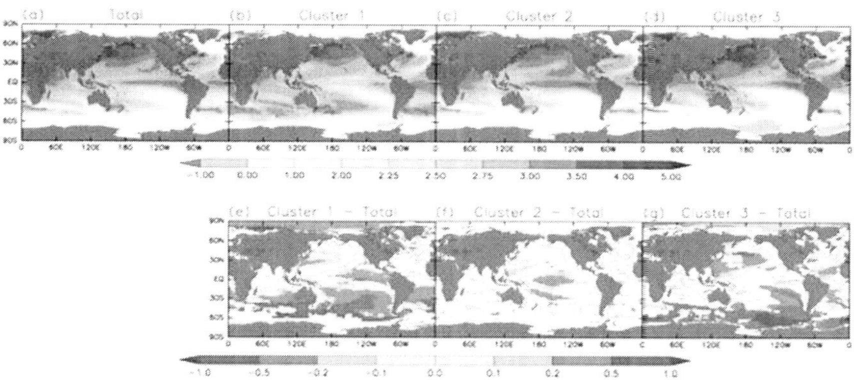

Figure 1. Annual mean sea surface temperature (SST) change (K) from the present-day (1979–2003, historical simulation) and the future (2075–2099, RCP8.5 scenario). (**a**) The composite of a total 28 models (C0); (**b**) the composite of the cluster C1; (**c**) C2; (**d**) C3; (**e–g**) differences for each cluster from the total mean (C0). The regions where over 75% of the models agree with the sign of the difference are colored. Contours denote zero. The change is normalized by the tropical (30° S–30° N) mean for each model before making the composition and then multiplied by the 28 models' mean tropical SST change (2.74 K). From [31].

3. PRESENT-DAY CLIMATE

3.1. Observation for Model Verification

To validate the model, we used the One-Degree Daily data (1dd) of the Global Precipitation Climatology Project (GPCP) v1.2 compiled by [34] from 1997–2008 (12 years). The horizontal resolution of this dataset is 1.0° in longitude and latitude, corresponding to a grid spacing of roughly 90 km at 35° N. The quantitative model performances were evaluated against GPCP 1dd data. Since the 60-km models have relatively higher horizontal resolution, we need observational data with higher horizontal resolution for verification. However, the GPCP 1dd data do not cover the whole target period of simulations from 1983–2003. Pentad and monthly data are calculated from daily data. In the verification, model data were interpolated to the location of grid points of the GPCP 1dd data.

Table 1. Definition of simulation names.

Present-Day Climate: 1983–2003, 21 Years				
Cumulus Convection [a]	Sea Surface Temperature (SST): Observation by the HadISST1 [28]			
YS	HPYS			
AS	HPAS			
KF	HPKF			
Future Climate: 2079–2099, 21 years				
Cumulus Convection [a]	Sea surface Temperature (SST): Projections by the CMIP5 AOGCMs for the Emission Scenario RCP8.5			
	Cluster 0 MME	Cluster 1	Cluster 2	Cluster 3
YS	HFYSC0	HFYSC1	HFYSC2	HFYSC3
AS	HFASC0	HFASC1	HFASC2	HFASC3
KF	HFKFC0	HFKFC1	HFKFC2	HFKFC3

First character of simulation name denotes horizontal resolution: H = 60 km; Second character denotes the target period: P = present-day, F = Future; Third and fourth characters denote the type of cumulus convection scheme; [a] Yoshiumura (YS): [23]; Arakawa-Schubert (AS): [25]; Kain-Fritsch (KF): [26]; HadISST1: the Hadley Centre sea ice and SST data set version 1; CMIP5: The fifth phase of the Coupled Model Intercomparison Project; AOGCM: Atmosphere-Ocean General Circulation Model; RCP: Representative Concentration Pathway; MME: multi-model ensemble.

Observational data themselves have uncertainty of observation so that model skill depends on the selection of observational data [35]. The pentad and monthly data of GPCP v2.2 provided by [36] are also utilized from 1983–2003 (21 years). These data cover the whole target period of simulations from 1983–2003. The horizontal resolution of this dataset is 2.5° in longitude and latitude, corresponding to a grid spacing of roughly 210 km at 35° N.

Further, the pentad and monthly data of the Climate Prediction Center Merged Analysis of Precipitation (CMAP) v1201 compiled by [37] are used from 1983–2003 (21 years). The horizontal resolution of this dataset is 2.5° in longitude and latitude, which is the same as GPCP v2.2. Moreover, we used the Tropical Rainfall Measuring Mission (TRMM) 3B43 compiled by [38] from 1998–2010 (13 years). The horizontal resolution is 0.25° in longitude and latitude, corresponding to a grid spacing of about 25 km at 35° N. However, regional coverage is restricted to a global belt extending from 50° S–50° N. Table 2 summarizes the observations used for verification.

Table 2. Observational data of precipitation used in this study. GPCP: Global Precipitation Climatology Project. 1dd: the One-Degree Daily data. CMAP: Climate Prediction Center Merged Analysis of Precipitation, TRMM: the Tropical Rainfall Measuring Mission.

Name	Time Resolution	Spatial Resolution	Period	Region	Reference
GPCP 1ddv1.2	Day	1.0 deg	1997–2013, 17 years	Global	[34]
GPCP 1ddv1.2	Month	1.0 deg	1997–2013, 17 years	Global	[34]
GPCP v2.2	Month	2.5 deg	1981–2000, 20 years	Global	[36]
CMAP v1201	Month	2.5 deg	1981–2000, 20 years	Global	[37]
TRMM 3B43	Month	0.25 deg	1998–2013, 16 years	50° S–50° N	[38]

3.2. Global Distribution of Precipitation

The global distributions of annual average precipitation (PAV) simulated by the models are compared with observations (Figure 2). In the observation of GPCP 1dd (Figure 2a), the region of large precipitation extends over the Equator in the Indian Ocean, the Pacific Ocean, the Atlantic Ocean and the Amazon in South America. Furthermore, precipitation is large over the South Pacific Convergence Zone (SPCZ), which spreads over the MC and Papua New Guinea. In middle and high latitudes in the Northern Hemisphere, precipitation shows moderate maxima over the North West Pacific Ocean and the North West Atlantic Ocean. These regions correspond to the location of storm tracks owing to the frequent passage of extratropical cyclones. Convective rainfall driven by the underlying warm western boundary ocean currents also contributes in these regions [39]. The large-scale geographical distributions of precipitation by other observations (Figure 2b–d) are approximately similar to that of GPCP 1dd (Figure 2a) Local maximum of precipitation over the MC is larger in Figure 2c,d than in Figure 2a, while precipitation over the MC is smaller in Figure 2b than in Figure 2a (also see Figure S1).

The 60-km models well reproduce the large-scale distribution of precipitation (Figure 2e–g), but the precipitation over the maritime continent and the SPCZ is overestimated compared with observations (Figure 2a–d). The MME mean of the CMIP5 atmospheric models' AMIP runs (Table S1) forced with a similar SST as the 60-km models shows also a realistic precipitation distribution (Figure 2h). The grid size at 35° N of CMIP5 atmospheric models ranges from 28 km–342 km with an average of 171.5 km and a median 171 km, which is larger than that of the 60-km model (last column in Table S2). The 60-km model with the KF scheme (Figure 2g) shows the largest precipitation over the MC compared with the other models (Figure 2e,f,h and Figure S2), which can be also confirmed in Figure S2f.

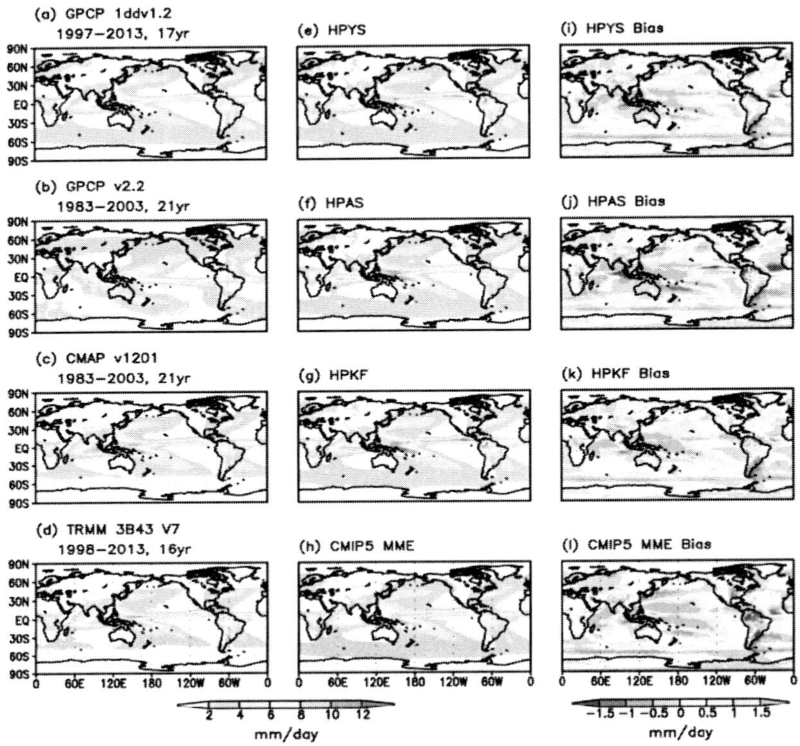

Figure 2. The climatology of annual average precipitation (PAV). Unit is mm day^{-1}. (**a**–**d**) Observation (Table 2); (**e**–**h**) simulations by models averaged for 1983–2003 (21 years); (**e**) HPYS; (**f**) HPAS; (**g**) HPKF; (**h**) MME mean of 24 CMIP5 atmospheric models' Atmospheric Model Intercomparison Project (AMIP) runs (Table S1); (**i**–**l**) bias against GPCP 1ddv1.2 (a). GPCP, Global Precipitation Climatology Project.

In terms of model biases against GPCP 1dd v1.2 (Figure 2i–l), all models show excessive precipitation over the MC and the central tropical Pacific Ocean.

3.3. Taylor Diagram

In order quantify the skill of models, we introduced the skill score *S* proposed by [40] against the GPCP 1dd data (Figure 2a). *S* is defined by:

$$S = \frac{4(1+R)}{(\sigma + 1/\sigma)^2 (1+R_0)}$$

where R is the spatial correlation coefficient between observation and simulation, σ is the spatial standard deviation of the simulation divided by that of the observations and R_0 is the maximum correlation attainable. Here, we assumed that $R_0 = 1$. S evaluates the spatial correlation coefficient, as well as the spatial standard deviation. S approaches unity in a perfect simulation.

Figure 3b displays the Taylor diagram [40] of S for the 60-km and CMIP5 atmospheric models with respect to PAV. Although S is widely used to verify model performance in many climate modelling studies, we must recognize that S cannot evaluate model bias because this method deals with deviation from the mean value of the target domain both for observation and simulations. Therefore, we also plotted the bias and root mean square error (RMSE) in Figure 3a. The target region is the global domain (Figure 2). In the Taylor diagram (Figure 3b), the radial distance from the origin is proportional to the standard deviation of a simulated spatial pattern normalized by the observed standard deviation. The spatial correlation coefficient between the observed and simulated fields is given by the angle from the Y-axis. The skills of the GPCP 2.5° data and CMAP 2.5° data against the GPCP 1dd data were also plotted to estimate the uncertainty of observations. This uncertainty originates from the differences of averaged period, horizontal resolution, original source of data and retrieval method. The difference among observations (green marks) is relatively small compared with the difference among all models (Figure 2a,b).

Figure 3a shows that all of the 60-km and all of the 24 CMIP5 models have positive biases mainly originating from overestimation of precipitation over the MC and the SPCZ. The RMSEs of the 60-km models (character H) are smaller than or comparable to those of the CMIP5 models (black X).

We calculated the average skill of the individual CMIP5 models (AVM, black square), as well as the skill of the spatial distribution of precipitation constructed from the MME mean (black circle) of CMIP5 models. In the calculation of the AVM, we first evaluated the skill of the geographical distribution of precipitation simulated by individual models, and then, we averaged all 24 skills. In the case of linear skill measures, such as average and bias, the MME mean is identical to the

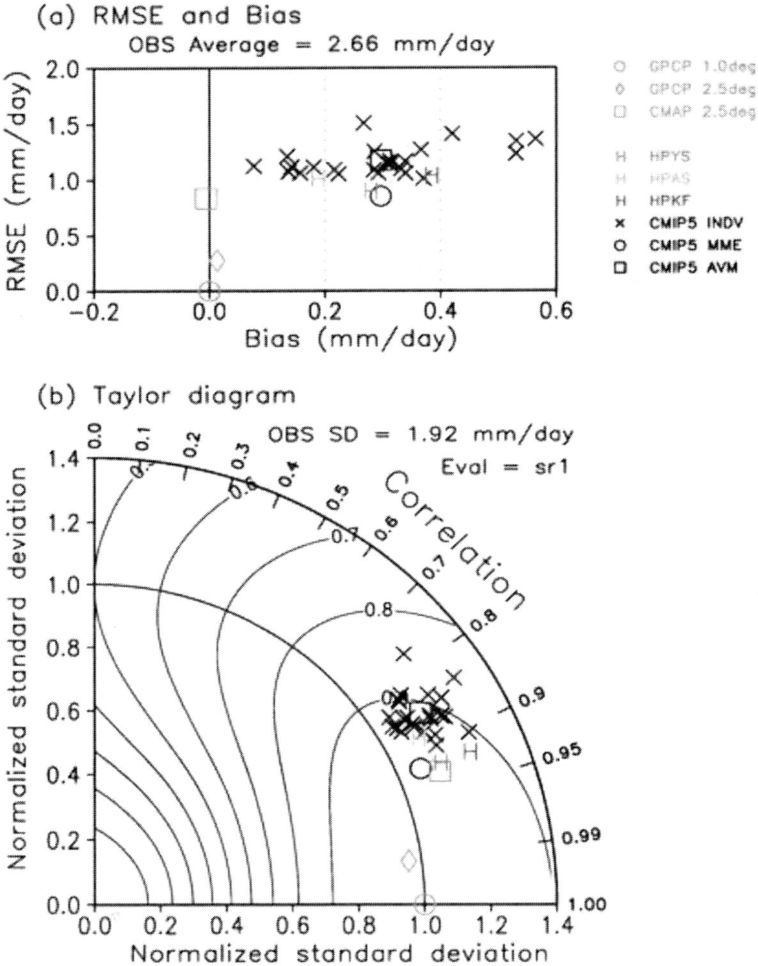

Figure 3. Skill of PAV simulated by models verified against the GPCP 1dd V1.2 data (green circle) for the global domain. Green marks denote other observations (Table 2). Color characters show the MRI-AGCM3.2H models. Red, orange and blue characters denote the YS, AS and KF schemes, respectively. Black marks X show the CMIP5 individual models. Black circles indicate the MME mean. Black squares indicate the average of the skill scores of all of the CMIP5 models (AVM). The target domain is the same as in Figure 2. (**a**) The root mean square error (RMSE) and bias (mm day^{-1}). The domain average of observation is shown above the panel. (**b**) Taylor diagram for displaying pattern statistics [40]. The standard deviation of the observation in the domain is shown above the panel. The contour shows the value of Taylor skill *S*.

AVM. In the case of nonlinear skill measures, such as RMSE, the correlation coefficient and Taylor skill score *S*, the MME mean and the AVM differ. In Figure 3a, the RMSE of MME (black circle) is smaller than that of the AVM (black square). This is consistent with previous findings that the MME mean average can be expected to outperform individual models in climate simulations [41,42,43]. The 60-km model with YS (red H) shows the smallest RMSE of all models.

In the Taylor diagram (Figure 3b), the Taylor skill scores *S* (contour plot) of the 60-km models (H) are also closer to the GPCP 1dd observation (green circle) than those of the CMIP5 models (black X). The MME mean (black circle) is closer to the observation (green circle) than the AVM (black square) as in Figure 3a. All of the models are plotted outside the quadrant of radius one. This means that the simulated precipitation distribution has larger spatial variability than observation. The spatial variability of the 60-km models is nearly comparable to that of the CMIP5 models, but the spatial pattern simulated by the 60-km models is better than that by most CMIP5 models. The 60-km model with YS (red H) is the closest to the observation (green circle) of all models.

In summary, the 60-km models perform better than or equal to the CMIP5 models in terms of bias, RMSE, spatial pattern and Taylor skill score *S*, although the spatial variability of the 60-km model is almost equivalent to that of the CMIP5 models.

3.4. Extreme Precipitation Events

We adopted four extreme precipitation indices (Table 3) from those proposed by [44]. PAV is also included for comparison. These are indices measuring precipitation intensity, except for consecutive dry days (CDD), which is a measure for dryness and the possibility of drought. These indices are based on annual statistics. The Simple Daily precipitation Intensity Index (SDII) is widely adopted in model studies, such as [45]. The maximum 5-day precipitation total (R5d) is a relevant measure to estimate the possibility of natural disasters, such as flood and land slide. The indices SDII, R5d and CDD are also adopted in Figure 9.37 of [46]. In the previous studies dealing with extreme precipitation events, notations Rx5day and Rx1day are widely used to define R5d and The maximum 1-day precipitation total (PMAX), respectively.

Table 3. Indices of precipitation.

Index	Name	Definition	Unit
PAV	Annual average precipitation	Annual average precipitation	mm day^{-1}
SDII	Simple daily precipitation intensity index	Total annual precipitation divided by the number of rainy days (precipitation ≥ 1 mm)	mm day^{-1}
R5d	Maximum 5-day precipitation total	Annual maximum of consecutive 5-day precipitation	mm
PMAX	Maximum 1-day precipitation total	Annual maximum of daily precipitation	mm
CDD	Consecutive dry days	Annual maximum number of consecutive dry days (precipitation < 1 mm)	day

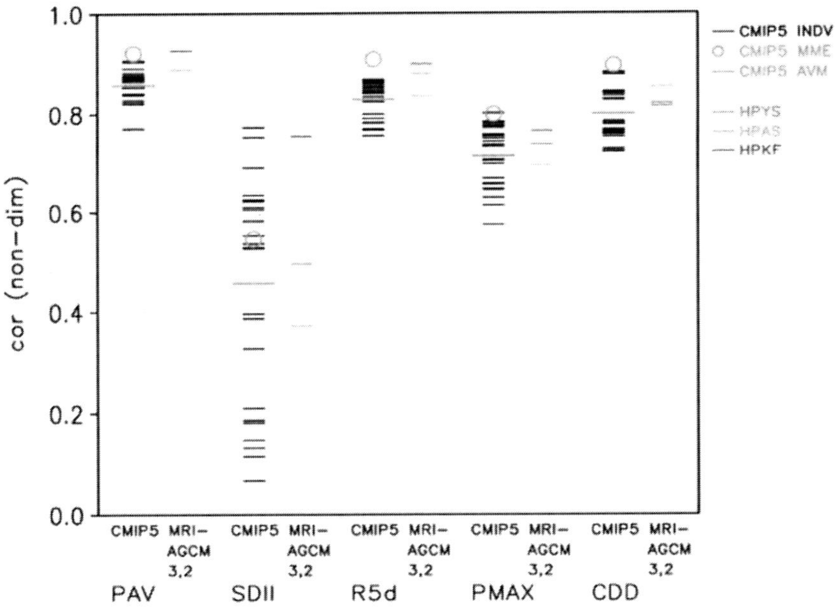

Figure 4. Spatial correlation coefficients of the global distribution of precipitation indices (Table 3) between observations by GPCP 1ddv1.2 and model simulations. Red, orange and blue bars denote the 60-km models with the YS, AS and KF schemes, respectively. Black bars show the CMIP5 individual models. Green circles indicate the MME mean of CMIP5 models. Green bars indicate the AVM of CMIP5 models. CDD, consecutive dry days.

Figure 4 compares the spatial correlation coefficients of the 60-km models against GPCP 1dd with those of CMIP5 models with respect to the global distributions of extreme precipitation indices (Table 3). In the case of PAV, the skills of the 60-km models are higher than the average skill of CMIP5 models (AVM, green line). The skills of the 60-km models with YS (red line) and with the KF scheme (blue line) are even higher than the MME mean of CMIP5 models (green circle). In the case of SDII, R5d, PMAX and CDD, the skills of the 60-km models are higher or comparable to those of CMIP5 models. However, the models with the AS scheme show the lowest correlation coefficient among the 60-km models. This is due to the under-estimation of precipitation intensity by the 60-km model with the AS scheme compared with models with the other cumulus scheme. See Figure S3 for the R5d case.

In terms of RMSE, the skills of the 60-km models are also higher or comparable to those of CMIP5 models, except for CDD (Figure S4). The 60-km models have relatively larger positive bias of CDD compared to those of CMIP5 models (Figure S5). This leads to the relatively larger RMSE of the 60-km models compared to those of CMIP5 models.

The 60-km model with the KF scheme shows a higher spatial correlation coefficient than the 60-km models with the YS and AS schemes except for CDD (Figure 4). However, the RMSE (Figure S4) and bias (Figure S5) of the 60-km model with the KF scheme are generally larger than those of the YS and AS schemes. Larger bias (Figure S5), as well as larger spatial variability by the KF scheme (Figure 3b) might lead to larger RMSE (Figure S4) in spite of a higher spatial correlation coefficient (Figure 4). The larger spatial variability by the KF scheme might originate from the tendency to produce excessive local orographic rainfall.

We speculate that the higher skill of the 60-km model originates in the higher horizontal resolution of the 60-km model as compared to that of the CMIP5 models (average 171 km). As for PAV, we calculated the correlation coefficient between the model skill measured by the spatial correlation coefficient and the grid size of the models. We used 31 models, including seven MRI-AGCMs with different grid sizes and cumulus convection schemes, as well as 24 CMIP5 models (Figure S6). We found a statistically-significant negative correlation between model skill and grid size, suggesting that higher horizontal resolution is favorable for higher skill. This tendency is also confirmed with R5d, PMAX and CDD (Figure S7). However, in the case of RMSE, the advantage of higher horizontal resolution was not clear. As for the precipitation over East Asa, the cumulus convection scheme, as well as higher horizontal resolution contributes to higher skill of the MRI-AGCM3.2 models [11].

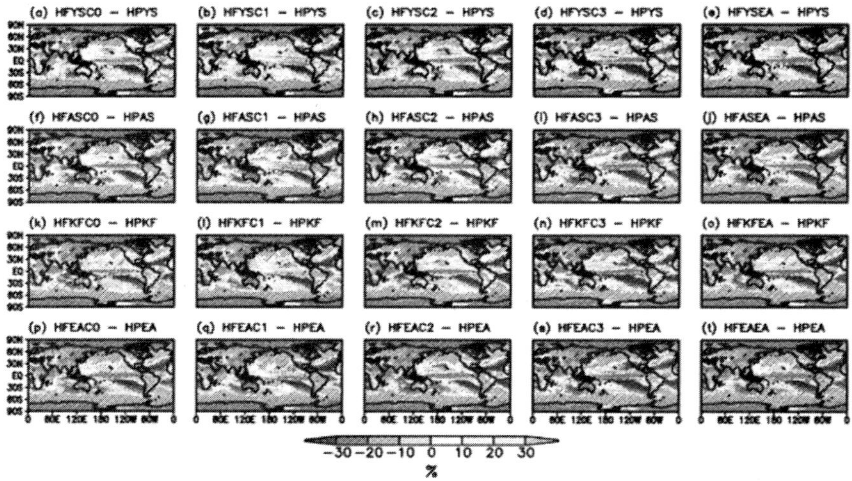

Figure 5. Future changes (2079–2099) in annual average precipitation (%) from the present-day climatology (1983–2003). Change is normalized by the present-day climatology. Hatched regions show changes above the 95% significance level based on Student's *t*-test. (First row; (**a**–**e**)) The 60-km model with YS. Changes are calculated from the present-day climate simulations with YS. (Second row; (**f**–**i**)) The 60-km model with the AS scheme. Changes are calculated from the present-day climate simulations with the AS scheme. (Third row; (**k**–**o**)) The 60-km model with the KF scheme. Changes are calculated from the present-day climate simulations with the KF scheme. (Fourth row; (**p**–**t**)) The ensemble average of 60-km model with the three convection schemes (first–third row). (First column; (**a**,**f**,**k**,**p**)) Simulations with the SST cluster C0. (Second column) Simulations with C1. (Third column) Simulations with C2. (Fourth column) Simulations with C3. (Fifth column) The ensemble average of simulations with the four SSTs (first–fourth column).

4. FUTURE CLIMATE

4.1. Changes in Annual Average Precipitation

Future changes in PAV are illustrated in Figure 5. In the case of the 60-km models with YS (Figure 5a–d), the present-day simulation with the same YS of HPYS (1983–2003, 21 years) is subtracted from future simulations of HFYSC0, HFYSC1, HFYSC2 and HFYSC3 (2079–2099, 21 years) to obtain future change, which is converted into the ratio (%) to the present-day climatology HPYS. Similarly, future changes by the 60-km model with the AS and KF schemes are calculated from the present-day simulations with the corresponding same cumulus convection schemes. The statistical significance of changes is evaluated by Student's *t*-test based on variances of year-to-year variability.

In the case of HFYSC0 (Figure 5a), precipitation increases over the equatorial regions of the Pacific Ocean, the Arabian Sea and the high latitudes of the Northern Hemisphere. These regions correspond to the larger warming of SST (Figure 1a). Precipitation also increases in the high latitudes of the Southern Hemisphere, but the warming of SST is relatively small in these regions (Figure 1a). This suggests that the precipitation increase in the high latitudes of the Southern Hemisphere is caused by large-scale circulation, rather than by the local SST effect. Precipitation decreases over the eastern Southern Pacific Ocean in the subtropical region around 30° S, the ocean to the west of Australia and the Atlantic Ocean in the subtropical region of the both hemispheres. These regions with decreased precipitation roughly correspond to the region with relatively small SST warming (Figure 1a). If we compare precipitation changes in Figure 5a–d, which are simulated by the same YS forced with different SSTs, the spatial structure is almost similar from the large-scale perspective, but the precipitation changes over the western tropical Pacific Ocean around Philippines differ among simulations. In the case of the 60-km models with the AS scheme (Figure 5f–i; second row) and with the KF scheme (Figure 5k–n; third row), the large-scale spatial structure is similar to those by the 60-km models with YS (Figure 5a–d; first row). However, precipitation decrease over the western tropical Pacific Ocean by the 60-km models with the KF scheme (Figure 5k–n; third row) is much more evident than the 60-km models with YS (Figure 5a–d; first row) and with the AS scheme (Figure 5f–i; second row). The dependence of precipitation change on the convection scheme is summarized by the ensemble averages (Figure 5e,j,o; last column). Precipitation increases over the western tropical Pacific Ocean by the 60-km models with YS (Figure 5e) and with the AS scheme (Figure 5j), while precipitation decreases over there by the 60-km models with the KF scheme (Figure 5o). The reason why the KF scheme behaves different from the YS and AS schemes might relate to the sensitive responses of the KF cumulus convection schemes over the MC (Figure S2), but the physical interpretation of the difference among future precipitation changes is not straightforward.

The dependence of precipitation change on SST is summarized by the ensemble averages (Figure 5p–s; bottom row). Precipitation increases over the equatorial regions of the Pacific Ocean in simulations with SST clusters C0 (Figure 5p), C2 (Figure 5r) and C3 (Figure 5s). On the other hand, in the case of C1 (Figure 5q), the region where precipitation increases over the equatorial regions of the Pacific Ocean is confined to the eastern side of the ocean and does not extend westward. This might be partly due to relatively small warming of SST in cluster C1 over the tropical Pacific Ocean compared with C0, C2 and C3 (Figure 1e–g). We will further discuss the dependence of precipitation change on convection scheme and SST in the later Section 4.3.

The ensemble average of all 12 simulations (Figure 5a–d,f–i,k–n) is displayed in Figure 5t. In general, precipitation increases globally, except some regions over the subtropical area. This suggests that global average precipitation will increase in the future. We will discuss this point from a global perspective in the later Section 4.4. The area of the statistically-significant region in the ensemble average (Figure 5t) is larger than that of individual simulations (Figure 5a–d,f–i,k–n). This is one of advantages of ensemble simulations and is mainly caused by the larger sample size (21 years/run × 12 runs = 252 years) as compared to individual simulation (21 years). The geographical distribution of precipitation change is qualitatively consistent with the Figure 12.10 in [46] based on the MME mean of CMIP5 AOGCMs.

4.2. Extreme Precipitation Events

Figure 6 displays the future change in extreme precipitation events (Table 3) including PAV for comparison. Each panel is calculated from the ensemble average of all 12 simulations. The area of precipitation increase in SDII (Figure 6b) is larger than that in PAV (Figure 6a). In PAV, precipitation decreases over the Mediterranean, the southern part of Africa, the ocean to the west of Australia and in the subtropics of the South Atlantic Ocean, but in SDII, precipitation increases over these regions. Another evident difference is that precipitation decreases over the western tropical Pacific Ocean and Antarctica in SDII as opposed to the precipitation increase over these regions in PAV. In the case of R5d (Figure 6c), the distribution of change is almost similar to SDII with some differences, but the magnitude of the positive and negative change in R5d is larger than those in SDII. Unlike SDII, R5d increases over Antarctica. This inconsistency might be originating in the difference of the index definition. In the case of PMAX (Figure 6d), the distribution of change is almost similar to R5d. Unlike the precipitation indices of PAV, SDII, R5d and PMAX, the index CDD is a measure of dryness and drought possibility. Note that the color bar in CDD (Figure 6e) is reversed in contrast to the other color bars (Figure 6a–d), such that brown color means dryness. CDD increases over the Mediterranean, the subtropical regions of eastern North Pacific Ocean, the United States, the Caribbean countries, the North Atlantic Ocean and the subtropical regions in the Southern Hemisphere. Since most of these regions are dry areas in the present-day observed climate condition, the drier region will get drier in the future. Over Europe, Africa, Australia, North America and South America, both CDD (Figure 6e) and extreme precipitation events (Figure 6b–d) will increase. This suggests that rainfall events in these regions will concentrate to a short period of time, presumably leading to the increase of natural disasters, such as flood and landslide.

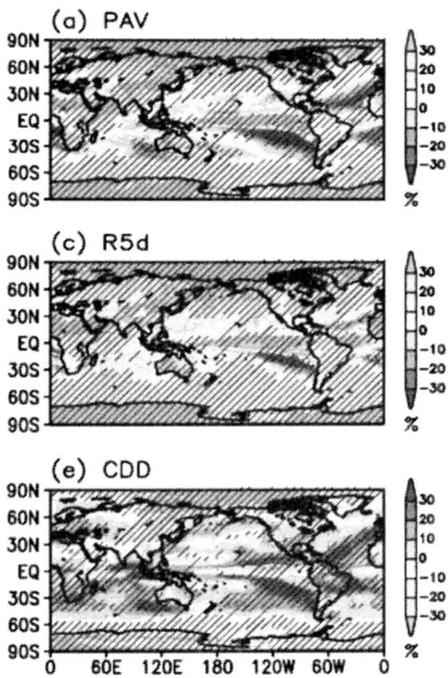

Figure 6. Future changes in precipitation indices (Table 3) by averaging all 12 simulations. Change is normalized by the present-day climatology. Unit is %. Hatched regions show changes above the 95% significance level. (**a**) PAV; (**b**) SDII; (**c**) R5d; (**d**) PMAX; (**e**) CDD. (**a**) is the same as Figure 5t.

Our finding that the area and rate of precipitation increase for heavy rainfall (Figure 6c,d) are larger than that for mean precipitation (Figure 6a) agrees with previous studies by [47,48,49,50,51]. This is consistent with the fact that changes in the extreme precipitation follow more closely the Clausius–Clapeyron (C-C) relationship than changes in the mean precipitation [51]. We will further discuss this point in Section 4.4.

Future R5d and CDD changes over land projected by the MME mean of CMIP5 AOGCMs are illustrated in Figure 12.26 of [46]. In this figure, R5d increases almost all land area, while CDD increases over the Mediterranean, the USA, South America and the southern part of Africa. Our results are basically consistent with these changes.

4.3. Which Influences Precipitation Change, the Cumulus Convection Scheme or SST?

We evaluate the dependence of future precipitation change on the cumulus convection scheme and SST. For this purpose, we have applied a two-way of analysis of variance (ANOVA; [52]) to future precipitation changes by the ensemble simulations with respect to three different cumulus convection schemes and four different SST distributions (Figure 5a–d,f–i,k–n). The ANOVA is able to quantitatively separate the relative contributions of the variances originating from differences in cumulus convection schemes and from differences in SSTs to the total variance. For technical details, see the Supplemental Materials. Figure 7a displays the total variance of future precipitation changes among all 12 simulations (Figure 5a–d,f–i,k–n), which can be decomposed into contributions by cumulus convection (Figure 7b) and SST (Figure 7c). The rest of the variance is named the residual (Figure 7d), which can be interpreted as the component not explained by cumulus convection and SST or as a non-linear interaction between cumulus convection and SST. In addition, internal natural variability of the atmosphere [53] may contribute to this residual part. Normalized variances by the total variance (Figure 7a) are also shown in Figure 7e,f. Finally, the relative magnitudes of the contribution from cumulus convection and SST are illustrated in Figure 7g. In red regions, cumulus convection affects precipitation change. On the other hand, in the blue region, SST affects precipitation change. In the tropics where convection is relatively active, cumulus convection influences precipitation. As for the Pacific Ocean, SST influences precipitation on some regions in the subtropics and middle latitudes, as well as the tropics. This is partly consistent with [19] in which they stressed that the SST distribution is a dominant source of uncertainty in future precipitation change. However, as for the Atlantic Ocean, SST influences precipitation only in the northern part.

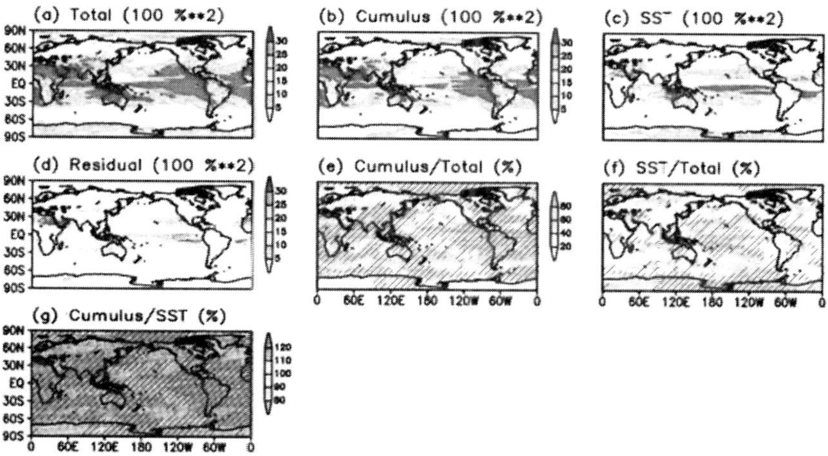

Figure 7. A two-way analysis of variance (ANOVA; [52]) applied to future annual average precipitation changes by all 12 ensemble simulations with respect to three different cumulus convection schemes and four different SST distributions (Figure 6a–d,f–i,k–n). (**a**) Total variance ($100\%^2$); (**b**) variance due to convection scheme ($100\%^2$); (**c**) variance due to SST ($100\%^2$); (**d**) residual of variance ($100\%^2$); (**e**) relative contribution of cumulus convection scheme as the ratio to the total variance (%); (**f**) relative contribution of SST as the ratio to the total variance (%); (**g**) ratio (%) of the variance by the cumulus convection scheme (b) to the variance by SST (c). Hatches show regions above the 95% significance level based on the *F*-test.

Total variance of changes in PAV is small over land at high latitudes of the Northern Hemisphere (Figure 7a). This is the outcome of the result that spatial patterns of precipitation change over these regions are almost same (Figure 5a–d,f–i,k–n), indicating that change over this region is not sensitive to the convection scheme and future SST distribution. This is caused by robust circulation changes at the 850 hPa height (Figure S8) where the spatial pattern of the height change over these regions is almost same.

We applied the ANOVA to extreme precipitation indices (Figure 8). In the case of SDII (Figure 8b), cumulus convection affects most part of the globe. As for R5d and PMAX, cumulus convection also affects most part of the globe, but SST influences some regions in the Pacific Ocean and the Atlantic Ocean. Figure 8a–c indicates that future changes in intense precipitation represented by SDII, R5D and PMAX (Figure 8b–d) are much more sensitive to the cumulus convection scheme than those by PAV (Figure 8a).

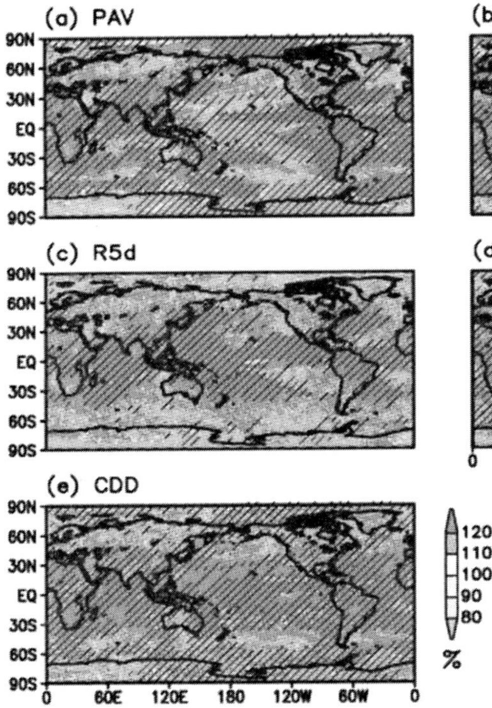

Figure 8. Ratio (%) of the variance by the cumulus convection scheme to the variance by SST. The ANOVA is applied for future changes in precipitation indices (Table 3). Hatched regions show the ratio above the 95% significance level. (**a**) PAV; (**b**) SDII; (**c**) R5d; (**d**) PMAX; (**e**) CDD. (a) is the same as Figure 7g.

The authors in [9] have conducted similar ensemble simulations for different cumulus convections and different SSTs using the MRI-AGCM3.2S and MRI-AGCM3.2H. They used the SSTs of CMIP3 AOGCMs with the A1B scenario, whereas we used the SSTs of CMIP5 AOGCMs with RCP8.5 in this paper. They reported that the precipitation change in PAV over the tropical Pacific Ocean is much affected by SST, but that the contribution of the cumulus convection scheme to the heavy precipitation over the tropical Pacific Ocean is larger than the contribution to PAV. Our results are consistent with their results, although both results cannot be directly compared due to some differences in experimental design.

4.4. Precipitation Efficiency

In principle, the increase in annual average precipitation and precipitation intensity can be attributed to the increased availability of water vapor caused by the warming air temperature of the atmosphere [46,54]. Figure 9a shows the future changes in annual average surface air temperature (SAT) projected by HPYS and HFYSC0. Warming over land is larger than over ocean because the heat capacity of land is smaller than that of ocean. Warming in the Northern Hemisphere is larger than in the Southern Hemisphere because the area of land in the Northern Hemisphere is larger than that in the Southern Hemisphere. Especially, warming over the high latitudes in the Norther Hemisphere is conspicuous. This is mainly caused by the surface albedo feedback mechanism. The reduction in snow over land and the extent of sea ice over the Arctic Ocean leads to the decrease of albedo at the surface, which enhances the absorption of the short wave at the surface to warm up SAT. The future change in precipitation projected by HPYS and HFYSC0 (Figure 9b) has some common features with changes in SAT (Figure 9a) in that precipitation change is generally larger over land than over ocean and is larger in the Northern Hemisphere than in the Southern Hemisphere.

Figure 9c shows the ratio of PAV change to SAT change at each grid point using the simulations of HPYS and HFYSC0. This quantity, which is the rate of precipitation change per one degree rise in SAT, can be interpreted as a kind of 'precipitation efficiency', which measures the conversion rate from water vapor to precipitation. This precipitation efficiency is referred to as the 'hydrological sensitivity' in Section 9.3.4.1 of [55]. Hereafter, we consistently use the terminology 'precipitation efficiency'. Figure 9c shows that the distribution of precipitation efficiency basically resembles that of precipitation change (Figure 9b), but the magnitudes of precipitation efficiency over the high latitudes in the Northern Hemisphere are reduced due to the large increase of temperature there. In Figure 9c, hatched regions denote that the precipitation efficiency exceeding the value 7.5%/°C theoretically expected from the C-C relationship [21]. Precipitation efficiency is larger than the theoretical value over the equatorial region of the Pacific Ocean, the Arabian Sea and high latitudes in the Southern Hemisphere, but precipitation efficiency is smaller than the theoretical value in most parts of the globe. This implies that the dynamical effect of atmospheric circulation is reducing the local thermodynamical effect of precipitation change in most regions. Our results are consistent with the findings by [21] in which they show that the precipitation efficiency of the MME mean of CMIP3 AOGCMs exceeds the theoretical value only over the central and eastern equatorial Pacific and the western Indian Ocean. The

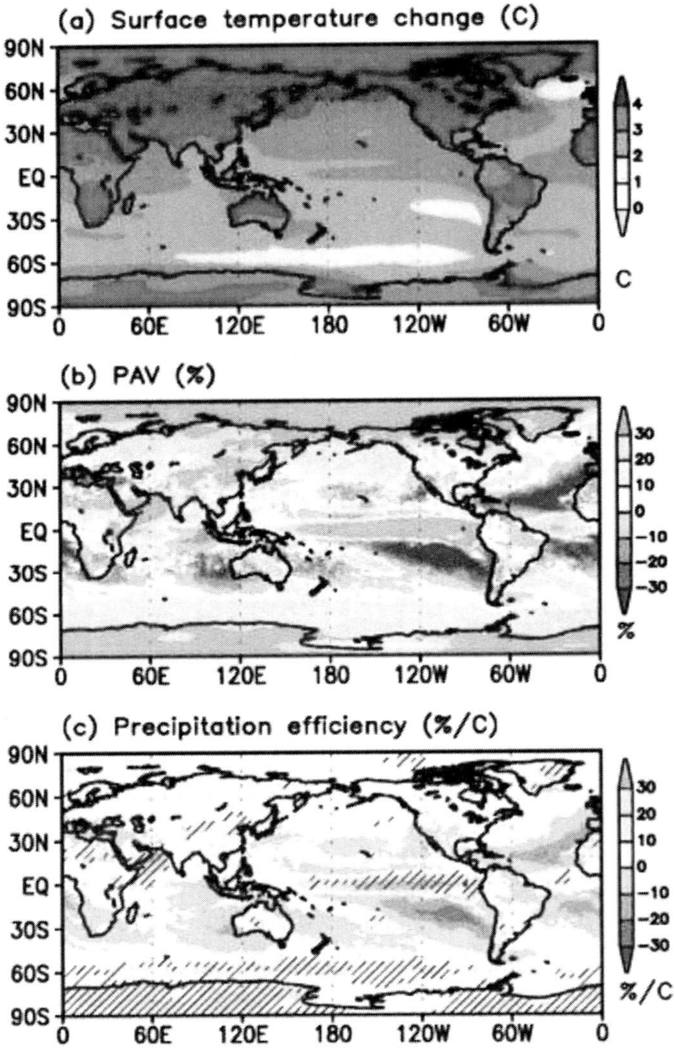

Figure 9. Precipitation efficiency calculated at each grid point. (**a**) Change in annual mean surface air temperature (°C) by HFYSC0 (2079–2099) from the HPYS (1983–2003); (**b**) PAV change (%) by HFYSC0 relative to HPYS; (**c**) precipitation efficiency (%/°C) defined as PAV change (**b**) divided by surface air temperature change (a). Hatched regions indicate that precipitation efficiency is larger than the rate of precipitation increase (7.5%/°C) expected from the Clausius–Clapeyron (C-C) relationship.

authors in [20] have revealed that precipitation is supposed to increase slower than the C-C relationship, and the models actually match the anticipation. Our results are consistent with their findings.

We further extended our analysis to intense precipitation. Figure 10 illustrates the precipitation efficiency of R5d and PMAX, as well as PAV for comparison, using all 12 simulations. The geographical distribution of precipitation efficiency by R5d and PMAX generally resembles that by PAV. However, the area of positive regions (blue) by R5d and PMAX is larger than that by PAV, which is directly caused by the distribution of precipitation change by R5d (Figure 6c) and PMAX (Figure 6d). Another noteworthy feature is that the magnitude of positive precipitation efficiency (blue regions) by R5d and PMAX is larger than that by PAV, especially over the northern Pacific Ocean.

From Figure 9, we expect that the global average precipitation efficiency by R5d and PMAX might be larger than that by PAV. From a global perspective, we firstly calculated the global average of precipitation changes and SAT changes and then evaluated precipitation efficiency. This procedure smooths out the strong locality in Figure 9c and Figure 10 in which precipitation efficiency is derived at each grid point. Another intention here is to investigate the dependence of precipitation efficiency on the cumulus convection scheme and SST, because Figure 9 and Figure 10 are based on the ensemble average of all 12 simulations. Figure 11a shows the relationship between PAV changes and SAT changes for all 12 simulations. The characters denote the convection schemes, and colors denote SST clusters. Precipitation changes are mainly determined by the cumulus convection scheme (character) and are not sensitive to SST (color). For instance, precipitation changes by the AS scheme (character A) are almost the same value of about 8.4% regardless of SAT increase ranging from about 3.1–3.6 °C. Conversely, SAT changes are mainly determined by SST and are less sensitive to the cumulus convection scheme. For instance, SAT changes by SST cluster C0 (black) are almost the same value of between 3.5 and 3.6 °C, regardless of large precipitation changes. As for precipitation change, the AS scheme gives the largest changes, whereas YS gives the smallest changes. As for SAT, SST cluster C1 gives the largest warming, whereas the SST cluster C3 gives the smallest warming. HFASC3 (blue A) gives the largest precipitation efficiency 2.46%/°C = 8.40%/3.42 °C, whereas HFYSC1 (red Y) gives the smallest precipitation efficiency 2.09%/°C = 7.59%/3.64 °C. The range of precipitation efficiency of PAV is small due to the narrow range of SAT change from about 3.4–3.7 °C.

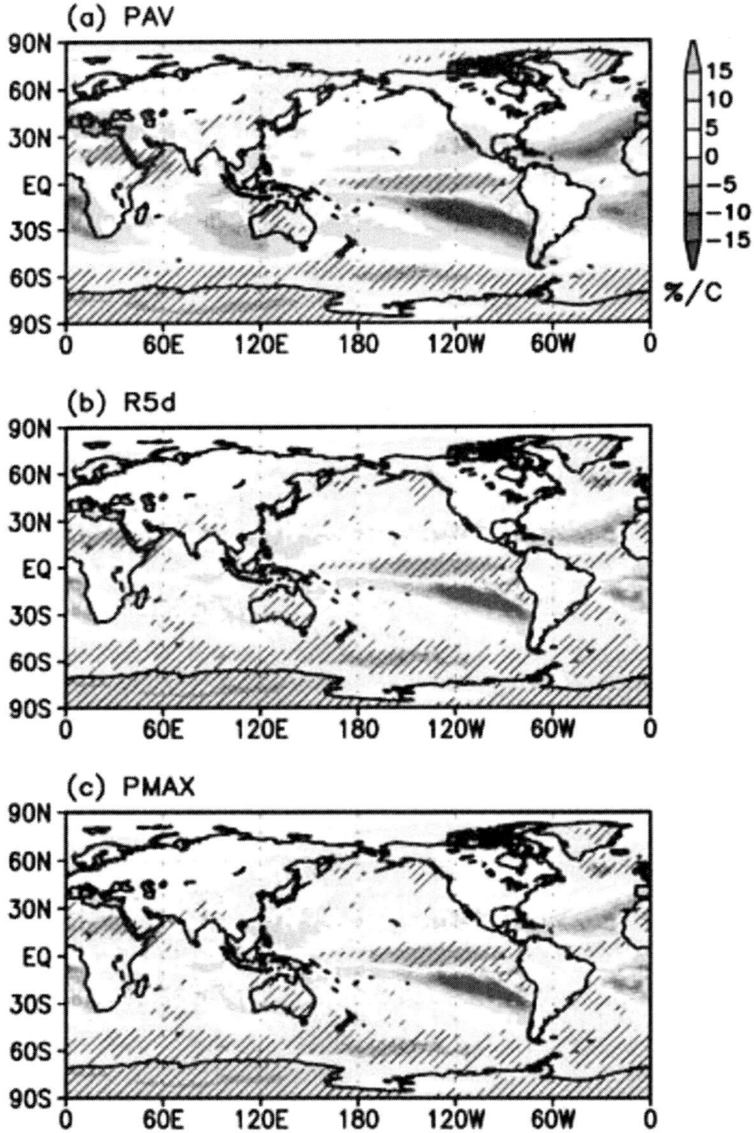

Figure 10. The averages of the precipitation efficiency (%/°C) of all 12 simulations. (**a**) PAV; (**b**) R5d; (**c**) PMAX. Hatched regions indicate that precipitation efficiency is larger than the rate of precipitation increase (7.5%/°C) expected from the C-C relationship.

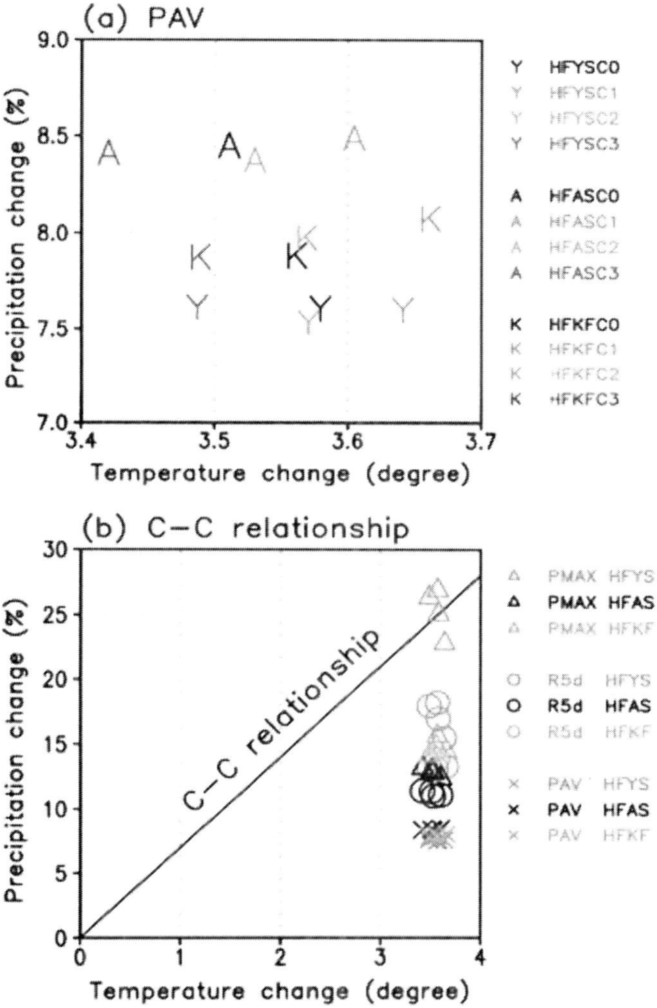

Figure 11. Dependence of precipitation change (%) on global average annual mean surface temperature change (°C). (**a**) Dependence of PAV changes on the cumulus convection scheme and SST. Characters Y, A and K mean the YS, AS and KF schemes, respectively. Black, red, green and blue colors mean SST cluster C0, C1, C2 and C3, respectively. (**b**) Dependence of precipitation change on different indices. The black slanted line denotes the rate of precipitation increase (7.5%/°C) expected from the Clausius–Clapeyron relationship. The X mark, circle and triangle mean PAV, R5d and PMAX, respectively. Red, black and green colors mean the YS, AS and KF schemes, respectively.

Figure 11b shows the relationship between precipitation changes and SAT changes for R5d and PMAX, as well as PAV for comparison. Note that the ranges in vertical and horizontal axes of Figure 11b are wider than those of Figure 11a. In Figure 11b, marks denote variables and colors denote the cumulus convection schemes. The differences of SST clusters are not discriminated in Figure 11b. In reference to the diagonal line indicating the theoretically-expected value of 7.5%/°C, precipitation efficiency of PAV (X mark) is far less than this value. If we focus only on the YS cumulus scheme (red), precipitation efficiency becomes larger in the following order: PAV (mark X), R5d (circle) and PMAX (triangle). The same relationship holds true for the AS scheme (black) and KF scheme (green). Since the indices PAV, R5d and PMAX represent a much more intense rainfall event in this order, Figure 11b implies precipitation changes associated with heavier rainfall events are larger than those by moderate or weak rainfall events. It is noteworthy that the precipitation efficiency of PMAX almost reaches the theoretical value of 7.5%/°C. In other words, it is reasonable that precipitation changes in heavy rainfall events are large because the conversion from water vapor into precipitation is much more effective than moderate or weak rainfall events. This is consistent with previous studies [48,49,50,51,56] in which the change of extreme precipitation is larger than that of mean precipitation.

The authors in [10] have investigated the future precipitation over East Asia using the MRI-AGCM3.2H with SSTs of CMIP3 AOGCMs assuming the A1B scenario and found that the precipitation efficiency of heavy precipitation is larger than that of mean precipitation. Our results are also consistent with their results. Using the same experiment by the study of [10], the authors in [57] have calculated precipitation efficiency over the Arctic and have revealed that precipitation efficiency of heavy precipitation is comparable or less than that of mean precipitation. This is presumably because rainfall over the Arctic is mainly caused by horizontal transport of moisture rather than cumulus convection.

5. CONCLUSIONS

In the present-day climate simulation for 1983–2003 (21 years) by the MRI-AGCM3.2H, we confirmed that the model well reproduces the global distributions of annual average precipitation and precipitation intensity. The performance of the MRI-AGCM3.2H is higher or equal to those by the CMIP5 atmospheric global models. This can be partly attributed to the higher horizontal resolution of the MRI-AGCM3.2H as compared with CMIP5 atmospheric models, because we found a statistically-significant correlation between model skill and grid size with respect to annual average precipitation and precipitation extremes.

In the future climate simulation for 2079–2099 (21 years), we have conducted the 12 ensemble simulations for three different convection schemes and four different future SST distributions, assuming the emission scenario RCP8.5. Annual average precipitation will increase over the equatorial regions and decrease over the subtropical regions. The area of precipitation increase by intense precipitation is larger than that by annual average precipitation. The differences of precipitation change among all 12 simulations are generally sensitive to the cumulus convection scheme, but changes are influenced by the SST over the some part of the Pacific Ocean. This is almost consistent with the previous work by [9], although their SST distributions are different from those used in this paper.

The precipitation efficiency defined as precipitation change per 1° surface temperature warming is evaluated. The global average of precipitation efficiency for annual precipitation was less than the maximum value expected by thermodynamical theory, indicating that dynamical atmospheric circulation is acting to reduce the conversion efficiency from water vapor to precipitation. The precipitation efficiency by intense precipitation is larger than that by moderate and weak precipitation. This agrees with previous works by [48,49,50,51,56].

Our results suggest that the selection of a convection scheme is crucial in climate simulation. Furthermore, implementing multiple cumulus convection schemes might be favorable to evaluate the spread of future precipitation change and enhance the reliability of future projection, if computer resources are rich and available. However, uncertainty in future precipitation change is often attributed to the different SST patterns among CMIP5 AOGCMs. We still recognize future projection is influenced by SST distributions, because future projected SST has also large uncertainties.

One of the major caveats in our simulations is the lack of air-sea interactions. In order to evaluate the effect of air-sea coupling process, we have already finished the future climate projection using the MRI-AGCM3.2H coupled with a full ocean model [18]. Figure S9 shows the difference between future precipitation changes projected by an atmospheric model and a coupled model. As is expected, differences are large over the eastern tropical Pacific Ocean. The reason and mechanism of the difference is not yet fully analyzed and should be explored in future studies.

SUPPLEMENTARY MATERIALS

The following are available online at www.mdpi.com/2073-4433/8/5/91/s1: Table S1: Previous studies on precipitation using the MRI-AGCM3, Table S2: Features of 24 CMIP5 models used in this study, Figure S1: Comparison among observations for PAV, Figure S2: Comparison among simulations for PAV, Figure S3: Same as Figure S2 but for R5d, Figure S4: Root mean square errors of global distribution of precipitation indices between observations by GPCP 1ddv1.2 and model simulations, Figure S5: Same as Figure S4 but for bias, Figure S6: Dependence of model skill on the grid spacing of 31 models including 7 MRI-AGCM3.2 models and 24 CMIP5 atmospheric models, Figure S7: Correlation coefficients between the skill and grid size of 31 models, Figure S8: Same as Figure 5 but for 850 hPa height, Figure S9: Comparison between future changes in PAV by atmospheric model and AOGCM, ANalysis Of VAriance (ANOVA): Technical details.

ACKNOWLEDGMENTS

This work was conducted under the framework of "the Integrated Research Program for Advanced Climate Modeling" supported by the TOUGOU Program of the Ministry of Education, Culture, Sports, Science, and Technology (MEXT) of Japan. We also thank the anonymous reviewers whose valuable comments and suggestions greatly improved the manuscript.

CONFLICTS OF INTEREST

The authors declare no conflict of interest.

REFERENCES

1. Cubasch, U.; Wuebbles, D.; Chen, D.; Facchini, M.C.; Frame, D.; Mahowald, N.; Winther, J.-G. Introduction. In Climate Change 2013: The Physical Science Basis; Contribution of Working Group I to the Fifth Assessment Report of the Intergovernmental Panel on Climate Change; Stocker, T.F., Qin, D., Plattner, G.-K., Tignor, M., Allen, S.K., Boschung, J., Nauels, A., Xia, Y., Bex, V., Midgley, P.M., Eds.; Cambridge University Press: Cambridge, UK; New York, NY, USA, 2013.

2. Shiogama, H.; Watanabe, M.; Ogura, T.; Yokohata, T.; Kimoto, M. Multi-parameter multi-physics ensemble (MPMPE): A new approach exploring the uncertainties of climate sensitivity. Atmos. Sci. Lett. 2014, 15, 97–102.
3. Kusunoki, S.; Yoshimura, J.; Yoshimura, H.; Noda, A.; Oouchi, K.; Mizuta, R. Change of Baiu rain band in global warming projection by an atmospheric general circulation model with a 20-km grid size. J. Meteorol. Soc. Jpn. 2006, 84, 581–611.
4. Kitoh, A.; Kusunoki, S. East Asian summer monsoon simulation by a 20-km mesh AGCM. Clim. Dyn. 2007, 31, 389–401.
5. Kusunoki, S. Is the global atmospheric model MRI-AGCM3.2 better than the CMIP5 atmospheric models in simulating precipitation over East Asia? Clim. Dyn. 2016, 1–22.
6. Kusunoki, S.; Mizuta, R. Future Changes in the Baiu Rain Band Projected by a 20-km Mesh Global Atmospheric Model: Sea Surface Temperature Dependence. Sci. Online Lett. Atmos. 2008, 4, 85–88.
7. Kusunoki, S.; Mizuta, R.; Matsueda, M. Future changes in the East Asian rain band projected by global atmospheric models with 20-km and 60-km grid size. Clim. Dyn. 2011, 37, 2481–2493.
8. Kusunoki, S.; Mizuta, R. Comparison of near future (2015–2039) changes in the East Asian rain band with future (2075–2099) changes projected by global atmospheric models with 20-km and 60-km grid size. Sci. Online Lett. Atmos. 2012, 8, 73–76.
9. Endo, H.; Kitoh, A.; Ose, T.; Mizuta, R.; Kusunoki, S. Future changes and uncertainties in Asian precipitation simulated by multiphysics and multi-sea surface temperature ensemble experiments with high-resolution Meteorological Research Institute atmospheric general circulation models (MRI-AGCMs). J. Geophys. Res. 2012, 117, D16118.
10. Kusunoki, S.; Mizuta, R. Changes in precipitation intensity over East Asia during the 20th and 21st centuries simulated by a global atmospheric model with a 60 km grid size. J. Geophys. Res. Atmos. 2013, 118, 11007–11016.
11. Kitoh, A.; Endo, H. Changes in precipitation extremes projected by a 20-km mesh global atmospheric model. Weather Clim. Extremes 2016, 11, 41–52.
12. Mizuta, R.; Murata, A.; Ishii, M.; Shiogama, H.; Hibino, K.; Mori, N.; Arakawa, O.; Imada, Y.; Yoshida, K.; Aoyagi, T.; et al. Over 5000 years of ensemble future climate simulations by 60 km global and 20 km regional atmospheric models. Bull. Am. Meteorol. Soc. 2016.
13. Endo, H.; Kitoh, A.; Mizuta, R.; Ishii, M. Future changes in precipitation extremes in East Asia and their uncertainty based on large ensemble simulations with a high resolution AGCM. Sci. Online Lett. Atmos. 2017, 13, 7–12.
14. Kusunoki, S. Future changes in precipitation over East Asia projected by the global atmospheric model MRI-AGCM3.2. Clim. Dyn. 2017, 1–17.
15. Bengtsson, L.; Hodges, K.I.; Keenlyside, N. Will extratropical storms intensify in a warmer climate? J. Clim. 2009, 22, 2276–2301.
16. Douville, H. Impact of regional SST anomalies on the Indian Monsoon response to global waming in the CNRM climate model. J. Clim. 2005, 18, 2008–2024.
17. Hasegawa, A.; Emori, S. Effect of air-sea coupling in the assessment of CO2-induced intensification of tropical cyclone activity. Geophys. Res. Lett. 2007, 34, L05701.
18. Ogata, T.; Mizuta, R.; Adachi, Y.; Murakami, H.; Ose, T. Effect of air-sea coupling on the frequency distribution of intense tropical cyclones over the northwestern Pacific. Geophys. Res. Lett. 2015, 42, 10415–10421.
19. Ma, J.; Xie, S.-P. Regional Patterns of Sea Surface Temperature Change: A Source of Uncertainty in Future Projections of Precipitation and Atmospheric Circulation. J. Clim. 2013, 26, 2482–2501.
20. Held, I.M.; Soden, B.J. Robust Responses of the Hydrological Cycle to Global Warming. J. Clim. 2006, 19, 5686–5699.
21. Vecchi, G.A.; Soden, B.J. Global warming and the weakening of the tropical circulation. J. Clim. 2007, 20, 4316–4340.

22. Mizuta, R.; Yoshimura, H.; Murakami, H.; Matsueda, M.; Endo, H.; Ose, T.; Kamiguchi, K.; Hosaka, M.; Sugi, M.; Yukimoto, S.; et al. Climate simulations using MRI-AGCM3.2 with 20-km grid. J. Meteorol. Soc. Jpn. 2012, 90A, 233–258.
23. Yoshimura, H.; Mizuta, R.; Murakami, H. A spectral cumulus parameterization scheme interpolating between two convective updrafts with semi-lagrangian calculation of transport by compensatory subsidence. Mon. Weather Rev. 2015, 143, 597–621.
24. Tiedtke, M. A comprehensive mass flux scheme for cumulus parameterization in large-scale models. Mon. Weather Rev. 1989, 117, 1779–1800.
25. Randall, D.A.; Pan, D.M. Implementation of the Arakawa-Schubert cumulus parameterization with a prognostic closure. In The Representation of Cumulus Convection in Numerical Models; American Meteorological Society: Boston, MA, USA, 1993; Volume 24, Chapter 11, No. 46; pp. 137–147.
26. Kain, J.S.; Fritsch, J.M. A one-dimensional entraining/detraining plume model and its application in convective parameterization. J. Atmos. Sci. 1990, 47, 2784–2802.
27. Yukimoto, S.; Yoshimura, H.; Hosaka, M.; Sakami, T.; Tsujino, H.; Hirabara, M.; Tanaka, T.; Deushi, M.; Obata, A.; Nakano, H.; et al. Meteorological Research Institute—Earth System Model Version 1 (MRI-ESM1)—Model Description; Technical Reports of the Meteorological Research Institute; Meteorological Research Institute: Tsukuba, Japan, 2011; Volume 64, p. 88.
28. Rayner, N.A.; Parker, D.E.; Horton, E.B.; Folland, C.K.; Alexander, L.V.; Rowell, D.P.; Kent, E.C.; Kaplan, A. Global analyses of sea surface temperature, sea ice, and night marine air temperature since the late nineteenth century. J. Geophys. Res. 2003, 108, 4407–4437.
29. Collins, M.; Knutti, R.; Arblaster, J.; Dufresne, J.-L.; Fichefet, T.; Friedlingstein, P.; Gao, X.; Gutowski, W.J.; Johns, T.; Krinner, G.; et al. Section 12; Long-term Climate Change: Projections, Commitments and Irreversibility. In Climate Change 2013: The Physical Science Basis; Contribution of Working Group I to the Fifth Assessment Report of the Intergovernmental Panel on Climate Change; Stocker, T.F., Qin, D., Plattner, G.-K., Tignor, M., Allen, S.K., Boschung, J., Nauels, A., Xia, Y., Bex, V., Midgley, P.M., Eds.; Cambridge University Press: Cambridge, UK; New York, NY, USA, 2013.
30. Mizuta, R.; Adachi, Y.; Yukimoto, S.; Kusunoki, S. Estimation of the Future Distribution of Sea Surface Temperature and Sea Ice Using the CMIP3 Multi-Model Ensemble Mean; Technical Reports of the Meteorological Research Institute; Meteorological Research Institute: Tsukuba, Japan, 2008; Volume 56, p. 28.
31. Mizuta, R.; Arakawa, O.; Ose, T.; Kusunoki, S.; Endo, H.; Kitoh, A. Classification of CMIP5 future climate responses by the tropical sea surface temperature changes. SOLA 2014, 10, 167–171.
32. Intergovermental Panel on Climate Change (IPCC). Special Report on Emissions Scenarios: A Special Report of Working Group III of the Intergovernmental Panel on Climate Change; Nakicenovic, N., Alcamo, J., Davis, G., de Vries, B., Fenhann, J., Gaffin, S., Gregory, K., Grübler, A., Jung, T.Y., Kram, T., et al., Eds.; Cambridge University Press: Cambridge, UK, 2000; p. 599.
33. Shibata, K.; Deushi, M.; Sekiyama, T.T.; Yoshimura, H. Development of an MRI chemical transport model for the study of stratospheric chemistry. Pap. Meteorol. Geophys. 2004, 55, 75–119.
34. Huffman, G.J.; Adler, R.F.; Morrissey, M.M.; Bolvin, D.T.; Curtis, S.; Joyce, R.; McGavock, B.; Susskind, J. Global precipitation at one-degree daily resolution from multisatellite observations. J. Hydrometeorol. 2001, 2, 36–50.
35. Sperber, K.R.; Annamalai, H.; Kang, I.S.; Kitoh, A.; Moise, A.; Turner, A.G.; Wang, B.; Zhou, T. The Asian summer monsoon: An intercomparison of CMIP5 vs. CMIP3 simulations of the late 20th century. Clim. Dyn. 2013, 41, 2711–2744.
36. Adler, R.F.; Huffman, G.J.; Chang, A.; Ferrano, R.; Xie, P.P.; Janowiak, J.; Rudolf, B.; Schneider, U.; Curtis, S.; Bolvin, D.; et al. The Version-2 Global Precipitation Climatology Preject (GPCP) monthly precipitation analysis (1979–Present). J. Hydrometeorol. 2003, 4, 1147–1167.

37. Xie, P.; Arkin, P. Global precipitation: A 17-year monthly analysis based on gauge observations, satellite estimates and numerical model outputs. Bull. Am. Meteorol. Soc. 1997, 78, 2539–2558.
38. Huffman, G.J.; Adler, R.F.; Bolvin, D.T.; Gu, G.; Nelkin, E.J.; Bowman, K.P.; Hong, Y.; Stocker, E.F.; Wolff, D.B. The TRMM Multisatellite Precipitation Analysis (TMPA): Quasi-global, multiyear, combined-sensor precipitation estimates at fine scales. J. Hydrometeorol. 2007, 8, 38–55.
39. Minobe, S.; Kuwano-Yoshida, A.; Komori, N.; Xie, S.-P.; Small, R J. Influence of the Gulf Stream on the troposphere. Nature2008, 452, 206–209.
40. Taylor, K.E. Summarizing multiple aspects of model performance in a single diagram. J. Geophys. Res. 2001, 106, 7183–7192.
41. Lambert, S.J.; Boer, G.J. CMIP1 evaluation and intercomparison of coupled climate models. Clim. Dyn. 2001, 17, 83–106.
42. Gleckler, P.J.; Taylor, K.E.; Doutriaux, C. Performance metrics for climate models. J. Geophys. Res. 2008, 113, D06104.
43. Reichler, T.; Kim, J. How well do coupled models simulate today's climate? Bull. Am. Meteorol. Soc. 2008, 89, 303–311.
44. Frich, P.; Alexander, L.V.; Della-Marta, P.; Gleason, B.; Haylock, M.; Klein Tank, A.M.G.; Peterson, T. Observed coherent changes in climatic extremes during the second half of the twentieth century. Clim. Res. 2002, 19, 193–212.
45. Dai, A. Precipitation characteristics in eighteen coupled climate models. J. Clim. 2006, 19, 4605–4630.
46. Intergovernmental Panel on Climate Change (IPCC). Climate Change 2013: The Physical Science Basis; Contribution of Working Group I to the Fifth Assessment Report of the Intergovernmental Panel on Climate Change; Stocker, T.F., Qin, D., Plattner, G.K., Tignor, M., Allen, S.K., Boschung, J., Nauels, A., Xia, Y., Bex, V., Midgley, P.M., Eds.; Cambridge University Press: Cambridge, UK; New York, NY, USA, 2013; p. 1535.
47. Hennessy, K.J.; Gregory, J.M.; Mitchell, J.F.B. Changes in daily precipitation under enhanced greenhouse conditions. Clim. Dyn. 1997, 13, 667–680.
48. Kharin, V.V.; Zwiers, F.W. Changes in the extremes in an ensemble of transient climate simulations with a coupled atmosphere-ocean GCM. J. Clim. 2000, 13, 3760–3788.
49. Wehner, M.F. Predicted 21st century changes in seasonal extreme precipitation events in the Parallel Climate Model. J. Clim.2004, 17, 4281–4290.
50. Emori, S.; Brown, S.J. Dynamic and thermodynamic changes in mean and extreme precipitation under changed climate. Geophys. Res. Lett. 2005, 32, L17706.
51. Pall, P.; Allen, M.R.; Stone, D.A. Testing the Clausius-Clapeyron constraint on changes in extreme precipitation under CO2 warming. Clim. Dyn. 2007, 28, 351–363.
52. Storch, H.V.; Zwiers, F.W. Section 9 Analysis of Variance. In Statistical Analysis in Climate Research; Storch, H.V., Zwiers, F.W., Eds.; Cambridge University Press: Cambridge, UK, 1999; pp. 171–192.
53. Deser, C.; Knutti, R.; Solomon, S.; Phillips, A.S. Communication of the role of natural variability in future North American climate. Nat. Clim. Chang. 2012, 2, 775–779.
54. Intergovernmental Panel on Climate Change (IPCC). Climate Change 2007: The Physical Science Basis; Contribution of Working Group I to the Fourth Assessment Report of the Intergovernmental Panel on Climate Change; Solomon, S., Qin, D., Manning, M., Chen, Z., Marquis, M., Averyt, K.B., Tignor, M., Miller, H.L., Eds.; Cambridge University Press: Cambridge, UK; New York, NY, USA, 2007; p. 996.
55. Intergovernmental Panel on Climate Change (IPCC). Climate Change 2001: The Scientific Basis; Contribution of Working Group I to the Third Assessment Report of the Intergovernmental Panel on Climate Change; Houghton, J.T., Ding, Y., Griggs, D.J., Noguer, M., van der Linden, P.J., Dai, X., Maskell, K., Johnson, C.A., Eds.; Cambridge University Press: Cambridge, UK; New York, NY, USA, 2001; p. 881.

56. Hegerl, G.C.; Zwiers, F.W.; Stott, P.A.; Kharin, V.V. Detectability of anthropogenic changes in annual temperature and precipitation extremes. J. Clim. 2004, 17, 3683–3700.
57. Kusunoki, S.; Mizuta, R.; Hosaka, M. Future changes in precipitation intensity over the Arctic projected by a global atmospheric model with a 60-km grid size. Polar Sci. 2015, 9, 277–292.

Index